ORGANIC SYNTHESES

ORGANIC SYNTHESES

AN ANNUAL PUBLICATION OF SATISFACTORY
METHODS FOR THE PREPARATION OF
ORGANIC CHEMICALS
VOLUME 92
2015

VIRESH H. RAWAL
VOLUME EDITOR

The procedures in this article are intended for use only by persons with prior training in experimental organic chemistry. These procedures must be conducted at one's own risk. *Organic Syntheses, Inc.*, its Editors, and its Board of Directors do not warrant or guarantee the safety of individuals using these procedures and hereby disclaim any liability for any injuries or damages claimed to have resulted from or related in any way to the procedures herein.

For general information on our other products and services or for technical support, please contact our Customer Care Department within the United States at (800) 762-2974, outside the United States at (317) 572-3993 or fax (317) 572-4002.

Wiley also publishes its books in a variety of electronic formats. Some content that appears in print may not be available in electronic formats. For more information about Wiley products, visit our web site at www.wiley.com.

"John Wiley & Sons, Inc. is pleased to publish this volume of Organic Syntheses on behalf of Organic Syntheses, Inc. Although Organic Syntheses, Inc. has assured us that each preparation contained in this volume has been checked in an independent laboratory and that any hazards that were uncovered are clearly set forth in the write-up of each preparation, John Wiley & Sons, Inc. does not warrant the preparations against any safety hazards and assumes no liability with respect to the use of the preparations."

Library of Congress Catalog Card Number: 21-17747

ISBN: 978-1-119-28068-2

Printed in the United States of America

10 9 8 7 6 5 4 3 2 1

ORGANIC SYNTHESES

VOLUME	VOLUME EDITOR	PAGES
I*	† ROGER ADAMS	84
II*	† JAMES BRYANT CONANT	100
III*	† HANS THACHER CLARK	105
IV*	† OLIVER KAMM	89
V*	† CARL SHIPP MARVEL	110
VI*	† HENRY GILMAN	120
VII*	† FRANK C. WHITMORE	105
VIII*	† ROGER ADAMS	139
IX*	† JAMES BRYANT CONANT	108
Collective Vol. I	A revised edition of Annual Volumes I-IX	580
	†HENRY GILMAN, *Editor-in-Chief*	
	2nd Edition revised by † A. H. BLATT	
X*	† HANS THACHER CLARKE	119
XI*	† CARL SHIPP MARVEL	106
XII*	† FRANK C. WHITMORE	96
XIII*	† WALLACE H. CAROTHERS	119
XIV	† WILLIAM W. HARTMAN	100
XV*	† CARL R. NOLLER	104
XVI*	† JOHN R. JOHNSON	104
XVII*	† L. F. FIESER	112
XVIII*	† REYNOLD C. FUSON	103
XIX*	† JOHN R. JOHNSON	105
Collective Vol. II	A revised edition of Annual volumes X-XIX	654
	† A. H. BLATT, *Editor-in-Chief*	
20*	† CHARLES F. H. ALLEN	113
21*	† NATHAN L. DRAKE	120
22*	† LEE IRVIN SMITH	114
23*	† LEE IRVIN SMITH	124
24*	† NATHAN L. DRAKE	119
25 *	† WERNER E. BACHMANN	120
26*	† HOMER ADKINS	124
27*	† R. L. SHRINER	121
28*	† H. R. SNYDER	121
29*	† CLIFF S. HAMILTON	119
Collective Vol. III	A revised edition of Annual Volumes 20–29	890
	† E. C. HORNING, *Editor-in-Chief*	
30*	† ARTHUR C. COPE	115
31*	† R. S. SCHREIBER	122
32*	† RICHARD ARNOLD	119
33*	† CHARLES PRICE	115
34*	† WILLIAM S. JOHNSON	121
35*	† T. L. CAIRNS	122

Out of print.
†*Deceased.*

VOLUME	VOLUME EDITOR	PAGES
36*	† N. J. Leonard	120
37*	† James Cason	109
38*	† James C. Sheehan	120
39*	† Max Tishler	114
Collective Vol. IV	A revised edition of Annual Volumes 30–39	1036
	† Norman Rabjohn, *Editor-in-Chief*	
40*	† Melvin S. Newman	114
41*	John D. Roberts	118
42*	† Virgil Boekelheide	118
43*	† B. C. McKusick	124
44*	† William E. Parham	131
45*	† William G. Dauben	118
46*	E. J. Corey	146
47*	† William D. Emmons	140
48*	† Peter Yates	164
49*	Kenneth B. Wiberg	124
Collective Vol. V	A revised edition of Annual Volumes 40–49	1234
	† Henry E. Baumgarten, *Editor-in-Chief*	
Cumulative Indices to Collective Volumes, I, II, III, IV, V		
	† Ralph L. and † Rachel H. Shriner, *Editors*	
50*	Ronald Breslow	136
51*	† Richard E. Benson	209
52*	† Herbert O. House	192
53*	† Arnold Brossi	193
54*	† Robert E. Ireland	155
55 *	† Satoru Masamune	150
56*	† George H. Büchi	144
57*	Carl R. Johnson	135
58*	† William A. Sheppard	216
59*	Robert M. Coates	267
Collective Vol. VI	A revised edition of Annual Volumes 50–59	1208
	Wayland E. Noland, *Editor-in-Chief*	
60*	† Orville L. Chapman	140
61*	† Robert V. Stevens	165
62*	Martin F. Semmelhack	269
63*	† Gabriel Saucy	291
64*	Andrew S. Kende	308
Collective Vol. VII	A revised edition of Annual Volumes 60–64	602
	† Jeremiah P. Freeman, *Editor-in-Chief*	
65*	Edwin Vedejs	278
66*	Clayton H. Heathcock	265
67*	Bruce E. Smart	289
68*	James D. White	318
69*	Leo A. Paquette	328

Out of print.
†*Deceased.*

VOLUME	VOLUME EDITOR	PAGES
Reaction Guide to Collective Volumes I-VII and Annual Volumes 65–68		854
	DENNIS C. LIOTTA AND MARK VOLMER, *Editors*	
Collective Vol. VIII	A revised edition of Annual Volumes 65–69	696
	† JEREMIAH P. FREEMAN, *Editor-in-Chief*	
Cumulative Indices to Collective Volumes, I, II, III, IV, V, VI, VII, VIII		
70*	† ALBERT I. MEYERS	305
71*	LARRY E. OVERMAN	285
72*	† DAVID L. COFFEN	333
73*	ROBERT K. BOECKMAN, JR.	352
74*	ICHIRO SHINKAI	341
Collective Vol. IX	A revised edition of Annual Volumes 70–74	840
	† JEREMIAH P. FREEMAN, *Editor-in-Chief*	
75	AMOS B. SMITH, III	257
76	STEPHEN MARTIN	340
77	DAVID S. HART	312
78	WILLIAM R. ROUSH	326
79	LOUIS S. HEGEDUS	328
Collective Vol. X	A revised edition of Annual Volumes 75–79	810
	† JEREMIAH P. FREEMAN, *Editor-in-Chief*	
80	STEVEN WOLFF	259
81	RICK L. DANHEISER	296
82	EDWARD J. J. GRABOWSKI	195
83	DENNIS P. CURRAN	221
84	MARVIN J. MILLER	410
Collective Vol. XI	A revised edition of Annual Volumes 80–84	1138
	CHARLES K. ZERCHER, *Editor-in-Chief*	
85	SCOTT E. DENMARK	321
86	JOHN A. RAGAN	403
87	PETER WIPF	403
88	JONATHAN A. ELLMAN	472
89	MARK LAUTENS	561
90	DAVID L. HUGHES	376
Collective Vol. XII	A revised edition of Annual Volumes 85–90	1570
	CHARLES K. ZERCHER, *Editor-in-Chief*	
91	KAY M. BRUMMOND	365
92	VIRESH H. RAWAL	386

Out of print.
†*Deceased.*

NOTICE

Beginning with Volume 84, the Editors of *Organic Syntheses* initiated a new publication protocol, which is intended to shorten the time between submission of a procedure and its appearance as a publication. Immediately upon completion of the successful checking process, procedures are assigned volume and page numbers and are then posted on the Organic Syntheses website (www.orgsyn.org). The accumulated procedures from a single volume are assembled once a year and submitted for publication. The annual volume is published by John Wiley and Sons, Inc., and includes an index. The hard cover edition is available for purchase through the publisher. Incorporation of graphical abstracts into the Table of Contents began with Volume 77. Annual volumes 70–74, 75–, 80–84 and 85–89 have been incorporated into five-year versions of the collective volumes of *Organic Syntheses*. Collective Volumes IX, X, XI and XII are available for purchase in the traditional hard cover format from the publishers.

Beginning with Volume 88, a new type of article, referred to as Discussion Addenda, appeared. In these articles submitters are provided the opportunity to include updated discussion sections in which new understanding, further development, and additional application of the original method are described. Organic Syntheses intends for Discussion Addenda to become a regular feature of future volumes.

Organic Syntheses, Inc., joined the age of electronic publication in 2001 with the release of its free web site (www.orgsyn.org). The site is accessible through internet browsers using Macintosh and Windows operating systems, and the database can be searched by key words and sub-structure. John Wiley & Sons, Inc., and Accelrys, Inc., partnered with Organic Syntheses, Inc., to develop a database (onlinelibrary.wiley.com/book/10.1002/0471264229) that is available for license with internet solutions from John Wiley & Sons, Inc. and intranet solutions from Accelrys, Inc.

Both the commercial database and the free website contain all annual and collective volumes and indices of *Organic Syntheses*. Chemists can draw structural queries and combine structural or reaction transformation queries with full-text and bibliographic search terms, such as chemical name, reagents, molecular formula, apparatus, or even hazard warnings or phrases. The contents of individual or collective volumes can be browsed by lists of titles, submitters' names, and volume and page references, with or without structures.

The commercial database at onlinelibrary.wiley.com/book/10.1002/0471264229 also enables the user to choose his/her preferred chemical drawing package, or to utilize several freely available plug-ins for entering queries. The user is also able to cut and paste existing structures and reactions directly into the structure search query or their preferred chemistry editor, streamlining workflow. Additionally, this database contains links to the full text of primary literature references via CrossRef, ChemPort, Medline, and ISI Web of Science. Links to local holdings for institutions using open url technology can also be enabled. The database user can limit his/her search to, or order the search results by, such factors as reaction type, percentage yield, temperature, and publication date, and can create a customized table of reactions for comparison. Connections to other Wiley references are currently made via text search, with cross-product structure and reaction searching to be added in the near future. Incorporations of new preparations will occur as new material becomes available.

INFORMATION FOR AUTHORS OF PROCEDURES

Organic Syntheses welcomes and encourages submissions of experimental procedures that lead to compounds of wide interest or that illustrate important new developments in methodology. Proposals for *Organic Syntheses* procedures will be considered by the Editorial Board upon receipt of an outline proposal as described below. A full procedure will then be invited for those proposals determined to be of sufficient interest. These full procedures will be evaluated by the Editorial Board, and if approved, assigned to a member of the Board for checking. In order for a procedure to be accepted for publication, each reaction must be successfully repeated in the laboratory of a member of the Editorial Board at least twice, with similar yields (generally ±5%) and selectivity to that reported by the submitters.

Organic Syntheses Proposals

A cover sheet should be included providing full contact information for the principal author and including a scheme outlining the proposed reactions (an *Organic Syntheses* Proposal Cover Sheet can be downloaded at orgsyn.org). Attach an outline proposal describing the utility of the methodology and/or the usefulness of the product. Identify and reference the best current alternatives. For each step, indicate the proposed scale, yield, method of isolation and purification, and how the purity of the product is determined. Describe any unusual apparatus or techniques required, and any special hazards associated with the procedure. Identify the source of starting materials. Enclose copies of relevant publications (attach pdf files if an electronic submission is used).

Submit proposals by mail or as e-mail attachments to:

Professor Charles K. Zercher
Associate Editor, *Organic Syntheses*
Department of Chemistry
University of New Hampshire
23 Academic Way, Parsons Hall
Durham, NH 03824

For electronic submissions: *org.syn@unh.edu*

Submission of Procedures

Authors invited by the Editorial Board to submit full procedures should prepare their manuscripts in accord with the Instructions to Authors which are described below or may be downloaded at orgsyn.org. Submitters are also encouraged to consult this volume of *Organic Syntheses* for models with regard to style, format, and the level of experimental detail expected in *Organic Syntheses* procedures. Manuscripts should be submitted to the Associate Editor. Electronic submissions are encouraged; procedures will be accepted as e-mail attachments in the form of Microsoft Word files with all schemes and graphics also sent separately as ChemDraw files.

Procedures that do not conform to the Instructions to Authors with regard to experimental style and detail will be returned to authors for correction. Authors will be notified when their manuscript is approved for checking by the Editorial Board, and it is the goal of the Board to complete the checking of procedures within a period of no more than six months.

Additions, corrections, and improvements to the preparations previously published are welcomed; these should be directed to the Associate Editor. However, checking of such improvements will only be undertaken when new methodology is involved.

NOMENCLATURE

Both common and systematic names of compounds are used throughout this volume, depending on which the Volume Editor felt was more appropriate. The Chemical Abstracts indexing name for each title compound, if it differs from the title name, is given as a subtitle. Systematic Chemical Abstracts nomenclature, used in the Collective Indexes for the title compound and a selection of other compounds mentioned in the procedure, is provided in an appendix at the end of each preparation. Chemical Abstracts Registry numbers, which are useful in computer searching and identification, are also provided in these appendices.

ACKNOWLEDGMENT

Organic Syntheses wishes to acknowledge the contributions of Amgen, Inc. and Boehringer Ingelheim to the success of this enterprise through their support, in the form of time and expenses, of members of the Board of Editors.

INSTRUCTIONS FOR AUTHORS

All organic chemists have experienced frustration at one time or another when attempting to repeat reactions based on experimental procedures found in journal articles. To ensure reproducibility, *Organic Syntheses* requires experimental procedures written with considerably more detail as compared to the typical procedures found in other journals and in the "Supporting Information" sections of papers. In addition, each *Organic Syntheses* procedure is carefully "checked" for reproducibility in the laboratory of a member of the Board of Editors.

Even with these more detailed procedures, the experience of *Organic Syntheses* editors is that difficulties often arise in obtaining the results and yields reported by the submitters of procedures. To expedite the checking process and ensure success, we have prepared the following "Instructions for Authors" as well as a **Checklist for Authors** and **Characterization Checklist** to assist you in confirming that your procedure conforms to these requirements. Please include a completed Checklist together with your procedure at the time of submission. Procedures submitted to *Organic Syntheses* will be carefully reviewed upon receipt and procedures lacking any of the required information will be returned to the submitters for revision.

Scale and Optimization

The appropriate scale for procedures will vary widely depending on the nature of the chemistry and the compounds synthesized in the procedure. However, some general guidelines are possible. For procedures in which the principal goal is to illustrate a synthetic method or strategy, it is expected, in general, that the procedure should result in at least 5 g and no more than 50 g of the final product. In cases where the point of the procedure is to provide an efficient method for the preparation of a useful reagent or synthetic building block, the appropriate scale also should not exceed 50 g of final product. Exceptions to these guidelines may be granted in special circumstances. For example, procedures describing the preparation of reagents employed as catalysts will often be acceptable on a scale of less than 5 g.

In considering the scale for an *Organic Syntheses* procedure, authors should also take into account the cost of reagents and starting materials. In general, the Editors will not accept procedures for checking in which the cost of any one of the reactants exceeds **$500** for a single full-scale run. Authors are requested to identify the most expensive reagent or starting material on the procedure submission checklist and to estimate its cost per run of the procedure.

It is expected that all aspects of the procedure will have been optimized by the authors prior to submission, and it is required that each reaction will have been carried out at least twice on exactly the scale described in the procedure, and with the results reported in the manuscript.

It is appropriate to report the weight, yield, and purity of the product of each step in the procedure as a range. In any case where a reagent is employed in significant excess, a Note should be included explaining why an excess of that reagent is necessary. If possible, the Note should indicate the effect of using amounts of reagent less than that specified in the procedure.

The Checking Process

A unique feature of papers published in *Organic Syntheses* is that each procedure and all characterization data is carefully checked for reproducibility in the laboratory of a member of the Board of Editors. In the event that an editor finds it necessary to make any modifications in an experimental procedure, then the published article incorporates the modified procedure, with an explanation and mention of the original protocol often included in a Note. The yields reported in the published article are always those obtained by the checkers. In general, the characterization data in the published article also is that of the checkers, unless there are significant differences with the data obtained by the authors, in which case the author's data will also be reported in a Note.

Reaction Apparatus

Describe the size and type of flask (number of necks) and indicate how *every* neck is equipped.

"A 500-mL, three-necked, round-bottomed flask equipped with an 3-cm Teflon-coated magnetic stirbar, a 250-mL pressure-equalizing addition funnel fitted with an argon inlet, and a rubber septum is charged with ...."

Indicate how the reaction apparatus is dried and whether the reaction is conducted under an inert atmosphere. Note that balloons are not acceptable as a means of maintaining an inert atmosphere. The description of the reaction apparatus can be incorporated in the text of the procedure or included in a Note.

"The apparatus is flame-dried and maintained under an atmosphere of argon during the course of the reaction."

In the case of procedures involving unusual glassware or especially complicated reaction setups, authors are encouraged to include a photograph or drawing of the apparatus in the text or in a Note (for examples, see *Org. Syn.,* Vol. 82, 99 and Coll. Vol. X, pp 2, 3, 136, 201, 208, and 669).

Use of Gloveboxes

When a glovebox is employed in a procedure, justification must be provided in a Note and the consequences of carrying out the operation without using a glovebox should be discussed.

Reagents and Starting Materials

All chemicals employed in the procedure must be commercially available or described in an earlier *Organic Syntheses* or *Inorganic Syntheses* procedure. For other compounds, a procedure should be included either as one or more steps in the text or, in the case of relatively straightforward preparations of reagents, as a Note. In the latter case, all requirements with regard to characterization, style, and detail also apply. Authors are encouraged to consult with the Associate Editor if they have any question as to whether to include such steps as part of the text or as a Note.

Authors are encouraged to consider the use of "substitute solvents" in place of more hazardous alternatives. For example, the use of *t*-butyl methyl ether (MTBE) should be considered as a substitute for diethyl ether, particularly in large scale work. Authors are referred to the articles "Sanofi's Solvent Selection Guide: A Step Toward More Sustainable Processes" (Prat, D.; Pardigon, O.; Flemming, H.-W.; Letestu, S.; Ducandas, V.; Isnard, P.; Guntrum, E.; Senac, T.; Ruisseau, S.; Cruciani, P. Hosek, P. *Org. Process Res. Dev.* **2013**, *17*, 1517-1525) and "Solvent Replacement for Green Processing" (Sherman, J.; Chin, B.; Huibers, P. D. T.; Garcia-Valis, R.; Hatton, T. A. *Environ. Health Perspect.* **1998**, *106* (Supplement I, 253-271) as well as the references cited therein for discussions of this subject. In addition, a link to a "solvent selection guide" can be accessed via the American Chemical Society Green Chemistry website at http://www.acs.org/content/acs/en/greenchemistry/research-innovation/tools-for-green-chemistry.html.

In one or more Notes, indicate the purity or grade of each reagent, solvent, etc. It is highly desirable to also indicate the source (company the chemical was purchased from), particularly in the case of chemicals where it is suspected that the composition (trace impurities, etc.) may vary from one supplier to another. In cases where reagents are purified, dried, "activated" (e.g., Zn dust), etc., a detailed description of the procedure used should be included in a Note. In other cases, indicate that the chemical was "used as received".

"Diisopropylamine (99.5%) was obtained from Aldrich Chemical Co., Inc. and distilled under argon from calcium hydride before use. THF (99+%) was obtained from Mallinckrodt, Inc. and distilled from sodium benzophenone ketyl. Diethyl ether (99.9%) was purchased from Aldrich Chemical Co., Inc. and purified by pressure filtration under argon through activated alumina. Methyl iodide (99%) was obtained from Aldrich Chemical Co., Inc. and used as received."

The amount of each reactant must be provided in parentheses in the order mL, g, mmol, and equivalents with careful consideration to the correct number of **significant figures**. Avoid indicating amounts of reactants with more significant figures than makes sense. For example, "437 mL of THF" implies that the amount of solvent must be measured with a level of precision that is unlikely to affect the outcome of the reaction. Likewise, "5.00 equiv" implies that an amount of excess reagent must be controlled to a precision of 0.01 equiv.

The concentration of solutions should be expressed in terms of molarity or normality, and not percent (e.g., 1 N HCl, 6 M NaOH, not "10% HCl").

Reaction Procedure

Describe every aspect of the procedure clearly and explicitly. Indicate the order of addition and time for addition of all reagents and how each is added (via syringe, addition funnel, etc.).

Indicate the temperature of the reaction mixture (preferably internal temperature). Describe the type of cooling (e.g., "dry ice-acetone bath") and heating (e.g., oil bath, heating mantle) methods employed. Be careful to describe clearly all cooling and warming cycles, including initial and final temperatures and the time interval involved.

Describe the appearance of the reaction mixture (color, homogeneous or not, etc.) and describe all significant changes in appearance during the course of the reaction (color changes, gas evolution, appearance of solids, exotherms, etc.).

Indicate how the reaction can be monitored to determine the extent of conversion of reactants to products. In the case of reactions monitored by TLC, provide details in a Note, including eluent, R_f values, and method of visualization. For reactions followed by GC, HPLC, or NMR analysis, provide details on analysis conditions and relevant diagnostic peaks.

"The progress of the reaction was followed by TLC analysis on silica gel with 20% EtOAc-hexane as eluent and visualization with p-anisaldehyde. The ketone starting material has $R_f = 0.40$ (green) and the alcohol product has $R_f = 0.25$ (blue)."

Reaction Workup

Details should be provided for reactions in which a "quenching" process is involved. Describe the composition and volume of quenching agent, and time and temperature for addition. In cases where reaction mixtures are added to a quenching solution, be sure to also describe the setup employed.

"The resulting mixture was stirred at room temperature for 15 h, and then carefully poured over 10 min into a rapidly stirred, ice-cold aqueous solution of 1 N HCl in a 500-mL Erlenmeyer flask equipped with a magnetic stirbar."

For extractions, the number of washes and the volume of each should be indicated as well as the size of the separatory funnel.

For concentration of solutions after workup, indicate the method and pressure and temperature used.

"The reaction mixture is diluted with 200 mL of water and transferred to a 500-mL separatory funnel, and the aqueous phase is separated and extracted with three 100-mL portions of ether. The combined organic layers are washed with 75 mL of water and 75 mL of saturated NaCl solution, dried over 25 g of $MgSO_4$, filtered through a 250-mL medium porosity sintered glass funnel, and concentrated by rotary evaporation (25 °C, 20 mmHg) to afford 3.25 g of a yellow oil."

"The solution is transferred to a 250-mL, round-bottomed flask equipped with a magnetic stirbar and a 15-cm Vigreux column fitted with a short path distillation head, and then concentrated by careful distillation at 50 mmHg (bath temperature gradually increased from 25 to 75 °C)."

In cases where solid products are filtered, describe the type of filter funnel used and the amount and composition of solvents used for washes.

" … and the resulting pale yellow solid is collected by filtration on a Büchner funnel and washed with 100 mL of cold (0 °C) hexane."

When solid or liquid compounds are dried under vacuum, indicate the pressure employed (rather than stating "reduced pressure" or "dried *in vacuo*").

" …. and concentrated at room temperature by rotary evaporation (20 mmHg) and then at 0.01 mmHg to provide …. "

"The resulting colorless crystals are transferred to a 50-mL, round-bottomed flask and dried overnight in a 100 °C oil bath at 0.01 mmHg."

Purification: Distillation

Describe distillation apparatus including the size and type of distillation column. Indicate temperature (and pressure) at which all significant fractions are collected.

" .... and transferred to a 100-mL, round-bottomed flask equipped with a magnetic stirbar. The product is distilled under vacuum through a 12-cm, vacuum-jacketed column of glass helices (Note 16) topped with a Perkin triangle. A forerun (ca. 2 mL) is collected and discarded, and the desired product is then obtained, distilling at 50–55 °C (0.04–0.07 mmHg) .... "

Purification: Column Chromatography

Provide information on TLC analysis in a Note, including eluent, R_f values, and method of visualization.

Provide dimensions of column and amount of silica gel used; in a Note indicate source and type of silica gel.

Provide details on eluents used, and number and size of fractions.

"The product is charged on a column (5×10 cm) of 200 g of silica gel (Note 15) and eluted with 250 mL of hexane. At that point, fraction collection (25-mL fractions) is begun, and elution is continued with 300 mL of 2% EtOAc-hexane (49:1 hexanes:EtOAc) and then 500 mL of 5% EtOAc-hexane (19:1 hexanes:EtOAc). The desired product is obtained in fractions 24–30, which are concentrated by rotary evaporation (25 °C, 15 mmHg) ...."

Purification: Recrystallization

Describe procedure in detail. Indicate solvents used (and ratio of mixed solvent systems), amount of recrystallization solvents, and temperature protocol. Describe how crystals are isolated and what they are washed with. A photograph of the crystalline product is often valuable to indicate the form and color of the crystals.

"The solid is dissolved in 100 mL of hot diethyl ether (30 °C) and filtered through a Buchner funnel. The filtrate is allowed to cool to room temperature, and 20 mL of hexane is added. The solution is cooled at −20 °C overnight and the resulting crystals are collected by suction filtration on a Buchner funnel, washed with 50 mL of ice-cold hexane, and then transferred to a 50-mL, round-bottomed flask and dried overnight at 0.01 mmHg to provide .... "

Characterization

Physical properties of the product such as color, appearance, crystal forms, melting point, etc. should be included in the text of the procedure. Comments on the stability of the product to storage, etc. should be provided in a Note.

In a Note, provide data establishing the identity of the product. This will generally include IR, MS, ^{1}H-NMR, and ^{13}C-NMR data, and in some cases UV data. Copies of the proton and carbon NMR spectra for the products of each step in the procedure should be submitted showing integration for all resonances. Submission of copies of the NMR spectra for other nuclei are encouraged as appropriate.

In the same Note, provide analytical data establishing that the purity of the **isolated** product is at least 97%. **Note that this data should be obtained for the material on which the yield of the reaction is based**, not for a sample that has been subjected to additional purification by chromatography, distillation, or crystallization. Elemental analysis for carbon and hydrogen (and nitrogen if present) agreeing with calculated values within 0.4% is preferred. However, **quantitative** NMR, GC, or HPLC analyses involving measurements versus an internal standard will also be accepted. See *Instructions for Authors* at orgsyn.org for procedures for quantitative analysis of purity by NMR and chromatographic methods. Provide details on equipment and conditions for GC and HPLC analyses.

In procedures involving non-racemic, enantiomerically enriched products, optical rotations should generally be provided, but **enantiomeric purity must be determined by another method** such as chiral HPLC or GC analysis.

In cases where the product of one step is used without purification in the next step, a Note should be included describing how a sample of the product can be purified and providing characterization data for the pure material. Copies of the proton NMR spectra of both the product both *before* and *after* purification should be submitted.

Hazard Warnings

Particularly significant hazards should be indicated in a statement within a box at the beginning of the procedure in italicized, red type. Hazard warnings should only be included in the case of procedures that involve unusual hazards such as the use of pyrophoric or explosive substances, and substances with a high degree of acute or chronic toxicity. Instructions are provided in the Article Template. For other procedures, it is not necessary to include a special caution note that refers to standard operating procedures such as working in a hood, avoiding skin contact, etc., since this is referenced in the "Working with Hazardous Chemicals" statement within each article. Efforts should be made to avoid the use of toxic and hazardous solvents and reagents when less hazardous alternatives are available.

Discussion Section

The style and content of the discussion section will depend on the nature of the procedure.

For procedures that provide an improved method for the preparation of an important reagent or synthetic building block, the discussion should focus on the advantages of the new approach and should describe and reference all of the earlier methods used to prepare the title compound.

In the case of procedures that illustrate an important synthetic method or strategy, the discussion section should provide a mini-review on the new methodology. The scope and limitations of the method should be discussed, and it is generally desirable to include a table of examples. Please be sure each table is numbered and has a title. Competing methods for accomplishing the same overall transformation should be described and referenced. A brief discussion of mechanism may be included if this is useful for understanding the scope and limitations of the method.

Titles of Articles

In cases where the main thrust of the article is the illustration of a synthetic method of general utility, the title of the article should incorporate reference to that method. Inclusion of the name of the final product is acceptable but not required. In the case of articles where the objective is the preparation of a specific compound of importance (such as a chiral ligand), then the name of that compound should be part of the title.

Examples

Title without name of product:

"Stereoselective Synthesis of 3-Arylacrylates by Copper-Catalyzed Syn Hydroarylation" (*Org. Synth.* **2010**, *87*, 53).

Title including name of final product (note name of product is not required):

"Catalytic Enantioselective Borane Reduction of Benzyl Oximes: Preparation of (S)-1-Pyridin-3-yl-ethylamine Bis Hydrochloride" (*Org. Synth.* **2010**, *87*, 36).

Title where preparation of specific compound is the subject:

"Preparation of (S)-3,3'-Bis-Morpholinomethyl-5,5',6,6',7,7',8,8'-octa hydro-1,1'-bi-2-naphthol" (*Org. Synth.* **2010**, *87*, 59).

Style and Format for Text

Articles should follow the style guidelines used for organic chemistry articles published in the ACS journals such as *J. Am. Chem. Soc.*, *J. Org. Chem.*, *Org. Lett.*, etc. as described in the ACS Style Guide (3rd Ed.). The text of the procedure should be created using the Word template available on the *Organic Syntheses* website. Specific instructions with regard to the manuscript format (font, spacing, margins) is available on the website in the "Instructions for Article Template" and embedded within the Article Template itself.

Style and Format for Tables and Schemes

Chemical structures and schemes should be drawn using the standard ACS drawing parameters (in ChemDraw, the parameters are found in the "ACS Document 1996" option) with a maximum full size width of 15 cm (5.9 inches). The graphics files should then be pasted into the Word document at the correct location and the size reduced to 75% using "Format Picture" (Mac) or "Size and Position" (Windows). Graphics files must also be submitted separately. All Tables that include structures should be entirely prepared in the graphics (ChemDraw) program and inserted into the word processing file at the appropriate location. Tables that include multiple, separate graphics files prepared in the word processing program will require modification.

Tables and schemes should be numbered and should have titles. The title for a Table should be included within the ChemDraw graphic and placed immediately above the table. The title for a scheme should be included within the ChemDraw graphic and placed immediately below the scheme. Use 12 point Palatino Bold font in the ChemDraw file for all titles. For footnotes in Tables use Helvetica (or Arial) 9 point font and place these immediately below the Table.

Acknowledgments and Author's Contact Information

Contact information (institution where the work was carried out and mailing address for the principal author) should be included as footnote 1. This footnote should also include the email address for the principal author. Acknowledgment of financial support should be included in footnote 1.

Biographies and Photographs of Authors

Photographs and 100-word biographies of all authors should be submitted as separate files at the time of the submission of the procedure. The format of the biographies should be similar to those in the Volume 84 procedures found at the orgsyn.org website. Photographs can be accepted in a number of electronic formats, including tiff and jpeg formats.

DISPOSAL OF CHEMICAL WASTE

General Reference: *Prudent Practices in the Laboratory* National Academy Press, Washington, D.C. 2011.

Effluents from synthetic organic chemistry fall into the following categories:

1. **Gases**
    1a. Gaseous materials either used or generated in an organic reaction.
    1b. Solvent vapors generated in reactions swept with an inert gas and during solvent stripping operations.
    1c. Vapors from volatile reagents, intermediates and products.

2. **Liquids**
    2a. Waste solvents and solvent solutions of organic solids (see item 3b).
    2b. Aqueous layers from reaction work-up containing volatile organic solvents.
    2c. Aqueous waste containing non-volatile organic materials.
    2d. Aqueous waste containing inorganic materials.

3. **Solids**
    3a. Metal salts and other inorganic materials.
    3b. Organic residues (tars) and other unwanted organic materials.
    3c. Used silica gel, charcoal, filter aids, spent catalysts and the like.

The operation of industrial scale synthetic organic chemistry in an environmentally acceptable manner* requires that all these effluent categories be dealt with properly. In small scale operations in a research or academic setting, provision should be made for dealing with the more environmentally offensive categories.

1a. Gaseous materials that are toxic or noxious, e.g., halogens, hydrogen halides, hydrogen sulfide, ammonia, hydrogen cyanide, phosphine, nitrogen oxides, metal carbonyls, and the like.
1c. Vapors from noxious volatile organic compounds, e.g., mercaptans, sulfides, volatile amines, acrolein, acrylates, and the like.

---

*An environmentally acceptable manner may be defined as being both in compliance with all relevant state and federal environmental regulations *and* in accord with the common sense and good judgment of an environmentally aware professional.

2a. All waste solvents and solvent solutions of organic waste.

2c. Aqueous waste containing dissolved organic material known to be toxic.

2d. Aqueous waste containing dissolved inorganic material known to be toxic, particularly compounds of metals such as arsenic, beryllium, chromium, lead, manganese, mercury, nickel, and selenium.

3. All types of solid chemical waste.

Statutory procedures for waste and effluent management take precedence over any other methods. However, for operations in which compliance with statutory regulations is exempt or inapplicable because of scale or other circumstances, the following suggestions may be helpful.

Gases

Noxious gases and vapors from volatile compounds are best dealt with at the point of generation by "scrubbing" the effluent gas. The gas being swept from a reaction set-up is led through tubing to a large trap to prevent suck-back and into a sintered glass gas dispersion tube immersed in the scrubbing fluid. A bleach container can be conveniently used as a vessel for the scrubbing fluid. The nature of the effluent determines which of four common fluids should be used: dilute sulfuric acid, dilute alkali or sodium carbonate solution, laundry bleach when an oxidizing scrubber is needed, and sodium thiosulfate solution or diluted alkaline sodium borohydride when a reducing scrubber is needed. Ice should be added if an exotherm is anticipated.

Larger scale operations may require the use of a pH meter or starch/iodide test paper to ensure that the scrubbing capacity is not being exceeded.

When the operation is complete, the contents of the scrubber can be poured down the laboratory sink with a large excess (10-100 volumes) of water. If the solution is a large volume of dilute acid or base, it should be neutralized before being poured down the sink.

Liquids

Every laboratory should be equipped with a waste solvent container in which *all* waste organic solvents and solutions are collected. The contents of these containers should be periodically transferred to properly labeled waste solvent drums and arrangements made for contracted disposal in a regulated and licensed incineration facility.**

**If arrangements for incineration of waste solvent and disposal of solid chemical waste by licensed contract disposal services are not in place, a list of providers of such services should be available from a state or local office of environmental protection.

Aqueous waste containing dissolved toxic organic material should be decomposed *in situ*, when feasible, by adding acid, base, oxidant, or reductant. Otherwise, the material should be concentrated to a minimum volume and added to the contents of a waste solvent drum.

Aqueous waste containing dissolved toxic inorganic material should be evaporated to dryness and the residue handled as a solid chemical waste.

Solids

Soluble organic solid waste can usually be transferred into a waste solvent drum, provided near-term incineration of the contents is assured.

Inorganic solid wastes particularly those containing toxic metals and toxic metal compounds used Raney nickel manganese dioxide etc. should be placed in glass bottles or lined fiber drums sealed properly labeled and arrangements made for disposal in a secure landfill.** Used mercury is particularly pernicious and small amounts should first be amalgamated with zinc or combined with excess sulfur to solidify the material.

Other types of solid laboratory waste including used silica gel and charcoal should also be packed, labeled, and sent for disposal in a secure landfill.

Special Note

Since local ordinances may vary widely from one locale to another, one should always check with appropriate authorities. Also, professional disposal services differ in their requirements for segregating and packaging waste.

Gabriel Saucy
December 13, 1927 – November 8, 2014

Gabriel Saucy, 86, a noted synthetic organic chemist, died November 8, 2014 in Audubon, Pennsylvania. During a 30-year career at Hoffmann-La Roche in Basel, Switzerland and Nutley, New Jersey, his contributions were instrumental in the development of processes for the manufacture of vitamins and fine chemicals and resulted in 123 patents and 86 publications.

Born in Schaffhausen, Switzerland, Gaby took both his undergraduate and Ph. D. studies at the Swiss Federal Institute of Technology (ETH) in Zurich, the latter in the laboratory of Nobel Prize recipient Prof. Leopold Ruzicka. His doctoral work involved steroids and sesquiterpenes and set the stage for a career highlighted by a strong interest in the elucidation of novel and efficient synthetic methodology and stereochemical control.

In 1954 he joined Hoffmann-La Roche & Co. A.G. and, as part of a team including Dr. Otto Isler and other Swiss colleagues, devised innovative and practical methods for producing vitamin A and carotenoids as well as fragrance chemicals such as alpha-isomethyl ionone for then Roche subsidiary Givaudan. A notable discovery made during this period was the [3,3]-sigmatropic rearrangement of propargyl vinyl ethers yielding allenic aldehydes and ketones, a transformation that came to be known as the Saucy-Marbet reaction.

As part of a scientist exchange program between Roche Basel and Roche Nutley, Gaby joined Dr. Leo Sternbach in Nutley where he worked on benzodiazepines, a project of great importance to the company at the time. After returning to Basel, Gaby was invited by Dr. Arnold Brossi to relocate permanently to the Nutley campus, which he did in 1964. He remained there until his retirement in 1985 as Director of Process Research and Development.

The research projects under his direction involved processes for producing vitamins C and E, contraceptive steroids, insect juvenile hormone analogs, peptides and retinoids having anti-cancer activity. He was in the forefront of employing microbiological transformations for obtaining key synthetic intermediates and asymmetric induction leading to enantiomerically pure products.

One of his early interests was finding a totally synthetic route to commercially important 19-norsteroids. Extensive studies in his laboratory led to the development of a novel condensation of chiral hydroxyl vinyl ketones with 2-methyl-cyclopentane-1,3-dione resulting predominantly in C,D-ring intermediates possessing the desired beta-methyl configuration at C-13. Another example of well-designed synthetic creativity was his utilization of chiral allylic alcohols in the Claisen rearrangement leading to alpha-tocopherol side chain synthons possessing the natural absolute configuration. He also directed synthetic work on retinoids that played a crucial role in the discovery of one of the retinoid X receptors.

After retiring from Roche in 1985, Gaby joined Bio-Mega, Inc. in Montreal, Quebec, Canada as Vice President for Research & Development overseeing the solid-phase synthesis of biologically active peptides. He then held a similar position at Harbor Branch Oceanographic Institute in Fort Pierce, Florida directing the isolation and characterization of pharmacologically active marine natural products. He was also a consultant for Bio-Technical Resources in Manitowoc, Wisconsin.

Gaby was a member of the American Chemical Society (emeritus), the Swiss Chemical Society, the New York Academy of Sciences and the Editorial Board of Organic Synthesis, Inc. where he served as Editor-in-Chief of *Organic Syntheses, Vol. 63* (1985). He helped organize the first *Asymmetric Synthesis* (now *Stereochemistry*) Gordon Conference in 1975, the *International Symposium on Carotenoids and Vitamin A* at the University of Wisconsin in 1978, and the initial *Chemistry as a Life Science Symposium* held at Rutgers University in Newark, New Jersey in 1982.

Gaby loved skiing and sailing, and was an accomplished painter and cabinet maker. His beautiful modern cabinets were made in his extensive workshop at his Essex Fells, New Jersey home.

Gabriel Saucy is survived by his widow Sonia, daughter Michele Mitchell, son Daniel and six grandchildren.

Noal Cohen
Percy Manchand

Ekkehard Winterfeldt
May 13, 1932 – October 11, 2014

Ekkehard Winterfeldt passed away on October 11, 2014 in Isernhagen, Germany. Recognized as a scholar and a gentleman, his passion, excitement, and enthusiasm about organic chemistry were on display to all those who had the pleasure to work with him.

Winterfeldt was born on May 13, 1932 in Danzig. Upon completion of the Second World War, he finished his high-school education in 1952 in Schleswig. He initiated studies in chemistry at the Technische Hochschule Braunschweig, and joined the research group of Ferdinand Bohlmann. He completed a diploma thesis on the synthesis of oenanthotoxin, and completed his Ph.D. dissertation on the synthesis of hydroxysparteines in 1958. He moved with Bohlmann to the Technische Universität (TU) Berlin where he completed his habilitation in 1962, working on the structural characterization and synthesis of lupin alkaloids. He accepted a position at the University of Hannover (now Leibniz Universität Hannover) in 1970 where he served as Director of the Institut für Organische Chemie until his retirement in 2000. He continued as Professor Emeritus until his death in 2014.

Early on at Berlin, he was interested in the synthesis of alkaloids, and completed the first total syntheses of a variety of indole alkaloids, such as akuammigine, tetrahydroalstonine, and ajmalicine. He also started working on stereoselective transformations of triple bonds, which resulted in a variety of practical protocols still employed today.

In Hannover he continued with his lifelong passion, the stereoselective syntheses of alkaloids. Syntheses of camptothecin, geissoschizine, and natural products of the indoloquinolizine family were followed by syntheses of the histrionicotoxins, toxins isolated from South American frogs, and a wide variety of cyclopentane-containing natural products. For syntheses of prostaglandins, carbacyclins, macrolides, triqinanes, and sesquiterpenes, a stereoselective approach based on 4-acetoxy-2-cyclopentene-1-one as the substitute for cyclopentadienone was developed in in his laboratories. During the 1980's, a widely used synthetic route to ansa-steroids was developed. His work using Diels–Alder reactions introduced the concept of using chiral dienes for the introduction of chirality into achiral molecules. Syntheses of natural products such as didemnones or clavularin A emanated from this protocol. His studies on the synthesis of unsymmetrical bissteroidal pyrazines laid the groundwork for the synthesis of the cephalostatins.

In addition to his scholarly contributions, Winterfeldt earned a reputation as a passionate, energetic, and dedicated educator. Almost 200 Ph.D. students were trained under his supervision, and he published more than 240 papers with these coworkers. His early morning lectures, presented with energy and a very human face, were particularly memorable.

He served as a co-Editor of several scientific journals, including *Organic Syntheses*, where he served on the Board of Editors from 1986–1991. During his long and distinguished career Ekkehard Winterfeldt received numerous awards including the Emil Fischer Medal (1990), the Adolf Windaus Medal (1993), and the Richard Kuhn Medal (1995), and an honorary doctorate from the University of Liege (1991). He also was extraordinary in his service to the field of Chemistry, a true ambassador for Chemistry. Among his achievements, he was elected President of the Gesellschaft Deutscher Chemiker (German Chemical Society; GDCh) 1996/1997. During his time as President he was responsible for the reorientation (Europeanization) of the GDCh journals and for establishing the "JungChemikerForum" which is now an established section of the GDCh. In 2011, he was appointed an Honorary Member of the GDCh for his contributions towards the advancement of chemistry and the goals of the GDCh. He was also a member of the Senate of the Deutsche Forschungsgemeinschaft and the Kuratorium of the Fonds der Chemischen Industrie.

With his service on the *Organic Syntheses* Board of Editors Ekkehard became well known personally to several of the present members of the *Organic Syntheses* family. Beginning in the late 1980's several of us including the authors had the good fortune to be hosted by Ekkehard at the Institute for Organic Chemistry in Hannover during a sabbatical leave with the generous support of the Alexander von Humboldt Stiftung. As a result of these visits, Ekkehard spent a period as a visiting professor at the University of California, Irvine in early 1990. It was always a delight to discuss chemistry with Ekkehard, as he was interested in all aspects of organic synthesis. He particularly liked chemistry that was practical, which made him a highly sought consultant in the pharmaceutical industry and an insightful member of the *Organic Syntheses* Board of Editors.

Winterfeldt was a regular presence at national and international chemistry meetings where his booming voice and thunderous laugh could easily be heard above the din of the crowd. Indeed, the halls and laboratories of the Institute regularly reverberated with his laugh when he was present as his students and colleagues will attest.

Ekkehard Winterfeldt was an innovative researcher, a true scholar, and an even finer human being. He was strongly supportive of his students and colleagues and many have gone on to prominent positions in Chemistry around the world. But more than that, he was a warm and generous man, who welcomed all with whom he had the occasion to interact both inside and outside chemistry. He cared deeply for his wife Marianne and his children, and loved good food, drink, and the company of friends. He enjoyed immensely the outdoors and his lovely wooden home in Isernhagen of which he was so proud.

Robert K. Boeckman Jr.
Rochester, NY
Larry E. Overman
Irvine, CA

PREFACE

Fashions and trends in everyday life represent fleeting fascinations with something new (or old) that some luminaries deem special and worthy. Such considerations also impact science, serving to direct attention to research areas of unproven merit. Fortunately, synthetic chemists are in general of a practical ilk, interested in problems that expand our capability to make molecules, through the development of new or better reagents and methods. Whether the new developments are truly useful is determined over time, through its examination by fellow chemists, whose successful use, especially repeated use, serves to put the stamp of approval on the new advance. So certified, the new development becomes part of the chemist's toolbox for synthesis. One of the main objectives of *Organic Syntheses* (OS) is to greatly accelerate the vetting process for chemical methods or procedures that are expected to be of use to synthetic chemists. The Board of Editors of OS, composed of prominent researchers with expertise in different facets of organic chemistry, identify procedures that demonstrate the synthesis of valuable compounds or the illustration of an important new method. After careful evaluation by the board members, procedures considered to be of value to the community are recommended for initial submission and validation.

A considerable amount of work has gone into the development of the different procedures in this volume. The transformation of a methodology that works on several milligram scale to one that is successful with 5 or more grams requires much development, and the many authors of this volume are to be congratulated for their valuable contributions to the synthesis community. What may not be as evident, however, is the enormous effort that goes into the checking of each procedure. While the term "checking" implies repeating the protocol to validate the procedure, the reality is often quite different. Even though the procedures as submitted are written with the clear objective of enabling repetition in a different laboratory, many procedure do not reproduce so easily. Indeed, it is not uncommon for a student in the checker's laboratory to repeat a procedure several times to work out the details of some of the operations. There are even instances where procedure have real issues that require modification of the experimental protocol, resulting in improved reproducibility or yield. In the course of 8 years on the Editorial Board of OS, I have observed many instances, a few in my own labs, where checkers have taken what was initially a problematic procedure and transformed it into one that was significantly better.

The present volume includes 28 useful procedures that, like all *Organic Syntheses* procedures, have been independently checked by researchers in the labs of one of the OS board members. Although the procedures are listed in the order that checking was completed, they could just as well be organized based on broad themes, similarities in the type of chemistry being demonstrated. Several areas of contemporary interest to synthetic chemists are represented through these procedures.

Transition metal-catalyzed processes have significantly impacted the development of synthetic methodology over the past few decades, and their importance is clearly reflected in the many procedures in which at least one step, generally the critical one, involves the use of a transition metal. Cross-coupling reactions are widely represented in the present volume, with five different examples illustrating the use of such processes. Millet and Baudoin, page 76, describe the preparation of a pyrrole-based phosphine ligand and illustrate its usefulness in cross-coupling through the synthesis of a 3-arylpiperidine. The procedure provides a nice demonstration of palladium-catalyzed migrative β-arylation of *N*-Boc-piperidines. Atmuri and Lubell, page 103, describe the Negishi cross-coupling between the zincate derived from alaninyl iodide and vinyl bromide, to give allyl glycine derivatives, which are building blocks for the synthesis of macrocyclic dipeptide β-turn mimics. Read, Wang, and Danheiser, page 156, describe the synthesis of phosphoryl ynamides via copper-catalyzed coupling of an alkynyl bromide with a phosphoramidate. A more typical example of transition metal-catalyzed C-N bond formation is demonstrated in the procedure by Wong, Choy, Yuen, So, and Kwong, page 195. The authors use aryl mesylates as substrates for Buchwald-Hartwig amination, as well as for Suzuki-Miyaura cross-coupling reactions. The procedure on page 237, by Busacca, Eriksson, Qu, Lee, Li, and Senanayake, describes the use of a bulky ferrocenyl phosphine to catalyze the aminocarbonylation of an arylbromide.

Nearly as popular as cross-coupling reactions are C-H functionalization processes. The definition of this class of chemistry remains unresolved, as many traditional transformations, such as Friedel-Crafts reactions and directed lithiations, are formally C-H functionalizations. The first of the C-H functionalization processes is shown on page 58, reported by Jin-Quan Yu, one of the leaders in this field, along with his co-authors Engle, Dastbaravardeh, Thuy-Boun, Wang, and Sather. The procedure demonstrates ligand-assisted ortho-palladation of an aryl acetic acid followed by a Heck reaction. Further into the volume, on page 131, is a procedure by Wang, Rauch, Lygin, Kozhushkov, and Ackermann, which illustrates the related ortho-alkenylation of an aryl acetamide-catalyzed by a ruthenium complex. The direct functionalization of the C7-position of indole is illustrated in the procedure by Amaike, Loach, and Movassaghi, page 373. The authors utilize an iridium complex to forge two carbon-boron bonds, at C2 and C7, on the indole portion of tryptophan, and then expunge the more fragile one to give the final C7-functionalized indole.

Transition metals also play a prominent role in asymmetric catalysis, as evidenced by the four procedures in the present volume. Hamilton, Sarlah, and Carreira start off the present volume with the synthesis of a binol–dibenzoazepine phosphoramidite, which is then used for the iridium-catalyzed enantioselective vinylation of allylic alcohols. A different binol-derived phosphoramidite is used by Beaton and Fillion, page 182, for the copper-catalyzed enantioselective conjugate addition of an ethyl group to alkylidene Meldrum's acid derivatives, the product of which after hydrolytic decarboxylation give chiral β-aryl-propionic acid derivatives. Another conjugate addition reaction is reported on page 247 by Holder, Shockley, Wiesenfeldt, Shimizu, and Stoltz. The authors describe the preparation of the chiral ligand (S)-tert-butylPyOx and then demonstrate its use for the 1,4-addition of an arylboronic acid to 3-methylcyclohexenone. Chen, Ding, He, and Fan, page 213, report the synthesis of a chiral iridium-diamine catalyst, which is then used for the asymmetric hydrogenation of 2-methylquinoline.

Two C-C bond-forming methodologies also benefit from transition metal catalysis. A combination of copper and gold are used as catalysts for the hydroalkylative cyclization of unactivated alkenes, as shown by Fang, Presset, Guérinot, Bour, Bezzenine-Lafollée, and Gandon, page 117. Arnold, Krainz, and Wipf, page 277, demonstrate the copper-catalyzed conjugate addition of an organozirconium species, wherein a potentially problematic alcohol is temporarily silenced through the formation of a boronate ester.

Transition metal-free chemistry comprises just under half of the procedures in the present volume. Four of these involve applications to asymmetric synthesis. Abbott, Allais, and Roush have contributed two procedures, pages 26 and 38, in this regard. The first of these describes a detailed procedure for the preparation of crystalline (diisopinocampheyl)borane. The second procedure has two parts, the first on the use of the pure reagent for a reductive syn-aldol reaction with acryloylmorpholine, and the second one on the use of the in situ generated reagent for the same reaction. Two back-to-back procedures by Boeckman, Tusch, and Biegasiewicz describe the synthesis and use of diphenylprolinol trimethylsilyl ether, the Hayashi-Jorgensen catalyst. The first procedure, page 309, details the synthesis of the catalyst, and the second one, page 320, describes its use for the enantioselective aldol reaction between isovaleraldehyde and formaldehyde.

A wide range of methodologies make up the remaining nearly dozen procedures. Deng, Wang, and Danheiser describe on page 13 the synthesis of the acetal of 3-butynal, a versatile bifunctional building block used by many synthetic chemists. On page 91, Kwon, Haussener, and Looper report the synthesis of mono-Cbz-protected guanidines. A procedure for the convenient synthesis of 3,5-dibromo-2-pyrone from coumalic acid is reported on page 148 by Cho and Cho. Kitamura and Murakami, page 171, present a convenient procedure for the synthesis of ADMP, a safe and stable diazo-transfer reagent for the conversion of primary amines to azides. A method is presented on page 227, reported by Tinnis, Lundberg, Kivijärvi, and Adolfsson, for the ZrCl$_4$-catalyzed conversion of carboxylic acids to amides. Zhao and Radosevich describe on page 267 a phosphorous (III)-mediated reductive amination of α-keto esters to amino esters. A copper-catalyzed protocol for peptide coupling is reported on page 296 by Suppo, Figueiredo, Campagne. A sulfone synthon having trioxabicyclo[2.2.2]octane orthoester group is reported on page 328 by Di Maso, St. Peter, and Shaw. A convenient procedure is reported on page 342 by Gardner, Kawamoto, and Curran for the preparation of the useful NHC-borane, 1,3-dimethylimidazoyl-2-ylidene borane. Finally, McDonald, Hendrick, Bitting, and Wang, on page 356, report the preparation of an O-benzoylhydroxylamine reagent and illustrate its use for amination of a heteroaryl-zinc, a transformation that effectively achieves the amination of an aromatic C-H bond.

I wish to thank all the submitters and checkers for the hard work and care that went into each procedure. Your dedicated effort will help our fellow chemists for years to come. I also want to thank all the board members with whom I have had the privilege of serving over the past eight years. I have learned much from the discussions and debates with regard to the merits of certain procedures or the tribulations associated with others, even when these deliberations protracted the meetings by several hours. Finally, and most importantly, I want to thank Rick Danheiser (Editor-in-Chief) and Chuck Zercher (Associate Editor) for all their help over the past eight years. Their immense dedication and tireless, behind-the-scenes work has been critical to the continuing success of *Organic Syntheses*.

VIRESH H. RAWAL
Chicago, Illinois

TABLE OF CONTENTS

Iridium-Catalyzed Enantioselective Allylic Vinylation with Potassium Alkenyltrifluoroborates   **1**
James Y. Hamilton, David Sarlah, and Erick M. Carreira*

Synthesis of 4,4-Dimethoxybut-1-yne   **13**
James Deng, Yu-Pu Wang, and Rick L. Danheiser*

Preparation of Crystalline (Diisopinocampheyl)borane   **26**
Jason R. Abbott, Christophe Allais, and William R. Roush*

Enantioselective Reductive *Syn*-Aldol Reactions of 4-Acryloylmorpholine: Preparation of (2*R*,3*S*)-3-Hydroxy-2-methyl-1-morpholino-5-phenylpentan-1-one

38

Jason R. Abbott, Christophe Allais, and William R. Roush*

Ligand-Accelerated *ortho*-C–H Olefination of Phenylacetic Acids

58

Keary M. Engle, Navid Dastbaravardeh, Peter S. Thuy-Boun, Dong-Hui Wang, Aaron C. Sather, and Jin-Quan Yu*

Palladium-catalyzed β-Selective C(sp³)–H Arylation of *N*-Boc-Piperidines

76

Anthony Millet and Olivier Baudoin*

Preparation of Mono-Cbz Protected Guanidines 91
Kihyeok Kwon, Travis J. Haussener, and Ryan E. Looper*

Preparation of *N*-(Boc)-Allylglycine Methyl Ester Using a 103
Zinc-mediated, Palladium-catalyzed Cross-coupling Reaction
N. D. Prasad Atmuri and William D. Lubell*

Copper(II) Triflate as Additive in Low Loading Au(I)- 117
Catalyzed Hydroalkylation of Unactivated Alkenes
Weizhen Fang, Marc Presset, Amandine Guérinot, Christophe Bour,
Sophie Bezzenine-Lafollée, and Vincent Gandon*

Ruthenium-Catalyzed Direct Oxidative Alkenylation of Arenes through Twofold C–H Bond Functionalization in Water: Synthesis of Ethyl (*E*)-3-(2-Acetamido-4-methylphenyl)acrylate

131

Lianhui Wang, Karsten Rauch, Alexander V. Lygin, Sergei I. Kozhushkov, and Lutz Ackermann*

Preparation of 3,5-Dibromo-2-pyrone from Coumalic Acid

148

Hyun-Kyu Cho and Cheon-Gyu Cho*

Synthesis of Phosphoryl Ynamides by Copper-Catalyzed Alkynylation of Phosphoramidates. Preparation of Diethyl Benzyl(oct-1-yn-1-yl)phosphoramidate

156

John M. Read, Yu-Pu Wang, and Rick L. Danheiser*

Synthesis of 2-Azido-1,3-dimethylimidazolinium   171
Hexafluorophosphate (ADMP)
Mitsuru Kitamura* and Kento Murakami

Asymmetric Synthesis of All-Carbon Benzylic Quaternary   182
Stereocenters via Conjugate Addition to Alkylidene
Meldrum's Acids
Eric Beaton and Eric Fillion*

Palladium-catalyzed Buchwald-Hartwig Amination and   195
Suzuki-Miyaura Cross-coupling Reaction of Aryl Mesylates
Shun Man Wong, Pui Ying Choy, On Ying Yuen, Chau Ming So,
and Fuk Yee Kwong*

Synthesis of Optically Active 1,2,3,4-Tetrahydroquinolines *via*   213
Asymmetric Hydrogenation Using Iridium-Diamine Catalyst
Fei Chen, Zi-Yuan Ding, Yan-Mei He, and Qing-Hua Fan*

Zirconium (IV) Chloride Catalyzed Amide Formation From   227
Carboxylic acid and Amine: (*S*)-*tert*-Butyl 2-(Benzylcarbamoyl)
pyrrolidine-1-carboxylate
Fredrik Tinnis, Helena Lundberg, Tove Kivijärvi, and Hans Adolfsson*

Aminocarbonylation Using Electron-rich Di-*tert*-butyl-   237
Phosphinoferrocene
Carl A. Busacca, Magnus C. Eriksson,* Bo Qu,* Heewon Lee, Zhibin Li, and
Chris H. Senanayake

Preparation of (*S*)-*tert*-ButylPyOx and Palladium-Catalyzed Asymmetric Conjugate Addition of Arylboronic Acids

247

Jeffrey C. Holder, Samantha E. Shockley, Mario P. Wiesenfeldt, Hideki Shimizu, and Brian M. Stoltz*

Phosphorus(III)-Mediated Reductive Condensation of α-Keto Esters and Protic Pronucleophiles

267

Wei Zhao and Alexander T. Radosevich*

One-pot Hydrozirconation/Copper-catalyzed Conjugate Addition of Alkylzirconocenes to Enones

277

David Arnold, Tanja Krainz, and Peter Wipf*

Dipeptide Syntheses via Activated α-Aminoesters

296

Jean-Simon Suppo, Renata Marcia de Figueiredo,* and Jean-Marc Campagne*

(S)-1,1-Diphenylprolinol Trimethylsilyl Ether 309

Robert K. Boeckman, Jr.*, Douglas J. Tusch, and Kyle F. Biegasiewicz

Organocatalyzed Direct Asymmetric α-Hydroxymethylation 320
of Aldehydes

Robert K. Boeckman, Jr.*, Douglas J. Tusch, and Kyle F. Biegasiewicz

4-Methyl-1-(2-(phenylsulfonyl)ethyl)-2,6,7- 328
trioxabicyclo[2.2.2]octane

Michael J. Di Maso, Michael A. St. Peter, and Jared T. Shaw*

1,3-Dimethylimidazoyl-2-ylidene Borane 342

Sean Gardner, Takuji Kawamoto, and Dennis P. Curran*

Copper-Catalyzed Electrophilic Amination of Heteroaromatic and Aromatic C–H Bonds via TMPZnCl•LiCl Mediated Metalation

Stacey L. McDonald, Charles E. Hendrick, Katie J. Bitting, and Qiu Wang*

Direct C7-Functionalization of Tryptophan. Synthesis of Methyl (S)-2-((*tert*-Butoxycarbonyl)amino)-3-(7-(4,4,5,5-tetramethyl-1,3,2-dioxaborolan-2-yl)-1*H*-indol-3-yl)propanoate

Kazuma Amaike, Richard P. Loach, and Mohammad Movassaghi*

Organic Syntheses

Iridium-Catalyzed Enantioselective Allylic Vinylation with Potassium Alkenyltrifluoroborates

James Y. Hamilton, David Sarlah, and Erick M. Carreira*

Eidgenössische Technische Hochschule Zürich, HCI H335, 8093 Zürich, Switzerland

Checked by Wen-bo (Boger) Liu, Seo-Jung Han, and Brian M. Stoltz

Procedure

A. *1-(Naphthalen-2-yl)prop-2-en-1-ol* (**1**). An oven-dried 250 mL round-bottomed flask, fitted with a rubber septum, is charged with a 2.5 cm Teflon-coated magnetic oval stir bar and 2-naphthaldehyde (4.69 g,

30.0 mmol, 1.0 equiv) (Note 1). The flask is flushed with nitrogen via a nitrogen line inlet and charged with anhydrous THF (60 mL) via a 60 mL syringe (Note 2). The stirred homogeneous solution (Note 3) is cooled in a dry ice-acetone bath (–78 °C bath temp), and vinylmagnesium chloride solution (20.6 mL, 33.0 mmol, 1.1 equiv) is added dropwise over 20 min via a syringe. The resulting clear yellow solution is stirred at –78 °C for 1 h and then at 0 °C in an ice bath for 30 min (Note 4). The reaction mixture is quenched with saturated aqueous NH$_4$Cl solution (30 mL) and diluted with H$_2$O (30 mL). After being warmed up to ambient temperature, the mixture is transferred to a 250 mL separatory funnel. Additional diethyl ether (20 mL) is used to assist transfer. The aqueous layer is separated and further extracted with diethyl ether (2 x 50 mL). The combined organic layers are washed with saturated aqueous NaCl solution (50 mL), dried over sodium sulfate (20 g), filtered through a 150 mL coarse porosity sintered glass funnel, and concentrated using a rotary evaporator (25 °C, 25 mmHg) to afford a yellow oil (Note 5). This crude product is purified by flash chromatography on silica gel to afford **1** (5.23–5.35 g, 95–97%) as a colorless oil (Note 6).

B. *(R)-(–)-(3,5-Dioxa-4-phospha-cyclohepta[2,1-a;3,4-a']dinaphthalen-4-yl)-dibenzo[b,f]-azepine* (**(R)-L**). An oven-dried 100 mL Schlenk flask, fitted with a rubber septum, is charged with a 2.5 cm Teflon-coated magnetic oval stir bar and *(R)-(+)-1,1□-bi(2-naphthol)* (2.29 g, 8.0 mmol, 1.0 equiv) (Note 7). The side arm of the flask, fitted with a glass stopcock, is connected to a vacuum/N$_2$ line. The flask is evacuated and refilled with nitrogen 3 times. Phosphorus trichloride (10.5 mL, 16.5 g, 15 equiv) and anhydrous *N,N*-dimethylformamide (19 µL, 18 mg, 0.24 mmol, 0.03 equiv) are added by syringes through the septum. The reaction mixture is stirred (Note 8) at 50 °C in an oil bath for 30 min, during which time it becomes a colorless homogeneous solution. Excess phosphorus trichloride is removed via vacuum distillation (Notes 9 and 10) and azeotropic removal with toluene (2 x 3 mL) under high vacuum (see photograph) (Notes 11 and 12) to afford the phosphochloridite as an oily foam. A separate oven-dried 250 mL round-bottomed flask is charged with a 2.5 cm Teflon-coated magnetic oval stir bar and 5*H*-dibenz[b,f]azepine (1.70 g, 8.80 mmol, 1.1 equiv). The flask is fitted with a rubber septum with a nitrogen line inlet and flushed with nitrogen. Anhydrous THF (47 mL) is added by syringe, and the stirred orange solution is cooled in a dry ice-acetone bath (–78 °C bath temp) (Note 13). *n*-Butyllithium (5.50 mL, 8.80 mmol, 1.1 equiv) is then added dropwise

over 10 min with a syringe, and the resulting dark blue solution is stirred at
−78 °C for 1 h. The phosphochloridite prepared as above is dissolved in
anhydrous THF (37 mL) and added to the deprotonated 5*H*-
dibenz[*b,f*]azepine solution at −78 °C dropwise via a cannula over 20 min.
Additional anhydrous THF (10 mL) is used to assist transfer. The dark blue
solution is stirred for 12 h while being gradually warmed to ambient
temperature (Note 14). Silica gel (20 g) is added to the orange reaction
mixture, and the resulting slurry is carefully concentrated by rotary
evaporation (35 °C, 25 mmHg). Flash chromatography on silica gel yields
(R)-L (2.96–3.07 g, 73–75%) as a white solid (Note 15).

  C. *(S,E)-2-(1-Phenylpenta-1,4-dien-3-yl)naphthalene* (**2**). A screw cap
100 mL cylindrical polyethylene bottle (diameter: 4.5 cm, height: 9.5 cm)
with a 4 cm Teflon-coated magnetic cylindrical stir bar is charged with
bis(1,5-cyclooctadiene)diiridium dichloride (0.725 g, 1.08 mmol, 0.04 equiv),
(R)-L (2.19 g, 4.32 mmol, 0.16 equiv) and 1,4-dioxane (34 mL) (Note 16). The
resulting dark brown solution (see photograph) is stirred for 15 min (Note
17). 1-(Naphthalen-2-yl)prop-2-en-1-ol (**1**) (4.97 g, 27.0 mmol, 1.0 equiv),
potassium *trans*-styryltrifluoroborate (8.51 g, 40.5 mmol, 1.5 equiv),
tetrabutylammonium bromide (0.87 g, 2.7 mmol, 0.1 equiv) and potassium
hydrogen difluoride (3.16 g, 40.5 mmol, 1.5 equiv) are sequentially added,

and additional 1,4-dioxane (2.0 mL) is added to rinse the reaction vessel wall. To the resulting stirred dark-yellow heterogeneous mixture is added trifluoroacetic acid (5.17 mL, 7.70 g, 67.5 mmol, 2.5 equiv) dropwise. The red heterogeneous mixture is vigorously stirred for 6 h, during which time it gradually turns light yellow (Note 18). Upon completion of the reaction, excess trifluoroacetic acid is quenched by addition of triethylamine (5 mL), and the heterogeneous mixture is filtered through a short silica pad (Note 19). The filtrate is concentrated by rotary evaporation (35 °C, 25 mmHg). The concentrate is purified by flash chromatography on silica gel to afford **2** (5.74 g, 79%) as a white solid (Notes 20 and 21).

Notes

1. The following reagents in this section were purchased from commercial sources and used without further purification: 2-naphthaldehyde (98%, Aldrich) and vinylmagnesium chloride solution (1.6 M in THF, Aldrich).
2. Anhydrous THF in all sections was obtained by passage over activated alumina under an atmosphere of argon (H_2O content <30 ppm, *Karl Fischer* titration).
3. The mixture is stirred at 900 rpm throughout the reaction.

4.   The reaction is monitored by TLC on Merck silica gel 60 F$_{254}$ TLC glass plates and visualized with UV light and KMnO$_4$ staining solution. R$_f$ (product): 0.23 (9:1 hexanes:EtOAc)

5.   Additional diethyl ether (2 x 10 mL) is used during filtration to assist transfer.

6.   The crude product is loaded onto a column (diameter: 5 cm, height: 16 cm) packed with silica gel (150 g) slurry in 9:1 hexanes:EtOAc. After 500 mL of initial elution, 100 mL fractions are collected. The desired product is obtained in fractions 3–12, which are concentrated by rotary evaporation (25 °C, 25 mmHg). The product has been characterized as follows: ^{1}H NMR (400 MHz, CDCl$_3$) δ: 2.04 (br s, 1 H), 5.25 (dt, J = 10.3, 1.4 Hz, 1 H), 5.35 (d, J = 6.0 Hz, 1 H), 5.41 (dt, J = 17.1, 1.5 Hz, 1 H), 6.13 (ddd, J = 17.1, 10.3, 6.0 Hz, 1 H), 7.47 – 7.52 (m, 3 H), 7.83 – 7.87 (m, 4 H); ^{13}C NMR (100 MHz, CDCl$_3$) δ: 75.5, 115.5, 124.6, 125.0, 126.1, 126.3, 127.8, 128.1, 128.4, 133.1, 133.4, 140.0, 140.2; IR (neat): 3358 (br), 3055, 1633, 1601, 1508, 1408, 1361, 1269, 1124, 1018, 988, 926, 819, 745 cm^{-1}; HRMS (EI): m/z calcd for C$_{13}$H$_{12}$O [M]$^+$ 184.0888, found 184.0860; Anal. calcd. for C$_{13}$H$_{12}$O: C, 84.75; H, 6.56; found: C, 84.31, H, 6.57.

7.   The following reagents in this section were purchased from commercial sources and used without further purification: (R)-(+)-1,1'-bi(2-naphthol) (98%, Combi-Blocks), phosphorus trichloride (99%, Sigma-Aldrich), anhydrous N,N-dimethylformamide (99.8%, Sigma-Aldrich), 5H-dibenz[b,f]azepine (97%, Aldrich) and n-butyllithium (1.6 M in hexane, Aldrich).

8.   The mixture is stirred at 300 rpm throughout the reaction.

9.   The reaction mixture is maintained at 50 °C throughout distillation and subsequent azeotropic removal of residual phosphorus trichloride with toluene.

10. The Schlenk flask is uncapped and quickly connected to a distillation apparatus (oven-dried and N$_2$ flushed). The receiving flask is cooled to −78 °C in a dry ice-acetone bath, and the vacuum line from the distillation head is connected to a liquid nitrogen cold trap. Upon reaching 375 mmHg (membrane pump), the pressure of the distillation apparatus is lowered carefully (50 mmHg per minute). Vigorous stirring (800 rpm) is maintained to keep the solution from rapid foaming. Upon reaching 150 mmHg, the distillation is maintained for 30 min.

11. After distillation, the system is refilled with N$_2$, and the distillation apparatus is quickly exchanged with a three-way stopcock and

connected to high vacuum line (0.9 mmHg) with liquid nitrogen cold trap. Toluene (3 mL) (≥99.7%, Fluka, ACS reagent) is added into the flask with a syringe through the three-way stopcock, followed by swirling of the mixture to completely dissolve the oily foam. Using the three-way stopcock and vigorous stirring (800 rpm), vacuum is applied carefully to avoid rapid foaming. After complete removal of toluene, another 3 mL of toluene is used to repeat the process. After the second azeotropic distillation, the vacuum is further maintained for 30 min.

12. Care should be taken for thorough removal of excess phosphorus trichloride and to avoid exposure of the air sensitive phosphochloridite to ambient atmosphere.

13. The mixture is stirred at 800 rpm throughout the reaction.

14. The reaction flask is kept in a dry ice-acetone bath that is gradually warmed to ambient temperature overnight.

15. Silica gel adsorbed with the crude product is dry-loaded onto a column (diameter: 5.5 cm, height: 22.5 cm) packed with silica gel (200 g) slurry in 2:1 hexanes:toluene (R_f(product): 0.31; visualized with UV light and KMnO$_4$ staining solution). After 500 mL of initial elution, 100 mL fractions are collected. The desired product is obtained in fractions 2–11, which are concentrated by rotary evaporation (35 °C, 25 mmHg). The purified product is stored under an inert atmosphere in the dark for long-term storage. The product has been characterized as follows: ^{1}H NMR (400 MHz, CDCl$_3$ (filtered through basic alumina)) δ: 6.58 (td, J = 7.5, 1.4 Hz, 1 H), 6.91 (d, J = 8.7 Hz, 1 H), 6.94 – 7.05 (m, 3 H), 7.12 – 7.20 (m, 1 H), 7.21 – 7.36 (m, 9 H), 7.38 – 7.46 (m, 2 H), 7.48 (d, J = 8.8 Hz, 1 H), 7.67 (d, J = 8.7 Hz, 1 H), 7.80 (dd, J = 8.2, 1.2 Hz, 1 H), 7.95 (d, J = 8.0 Hz, 1 H), 8.03 (d, J = 8.8 Hz, 1 H); ^{13}C NMR (100 MHz, CDCl$_3$) δ: 121.2 (d, J = 2.8 Hz), 121.6, 122.3 (d, J = 2.0 Hz), 124.4 (d, J = 5.3 Hz), 124.4, 124.9, 125.8, 126.2, 126.3, 126.8 (d, J = 1.4 Hz), 126.9, 127.2, 128.0, 128.4, 128.5, 128.7, 129.0, 129.10, 129.12, 129.2 (d, J = 2.0 Hz), 129.3, 130.3, 130.5, 131.5, 131.6, 131.7, 132.3 (d, J = 1.5 Hz), 133.0 (d, J = 1.6 Hz), 135.3, 136.6 (d, J = 3.6 Hz), 142.6, 143.0 (d, J = 24.0 Hz), 148.8 (d, J = 1.2 Hz), 150.0 (d, J = 8.0 Hz); ^{31}P NMR (162 MHz, CDCl$_3$) δ: 137.85; IR (neat): 3054, 3020, 1619, 1591, 1485, 1463, 1327, 1283, 1234, 1207, 1155, 1107, 1070, 982, 949, 866, 820, 802, 750 cm^{-1}; HRMS (ESI+): m/z calcd for C$_{34}$H$_{23}$NO$_2$P [M+H]$^+$ 508.1461, found 508.1464; [α]$^{20}_D$ = –325.4 (c = 1.04, CHCl$_3$); mp 249–250 °C. HPLC: >99% purity, t_R = 5.78 min (column: Eclipse Plus C8 2.1 x 50 mm, 1.8 micron; method: linear gradient of A (H$_2$O with 0.025% AcOH) and B (MeCN), flow rate of 1.0 mL/min,

254 nm detection, 25 °C column temperature, linear gradient: 40–95% B in 7 min).

16. The following reagents in this section were purchased from commercial sources and used without further purification: bis(1,5-cyclooctadiene)diiridium dichloride (97%, Combi-Blocks), potassium *trans*-styryltrifluoroborate (Sigma-Aldrich, or synthesized by the known procedure: Molander, G. A. *et al*, *J. Org. Chem.* **2002**, *67*, 8424.), tetrabutylammonium bromide (≥99%, Sigma-Aldrich), potassium hydrogen difluoride (≥99%, Sigma-Aldrich) and trifluoroacetic acid (≥99%, Sigma-Aldrich). 1,4-Dioxane (≥99.5%, Acros) was used as received.

17. The mixture is stirred at 400 rpm throughout the reaction. The reaction vessel is capped, and the reaction is performed under ambient atmosphere.

18. The reaction is monitored by TLC and visualized with UV light and KMnO$_4$ staining solution. R$_f$(SM): 0.42 (4:1 hexanes:EtOAc); R$_f$(product): 0.32 (19:1 hexane:CH$_2$Cl$_2$)

19. The reaction mixture is loaded onto a short silica pad column (diameter: 6.5 cm, height: 4.5 cm) packed with silica gel (70 g) slurry in hexanes. A mixture of 9:1 hexanes:EtOAc (2 L) is used as eluent. The filtrate is concentrated by rotary evaporation (25 °C, 25 mmHg) to yield a dark red oil.

20. The crude product is loaded onto a column (diameter: 5.5 cm, height: 22 cm) packed with silica gel (200 g) slurry in 19:1 hexanes:CH$_2$Cl$_2$. After 500 mL of initial elution, 50 mL fractions are collected. The desired product is obtained in fractions 6–41, which are concentrated by rotary evaporation (25 °C, 25 mmHg). The regioisomeric ratio is >50:1, determined by ^{1}H NMR. The product has been characterized as follows: ^{1}H NMR (400 MHz, CDCl$_3$) δ: 4.40 – 4.44 (m, 1 H), 5.21 (dt, J = 17.2, 1.5 Hz, 1 H), 5.26 (dt, J = 10.2, 1.4 Hz, 1 H), 6.22 (ddd, J = 17.0, 10.2, 6.7 Hz, 1 H), 6.44 – 6.57 (m, 2 H), 7.20 – 7.27 (m, 1 H), 7.28 – 7.35 (m, 2 H), 7.37 – 7.51 (m, 5 H), 7.72 (d, J = 1.7 Hz, 1 H), 7.78 – 7.87 (m, 3 H); ^{13}C NMR (100 MHz, CDCl$_3$) δ: 52.5, 116.1, 125.7, 126.2, 126.42, 126.44, 127.0, 127.4, 127.8, 127.9, 128.3, 128.7, 131.1, 131.8, 132.5, 133.8, 137.5, 140.1, 140.2; IR (neat): 3055, 3023, 1631, 1598, 1494, 1446, 1270, 995, 968, 914, 856, 817, 743 cm^{-1}; HRMS (EI): *m/z* calcd. for C$_{21}$H$_{18}$ [M]$^+$ 270.1408, found 270.1417; [α]$^{24}_D$ = –1.6 (c = 1.1, CHCl$_3$); mp 74–75 °C; SFC: Daicel Chiralcel OJ-H, 10% MeOH, 2.5 mL/min, 40 °C, 254 nm; >99% ee (t$_R$ (minor) = 23.5 min, t$_R$ (major) = 25.1 min). HPLC: >99%

purity, t_R = 7.41 min (column: Eclipse Plus C8 2.1 x 50 mm, 1.8 micron; method: linear gradient of A (H_2O with 0.025% AcOH) and B (MeCN), flow rate of 1.0 mL/min, 254 nm detection, 25 °C column temperature, linear gradient: 20–95% B in 10 min).

21. The reaction was also checked with half scale (13.5 mmol), and 2.44 g (67% yield) of (*S,E*)-2-(1-phenylpenta-1,4-dien-3-yl)naphthalene) was obtained with >99% ee.

Working with Hazardous Chemicals

The procedures in *Organic Syntheses* are intended for use only by persons with proper training in experimental organic chemistry. All hazardous materials should be handled using the standard procedures for work with chemicals described in references such as "Prudent Practices in the Laboratory" (The National Academies Press, Washington, D.C., 2011; the full text can be accessed free of charge at http://www.nap.edu/catalog.php?record_id=12654). All chemical waste should be disposed of in accordance with local regulations. For general guidelines for the management of chemical waste, see Chapter 8 of Prudent Practices.

In some articles in *Organic Syntheses*, chemical-specific hazards are highlighted in red "Caution Notes" within a procedure. It is important to recognize that the absence of a caution note does not imply that no significant hazards are associated with the chemicals involved in that procedure. Prior to performing a reaction, a thorough risk assessment should be carried out that includes a review of the potential hazards associated with each chemical and experimental operation on the scale that is planned for the procedure. Guidelines for carrying out a risk assessment and for analyzing the hazards associated with chemicals can be found in Chapter 4 of Prudent Practices.

The procedures described in *Organic Syntheses* are provided as published and are conducted at one's own risk. *Organic Syntheses, Inc.*, its Editors, and its Board of Directors do not warrant or guarantee the safety of individuals using these procedures and hereby disclaim any liability for any injuries or damages claimed to have resulted from or related in any way to the procedures herein.

Discussion

Transition metal-catalyzed asymmetric allylic substitution is one of the most powerful methods for enantioselective formation of carbon–heteroatom and carbon–carbon bonds.[2] Despite significant advances in this field, there are only a limited number of methods that offer high stereoselectivity and allow direct substitution of allylic alcohols without prior activation.[3]

Iridium-catalyzed allylic vinylation described above affords highly enantioenriched 1,4-dienes directly from allylic alcohols and potassium alkenyltrifluoroborates.[4] The reaction displays high regioselectivity under conditions that circumvent hazardous handling of hydrofluoric acid.[5] Furthermore, this catalytic enantioselective transformation can be conveniently performed without exclusion of air or moisture, using technical grade solvents. Alternatively, similar structural motifs can be accessed by Cu-catalyzed processes for allylic substitution of allylic phosphates with vinylaluminum and vinylboronic acid ester reagents, reported by Hoveyda and Hayashi, respectively.[6]

The described method is applicable to a range of aryl and heteroaryl allylic alcohols. Potassium alkenyltrifluoroborates as well as potassium alkynyltrifluoroborates of various substitution patterns can be successfully used under the described reaction conditions.[4,5]

References

1. Laboratorium für Organische Chemie, Eidgenössische Technische Hochschule Zürich, HCI H335, Wolfgang-Pauli-Strasse 10, CH-8093, Zürich (Switzerland). E-mail: carreira@org.chem.ethz.ch. We are grateful to the ETH Zürich and the Swiss National Science Foundation for financial support.

2. "Transition Metal Catalyzed Enantioselective Allylic Substitution in Organic Synthesis": *Topics in Organometallic Chemistry, Vol. 38* (Ed.: U. Kazmaier), Springer, Heidelberg, **2012**.

3. Sundararaju, B.; Achard, M.; Bruneau, C. *Chem. Soc. Rev.* **2012**, *41*, 4467–4483 and references therein.

4. Hamilton, J. Y.; Sarlah, D.; Carreira, E. M. *J. Am. Chem. Soc.* **2013**, *135*, 994–997.

5. Hamilton, J. Y.; Sarlah, D.; Carreira, E. M. *Angew. Chem. Int. Ed.* **2013**, *52*, 7532–7535.
6. (a) Akiyama, K.; Gao, F.; Hoveyda, A. H. *Angew. Chem., Int. Ed.* **2010**, *49*, 419–423. (b) Gao, F.; McGrath, K. P.; Lee, Y.; Hoveyda, A. H. *J. Am. Chem. Soc.* **2010**, *132*, 14315–14320. (c) Gao, F.; Carr, J. L.; Hoveyda, A. H. *Angew. Chem., Int. Ed.* **2012**, *51*, 6613–6617. (d) Shintani, R.; Takatsu, K.; Takeda, M.; Hayashi, T. *Angew. Chem., Int. Ed.* **2011**, *50*, 8656–8659.

Appendix
Chemical Abstracts Nomenclature (Registry Number)

1-(Naphthalen-2-yl)prop-2-en-1-ol: 2-Naphthalenemethanol, α-ethenyl-; (76635-88-6)

(*R*)-(–)-(3,5-Dioxa-4-phospha-cyclohepta[2,1-a;3,4-a']dinaphthalen-4-yl)-dibenzo[b,f]-azepine ((**R**)-**L**): 5*H*-Dibenz[*b,f*]azepine, 5-(11b*R*)-dinaphtho[2,1-*d*:1',2'-*f*][1,3,2]dioxaphosphepin-4-yl-; (1265884-98-7)

(*S*,*E*)-2-(1-Phenylpenta-1,4-dien-3-yl)naphthalene: Naphthalene, 2-[(1*S*,2*E*)-1-ethenyl-3-phenyl-2-propen-1-yl]-;

2-Naphthaldehyde: 2-Naphthalenecarboxaldehyde; (66-99-9)

Vinylmagnesium chloride: Magnesium, chloroethenyl-; (3536-96-7)

(*R*)-(+)-1,1'-Bi(2-naphthol): [1,1'-Binaphthalene]-2,2'-diol, (1*R*)-; (18531-94-7)

Phosphorous trichloride; (7719-12-2)

5*H*-Dibenz[*b,f*]azepine; (256-96-2)

n-Butyllithium: Lithium, butyl-; (109-72-8)

Bis(1,5-cyclooctadiene)diiridium dichloride: Iridium, di-μ-chlorobis[(1,2,5,6-η)-1,5-cyclooctadiene]di-; (12112-67-3)

Potassium *trans*-styryltrifluoroborate: Borate(1-), trifluoro[(1*E*)-2-phenylethenyl]-, potassium (1:1), (*T*-4)-; (201852-49-5)

Tetrabutylammonium bromide: 1-Butanaminium, *N,N,N*-tributyl-, bromide (1:1); (1643-19-2)

Potassium hydrogen difluoride: Potassium fluoride (K(HF$_2$)); (7789-29-9)

Trifluoroacetic acid: Acetic acid, 2,2,2-trifluoro-; (76-05-1)

Prof. Erick M. Carreira obtained a B.S. degree from the University of Illinois at Urbana-Champaign (1984) and a Ph.D. degree from Harvard University (1990). After carrying out postdoctoral work at the California Institute of Technology through late 1992, he joined the faculty at the same institution as an assistant professor of chemistry and was subsequently promoted to the rank of full professor. Since September 1998, he has been professor of chemistry at ETH Zürich. His research program focuses on asymmetric synthesis of complex natural products, the development of catalytic along with stoichiometric reactions for asymmetric stereocontrol, chemical biology, and medicinal chemistry.

James Y. Hamilton was born in South Korea. He received his Bachelor's degree in chemistry and biology from the University of Wyoming (2006) and Master's degree from the University of California, Berkeley (2008) under the supervision of Prof. Dirk Trauner. Since May 2011, he has been a doctoral student in the group of Prof. Erick M. Carreira, working on design and development of transition metal-catalyzed asymmetric reactions.

David Sarlah was born and raised in Slovenia, where he obtained his Bachelor's Degree in Chemistry (University of Ljubljana). He obtained his Ph.D. in chemistry in 2011 for research conducted under Professor K. C. Nicolaou involving the total synthesis of complex natural products. In the fall of 2011, he joined Professor Erick M. Carreira's group as a postdoctoral researcher where he is currently developing new stereoselective methods involving allylic substitution. His research interests encompass chemical synthesis, reaction design and their application to complex natural product synthesis and chemical biology.

Wen-Bo Liu was born in China and he received his Bachelor's Degree in Chemistry from the Nankai University in 2006. He obtained his Ph.D. in organic chemistry (2011) from the Shanghai Institute of Organic Chemistry (SIOC) under the supervision of Professor Li-Xin Dai and Professor Shu-Li You. Then He joined the Professor Brian M. Stoltz laboratory at Caltech as a postdoctoral scholar, working on asymmetric catalysis and sustainable chemistry.

Seo-Jung Han graduated with a B.S. in chemistry from Sogang University in 2008. She received her M.S. degree in 2010 from Sogang University under the direction of Professor Duck-Hyung Lee. She then moved to the California Institute of Technology and began her doctoral studies under the guidance of Professor Brian M. Stoltz. Her graduate research focuses on total synthesis of complex polycyclic natural products.

rganic
yntheses

Synthesis of 4,4-Dimethoxybut-1-yne

James Deng, Yu-Pu Wang, and Rick L. Danheiser*

Department of Chemistry, Massachusetts Institute of Technology,
Cambridge, MA 02139.

Checked by Yutaka Kobayashi and John L. Wood

$$
\text{≡}\diagdown\text{Br} \xrightarrow[\substack{\text{HC(OMe)}_3 \\ -78\,°C,\,15\,\text{min}}]{\substack{\text{Al, cat. HgCl}_2 \\ \text{Et}_2\text{O, reflux;}}} \text{≡}\diagdown\diagup\substack{\text{OMe} \\ \text{OMe}}
$$

1

Procedure

*4,4-Dimethoxybut-1-yne (**1**).* A 500-mL, three-necked, round-bottomed flask equipped with a 30 x 15 mm, Teflon-coated, oval magnetic stir bar, a 30-cm Friedrichs condenser fitted with a nitrogen inlet adapter, a 125-mL pressure-equalizing addition funnel fitted with a rubber septum, and a glass stopper (Note 1) is charged with aluminum powder (6.56 g, 243 mmol, 1.5 equiv) and $HgCl_2$ (0.220 g, 0.810 mmol, 0.5 mol%) (Note 2) by temporarily removing the glass stopper. The glass stopper is replaced with a rubber septum fitted with a thermocouple temperature probe, and Et_2O (40 mL) (Note 3) is added through the addition funnel. The mixture is heated at reflux (preheated oil bath, temperature 50 °C) for 10 min (Note 4) and then allowed to cool to room temperature by removing the oil bath. The addition funnel is charged with a solution of propargyl bromide (25.1 g, 211 mmol, 1.3 equiv) (Note 5) in 80 mL of Et_2O (Note 6). The propargyl bromide solution is cautiously added dropwise to the reaction flask (Note 7) over 1.5 h at a rate that maintains the reaction mixture at gentle reflux (see photo) (Note 8).

Org. Synth. **2015**, *92*, 13-25
DOI: 10.15227/orgsyn.92.0013

Published on the Web 02/12/2015
© 2015 Organic Syntheses, Inc.

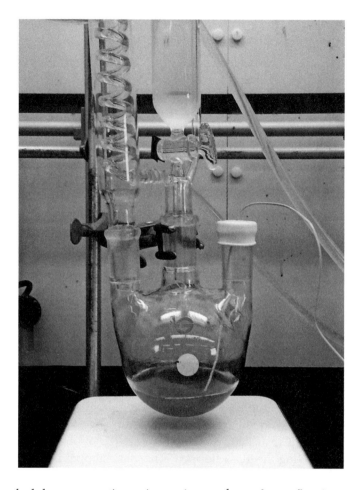

The dark brown reaction mixture is next heated at reflux in an oil bath for an additional hour and then allowed to cool to room temperature. Stirring is halted to allow the excess aluminum powder to settle and the brown supernatant solution is transferred via cannula to a 1-L, three-necked, round-bottomed flask equipped with a 30 x 15 mm, Teflon-coated, oval magnetic stir bar, a rubber septum, a 125-mL pressure-equalizing addition funnel fitted with a rubber septum, and a nitrogen inlet adapter (see photo). The solids in the original flask are rinsed with two 100-mL portions of Et$_2$O, which are transferred via cannula to the 1-L flask. The cannula is then removed from the rubber septum and replaced with a thermocouple temperature probe.

The dark green-brown solution (Note 9) is cooled to –73 °C (internal temperature) using an acetone/dry ice bath, and a solution of trimethyl orthoformate (HC(OMe)₃) (17.9 mL, 17.2 g, 162 mmol, 1.0 equiv) (Note 10) in 30 mL of Et₂O is added via the addition funnel over 50 min while the internal temperature of the reaction mixture is kept below –65 °C. The addition funnel is rinsed with 5 mL of Et₂O. The resulting gray suspension is stirred at –73 °C (internal temperature) for 15 min and 160 mL of deionized water is then added via the addition funnel over 45 min (Note 11). The acetone/dry ice bath is replaced by a room temperature water bath and the reaction mixture is allowed to warm to 20 °C over 30 min.

The contents of the flask are transferred to a 1-L separatory funnel. The aqueous layer is separated and extracted with Et₂O (4 x 50 mL). The combined organic layers are washed with 5 M aqueous NaOH solution (140 mL), deionized water (2 x 100 mL), and saturated NaCl solution (100 mL), dried over 20 g of K₂CO₃, and filtered through filter paper into a 1-L round-bottomed flask containing a 30 x 15 mm, Teflon-coated, oval magnetic stir bar. The solution is concentrated to a volume of approximately 40 mL by distillation at 34–35 °C through a distillation head with a 10-cm condenser side arm. The resulting yellow solution is transferred to a 100-

mL, round-bottomed flask containing a 13 x 8 mm Teflon-coated magnetic stir bar and equipped with a Perkin triangle. Distillation of the residual solvents is carried out at 20 °C, 300 mmHg and subsequently the pressure is reduced to 75 mmHg and the mixture heated in a 90 °C oil bath. A forerun of ca. 4 mL (bp 20–62 °C, 75 mmHg) and a main fraction containing the desired product of ca. 15 mL (bp 62–67 °C, 75 mmHg) are collected (see photo).

The distillation head is rinsed with 5 mL of pentane and the washings are combined with the forerun. The resulting solution and the residue in the original distillation flask (ca. 4 mL) are combined, transferred to a 25-mL, round-bottomed flask containing a 13 x 8 mm, Teflon-coated magnetic stir bar and equipped with a Perkin triangle, and distilled at 75 mmHg. After the volatile solvents have evaporated, a forerun of ca. 1 mL (bp 20-62 °C, 75 mmHg) and a main fraction containing the desired product of ca. 1 mL (bp 62-67 °C, 75 mmHg) are collected. The main fractions from each distillation are combined to provide a total yield of 14.3–14.6 g of 4,4-dimethoxybut-1-yne (1) (77–79%) as a colorless liquid (Note 12).

Notes

1. All glassware was flame-dried under vacuum (2 mmHg) and backfilled with nitrogen while hot, and then maintained under an atmosphere of nitrogen during the course of the reaction.
2. The checkers obtained aluminum (99.5%, -100+325 mesh powder) from Alfa Aesar and $HgCl_2$ (≥99.5%, ACS reagent ISO) from Sigma-Aldrich. The submitters obtained $HgCl_2$ (≥99.5%, ACS reagent) from Fluka. Both reactants were used as received. The submitters note that small amounts (ca. 3%) of unreacted propargyl bromide were observed when the reaction was carried out using old batches of aluminum powder containing visible clumps.
3. The checkers purchased diethyl ether from Fisher Scientific that was then purified employing a Glass Contour Solvent Purification System. The submitters note that diethyl ether (ultra low water) was purchased from J. T. Baker and purified by pressure filtration through activated alumina prior to use.
4. Heating with mercuric chloride serves to activate the aluminum powder. No visible change in the appearance of the reaction mixture is observed.
5. Propargyl bromide (>97%, stabilized with MgO) was purchased from TCI America and filtered under argon immediately prior to use through an inverted 16-gauge needle with the Luer end wrapped with filter paper and the needle end in a pre-dried 200-mL, round-bottomed flask

(see photos). Small amounts of residual suspended MgO in the filtered solution do not affect the reaction. The submitters note that alternatively, a commercial toluene solution of propargyl bromide can also be used, but the purification of final product by distillation then requires a fractionating column to remove toluene.

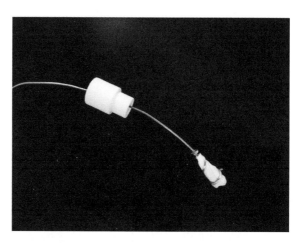

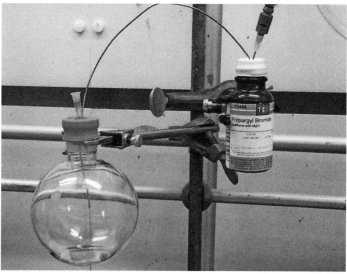

6. The propargyl bromide in the 200-mL round-bottomed flask is diluted with 70 mL of diethyl ether and transferred to the addition funnel with the aid of two 5-mL rinses of diethyl ether.

7. It is advisable to first add ca. 0.5 mL of solution dropwise and then halt the addition to ensure that there is not a significant initial exotherm.

8. The internal temperature increases to the boiling point of ether over 15 min and remains at that temperature for the duration of the addition. The addition should be temporarily stopped and the reaction flask should be placed in an ice/water bath if reflux becomes too vigorous.

9. A small amount of suspended aluminum powder may be present. This does not affect the reaction.

10. Trimethyl orthoformate (98%) was obtained from Sigma-Aldrich and distilled at 99–101 °C (760 mmHg) under argon prior to use.

11. The water that is added freezes to become a chunk of ice, which gradually melts and mixes with the reaction mixture upon warming to room temperature (see photo).

12. 4,4-Dimethoxybut-1-yne (**1**) has the following physical and spectroscopic properties: bp 62–67 °C (75 mmHg); ^1H NMR (CDCl$_3$, 600 MHz) δ 2.03 (t, J = 2.7 Hz, 1 H), 2.53 (dd, J = 5.6, 2.7 Hz, 2 H), 3.38 (s, 6 H), 4.55 (t, J = 5.6 Hz, 1 H). ^{13}C NMR (CDCl$_3$, 150 MHz) δ: 23.7, 53.4, 70.2, 79.3, 102.3. HRMS (ESI) m/z: Calcd. for C$_6$H$_{10}$O$_2$ [M+Na]$^+$: 137.0573, found: 137.0573. IR (cm^{-1}): 3289, 2938, 2834, 2124, 1447, 1423, 1362, 1240, 1192, 1120, 1064. Anal. calcd for C$_6$H$_{10}$O$_2$: C, 63.14; H, 8.83. Found: C, 62.86; H, 8.86.

Working with Hazardous Chemicals

The procedures in *Organic Syntheses* are intended for use only by persons with proper training in experimental organic chemistry. All hazardous materials should be handled using the standard procedures for work with chemicals described in references such as "Prudent Practices in the Laboratory" (The National Academies Press, Washington, D.C., 2011; the full text can be accessed free of charge at http://www.nap.edu/catalog.php?record_id=12654). All chemical waste should be disposed of in accordance with local regulations. For general guidelines for the management of chemical waste, see Chapter 8 of Prudent Practices.

In some articles in *Organic Syntheses*, chemical-specific hazards are highlighted in red "Caution Notes" within a procedure. It is important to recognize that the absence of a caution note does not imply that no significant hazards are associated with the chemicals involved in that procedure. Prior to performing a reaction, a thorough risk assessment should be carried out that includes a review of the potential hazards associated with each chemical and experimental operation on the scale that is planned for the procedure. Guidelines for carrying out a risk assessment and for analyzing the hazards associated with chemicals can be found in Chapter 4 of Prudent Practices.

The procedures described in *Organic Syntheses* are provided as published and are conducted at one's own risk. *Organic Syntheses, Inc.*, its Editors, and its Board of Directors do not warrant or guarantee the safety of individuals using these procedures and hereby disclaim any liability for any injuries or damages claimed to have resulted from or related in any way to the procedures herein.

Discussion

4,4-Dimethoxybut-1-yne (1) is a valuable difunctional synthetic building block which incorporates a masked aldehyde carbonyl group and terminal alkyne within a four-carbon chain. To date this versatile synthon has found application as a key intermediate in the synthesis of heterocyclic compounds, functionalized alkenes, and several classes of biomolecules including carbohydrates, polyketides, and lipids. For example, Jung and Gardiner employed dimethoxybutyne 1 in their approach to the stereoselective synthesis of carbohydrates,[2] and Tan and coworkers applied this compound in their stereocontrolled strategy for the construction of six-carbon polyketide fragments.[3] Vanderwal et al. utilized dimethoxybutyne 1 in their synthesis of chlorosulfolipids,[4] and Hénaff and Whiting have used this compound as an intermediate in a stereocontrolled route to complex polyenes.[5] 4,4-Dimethoxybut-1-yne was also employed as a key building block in the tandem benzannulation/cyclization strategy for the synthesis of highly substituted indoles developed in our laboratory.[6] The analogous *diethoxy* acetal derivative of butynal has found application in the total synthesis of several natural products and heterocyclic systems.[7]

Durand reported the first synthesis of 4,4-diethoxybutyne (2) in 1961.[8] As outlined in Scheme 1, alkylation of lithium acetylide with bromoacetaldehyde diethyl acetal provides the desired product in a single step. Unfortunately, the yield for this reaction is reported to be low and the product proved difficult to separate from unreacted bromoacetaldehyde acetal.

Scheme 1. Synthesis of 4,4-Diethoxybut-1-yne (Durand, 1961)

$$\text{Br}\!\!\bigwedge\!\!\substack{\text{OEt}\\\text{OEt}} \quad \xrightarrow[\text{then dioxane, reflux, 10 h}]{\substack{\text{1.25 equiv LiNH}_2\\\text{acetylene, liq NH}_3}} \quad \text{HC}\!\!\equiv\!\!\bigwedge\!\!\substack{\text{OEt}\\\text{OEt}}$$

Makin et al. reported a three-step synthesis of 4,4-dialkoxybut-1-ynes in 1981 that begins with the Lewis acid catalyzed self-condensation of alkyl vinyl ethers (Scheme 2).[9] Addition of molecular bromine and double dehydrobromination with excess sodamide in liquid ammonia then affords 1 and 2 in good overall yield.

Scheme 2. Synthesis of 4,4-Dialkoxybut-1-ynes (Makin, 1981)

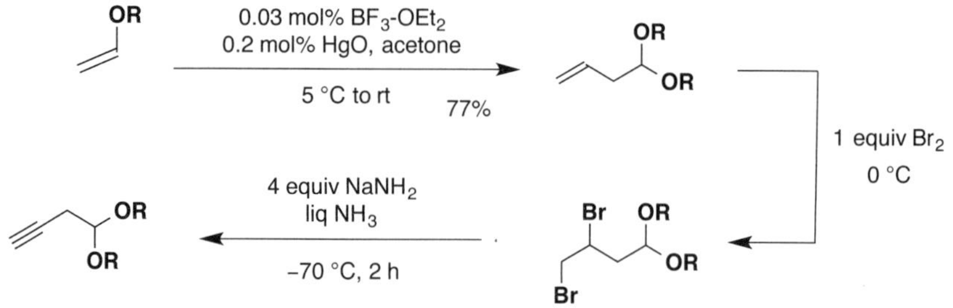

1 R = Me 62% over two steps
2 R = Et 63% over two steps

A third approach to the synthesis of 4,4-dialkoxybutynes is based on the substitution reaction of propargylic organometallic compounds with orthoformate esters. The reaction of Grignard reagents with orthoformates is a well established method for the preparation of acetals,[10] and several laboratories have described the application of this process to the synthesis of 4,4-dialkoxybutynes. In 1981, Miginiac et al. reported that the reaction of Grignard reagents with certain mixed orthoformates takes place at room temperature rather than at the elevated temperatures required for reactions with *trialkyl* orthoformates.[11] The synthesis of 4,4-diethoxybut-1-yne (**2**) was achieved in 55% yield by reaction of the Grignard derivative of propargyl bromide with phenyl diethyl orthoformate according to this method.[11] In a similar fashion, Quintard and coworkers prepared **2** in 74% yield by using acetyl diethyl orthoformate (AcOCH(OEt)$_2$).[12] In this case, however, the organoaluminum compound obtained from propargyl bromide was found to give superior results as compared to organomagnesium and organozinc compounds and was therefore employed for the reaction. This observation was consistent with earlier reports that organoaluminum compounds often react in higher yield with orthoesters as compared to the corresponding magnesium and zinc compounds. The rates of reaction of the organoaluminum compounds were also shown to be considerably faster, with the addition to orthoesters generally taking place rapidly at low temperature.

A drawback of the procedures outlined above is that the mixed orthoesters are ca. 50-100 times as expensive as simple *trialkyl* orthoformates. In 1986, Miginiac reported that the aluminum derivative of propargyl bromide reacts with inexpensive trimethyl orthoformate at low temperature to afford 4,4-dimethoxybut-1-yne in good yield.[13] In our view, this reaction provides the most economical and convenient method for the preparation of 4,4-dimethoxybut-1-yne and is thus the basis for the procedure described in the present *Organic Syntheses* article. This method involves a single step and begins with relatively inexpensive starting materials. The product acetal can be purified by distillation and can then be stored neat without detectable decomposition at 0 °C under inert atmosphere.

References

1. Department of Chemistry, Massachusetts Institute of Technology, Cambridge, MA 02139. Email: danheisr@mit.edu. We thank the National Science Foundation (CHE-1111567) for generous financial support.
2. Jung, M. E.; Gardiner, J. M. *Tetrahedron Lett.* **1994**, *35*, 6755–6758.
3. Shang, S.; Iwadare, H.; Macks, D. E.; Ambrosini, L. M.; Tan, D. S. *Org. Lett.* **2007**, *9*, 1895–1898.
4. Shibuya, G. M.; Kanady, J. S.; Vanderwal, C. D. *J. Am. Chem. Soc.* **2008**, *130*, 12514–12518.
5. Hénaff, N.; Whiting, A. *J. Chem. Soc. Perkin Trans. 1* **2000**, *3*, 395–400.
6. Lam, T. Y.; Wang, Y. P.; Danheiser, R. L. *J. Org. Chem.* **2013**, *78*, 9396–9414.
7. (a) Trost, B. M.; Weiss, A. H. *Angew. Chem., Int. Ed.* **2007**, *46*, 7664–7666. (b) Hénaff, N.; Whiting, A. *Tetrahedron* **2000**, *56*, 5193–5204. (c) Inack-Ngi, S.; Rahmani, R.; Commeiras, L.; Chouraqui, G.; Thibonnet, J.; Duchêne, A.; Abarbri, M.; Parrain, J. L. *Adv. Synth. Catal.* **2009**, *351*, 779–788. (d) Krafft, M. E.; Scott, I. L.; Romero, R. H.; Feibelmann, S.; Van Pelt, C. E. *J. Am. Chem. Soc.* **1993**, *115*, 7199–7207.
8. Durand, M. H. *Bull. Soc. Chim. Fr.* **1961**, *12*, 2396–2401.
9. Raifel'd, Yu. E.; Kusikova, I. P.; Zil'berg, L. L.; Arshava, B. M.; Limanova, O. V.; Makin, S. M. *J. Org. Chem. USSR (Engl. Transl.)* **1981**, *17*, 1425–1429; Raifel'd, Yu. E.; Kusikova, I. P.; Zil'berg, L. L., Arshava, B. M.; Limanova, O. V.; Makin, S. M. *Zh. Org. Khim.* **1981**, *17*, 1605–1609.

10. (a) Dewolfe, R. H. *Carboxylic Ortho Acid Derivatives*; Academic Press: New York, 1970; pp 224–230. (b) MacPherson, D. T.; Rami, K. R. In *Comprehensive Organic Functional Group Transformations*; Katritzky, A. R., Meth-Cohn, O., Rees, C. W., Eds.; Pergamon: Oxford, U.K., 1995; Vol. 4, p 192. (c) von Angerer, S.; Warriner, S. L. In *Science of Synthesis*; Warriner, S. L., Ed.; Thieme: Stuttgart, 2007; Vol. 29, pp 308–309.
11. (a) Barbot, F.; Miginiac, P. *J. Organomet. Chem.* **1981**, *222*, 1–15. (b) Barbot, F.; Poncini, L.; Randrianoelina, B.; Miginiac, P. *J. Chem. Res., Synop.* **1981**, *11*, 343.
12. Beaudet, I.; Duchêne, A.; Parrain, J. L.; Quintard, J. P. *J. Organomet. Chem.* **1992**, *427*, 201–212.
13. Picotin, G.; Miginiac, P. *Chem. Ber.* **1986**, *119*, 1725–1730.

Appendix
Chemical Abstracts Nomenclature (Registry Number)

4,4-Dimethoxybut-1-yne: 1-butyne, 4,4-dimethoxy-; 4,4-Dimethoxy-1-butyne; 4,4-Dimethoxybut-1-yne; (33639-45-1)

Trimethyl orthoformate: Methane, trimethoxy-; (149-73-5)

Propargyl bromide: 1-Propyne, 3-bromo-; (106-96-7)

Aluminum (powder) (7429-90-5)

Mercuric Chloride: Mercury dichloride; (7487-94-7)

Rick L. Danheiser received his undergraduate education at Columbia where he carried out research in the laboratory of Professor Gilbert Stork. He received his Ph.D. at Harvard in 1978 working under the direction of E. J. Corey on the total synthesis of gibberellic acid. Dr. Danheiser is the A. C. Cope Professor of Chemistry at MIT where his research focuses on the design and invention of new annulation and cycloaddition reactions, and their application in the total synthesis of biologically active compounds.

Organic
Syntheses

James Deng was born in New Haven, Connecticut and is currently an undergraduate chemistry major at the Massachusetts Institute of Technology and is expecting to graduate with a B.S. degree in chemistry in 2017. James joined the laboratory of Professor Danheiser as an undergraduate researcher in January 2014 and is currently investigating a synthetic route to bioactive polyacetylene natural products.

Yu-Pu Wang was born in Taipei, Taiwan. He received a B.S. degree in Chemistry in 2009 from Rice University working in the laboratory of Professor James M. Tour. He is currently pursuing a Ph.D. degree at the Massachusetts Institute of Technology in the research group of Professor Rick L. Danheiser and his work involves the development of new methods for the synthesis of highly substituted indoles and their application to the synthesis of natural products and polycyclic systems with interesting electronic properties.

Yutaka Kobayashi received a B.S. in chemistry from Kitasato University in 2008. He received a Ph.D. in Life Sciences from Kitasato University under the guidance of Professor Toshiaki Sunazuka. He is currently a post-doctoral fellow in the laboratory of Professor John L. Wood at Baylor University.

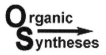

Preparation of Crystalline (Diisopinocampheyl)borane

Jason R. Abbott, Christophe Allais, and William R. Roush*[1]

Department of Chemistry, The Scripps Research Institute–Florida, 130 Scripps Way #3A2, Jupiter, FL 33458

Checked by Simon Krautwald, Simon Breitler, and Erick M. Carreira

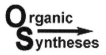

Procedure

A. *(+)-(Diisopinocampheyl)borane ((+)-(Ipc)₂BH) or ((ᴵIpc)₂BH) (2)*. A flame-dried 250-mL, two-necked, round-bottomed flask is equipped with a 4-cm Teflon-coated egg-shaped magnetic stir bar, a rubber septum, a thermometer (Note 1), and an argon line. The flask is charged with tetrahydrofuran (THF) (80 mL) and borane-methyl sulfide complex (Note 2) (8.2 mL, 6.47 g, 80.1 mmol, 1.00 equiv) is added via syringe. The mixture is cooled to 0 °C (Note 3) with an ice/water bath and (–)-(α)-pinene (1) (25.5 mL, 22.3 g, 160.2 mmol, 2.00 equiv) (Note 4) is added over 30 min using a syringe pump. Upon complete addition, the stirring is terminated, the thermometer replaced with a rubber septum, the argon line removed, and the septa are wrapped thoroughly with Parafilm®. The reaction flask is then placed in a 0 °C ice/water bath in a 4 °C cold room for 46 h (Note 5). After this time, the flask is allowed to warm to room temperature, the Parafilm® is discarded, and the supernatant is removed via cannula. Trituration of the residual chunks of (Ipc)₂BH is performed by introduction of diethyl ether (50 mL) via syringe and subsequent removal of the

supernatant by cannula. The trituration process is repeated two additional times before the cannula is removed and replaced with a needle attached to a vacuum line. The white crystals of (Ipc)₂BH are allowed to dry at <5 mmHg for 2 h. After this time, the flask is back-filled with argon, the septa are wrapped thoroughly with Parafilm®, and the flask is moved into a glovebox. Once inside, the chunks of (Ipc)₂BH are pulverized using a spatula and stored at –20 °C (Note 6). This procedure

provides 10.2–12.1 g (45–52%) of 97% pure (+)-(diisopinocampheyl)borane ((ᵈIpc)₂BH) (2) (Notes 7 and 8) as a fine white powder (Note 9). The enantiomeric purity of the crystalline (+)-(diisopinocampheyl)borane ((ᵈIpc)₂BH) (2) was determined to be 97% ee by Mosher ester analysis of the (+)-isopinocampheol produced by oxidation of 2 with sodium perborate (Note 10).

Notes

1. The submitters used a single-necked flask and monitored the internal temperature of the reaction mixture using an Oakton Instruments Temp JKT temperature meter with a Teflon-coated thermocouple probe (30.5 cm length, 3.2 mm outer diameter, temperature range –250 to 400 °C).

2. THF (HPLC Grade) and diethyl ether (Certified ACS, stabilized with BHT) were obtained from Fisher Scientific and purified by passage through activated alumina using a GlassContour solvent purification system.[2] Borane-methyl sulfide complex (94%) was obtained from Acros Organics and used as received. (–)-(α)-Pinene (1) (98%, ≥81% ee) and (+)-(α)-pinene (98%, ≥91% ee) were obtained from Aldrich Chemical Co., Inc. and used as received.

3. The internal temperature of the reaction mixture remained between 0.3 and 1.1 °C throughout the course of the reaction.

4. Due to the viscosity of (–)-(α)-pinene, it is recommended that a large-gauge (16-18) needle be used.

5. As reported by Brown and Singaram,[3] it is imperative that the crystallization be carried out at 0 °C. The submitters observed a significant decrease in yield (from 64–66% at 0 °C to 31% at –18.5 °C) with no discernable increase in reagent purity when the crystallization was carried out at –18.5 °C for 46 h. As shown in the companion article, the checkers found that the yield can vary between batches of borane-methylsulfide even from the same supplier (57-60% instead of 45-52%).

6. (+)-(Diisopinocampheyl)borane (($Ipc)_2BH$) stored in this way remains stable for periods of >1 year.

7. The sample for analysis was prepared as a solution in anhydrous, degassed d-THF in a J. Young NMR tube in a nitrogen-filled glovebox. It is important to use anhydrous NMR solvent as the presence of adventitious water will lead to hydrolysis of the borane. ^1H, ^{13}C and ^{11}B NMR spectra of the hydrolysis product, resulting from addition of water to a solution of 2 in anhydrous d-THF, are provided for the benefit of the user. The detailed appearance of the spectra of the hydrolysis product depends on the quantity of water added; representative spectra are included. Compound 2 exists as a mixture of species in anhydrous solution, and the utility of NMR spectroscopy in establishing the identity and purity of 2 is, therefore, limited.

Additional information on NMR spectroscopy of **2** can be found in the text of a report by Fürstner.[4] The signals corresponding to methyl groups in the ^{1}H NMR spectrum of **2** are tabulated as single peaks in the range of 0.85–1.27 ppm without assignment of their multiplicity: Crystalline (^{d}Ipc)$_2$BH (**2**) exhibits the following properties: mp 95–98 °C; ^{1}H NMR (500 MHz, d-THF) δ: 0.85, 0.87, 0.89, 0.91, 0.92, 0.93, 0.94, 0.96, 0.97, 0.99, 1.00, 1.02, 1.04, 1.05, 1.06, 1.07, 1.09, 1.12, 1.13, 1.14, 1.15, 1.17 (2), 1.19, 1.21, 1.23, 1.24, 1.27, 1.64 (m), 1.65–2.45 (m), 5.18 (m); ^{13}C NMR (125 MHz, d-THF) δ: 21.3, 22.6, 22.9, 23.1, 23.2, 23.3 (2), 23.4 (2), 23.7, 25.5, 26.8, 26.9, 27.6, 29.0, 29.2, 29.4, 30.1, 31.9, 32.2, 32.3, 32.6, 34.0, 34.7, 35.1, 35.6, 37.0, 38.9, 39.7, 39.8, 40.2, 40.3, 40.9, 41.6, 41.9, 42.6, 43.1 (2), 43.2, 43.3, 48.1, 49.6, 49.7, 49.9, 50.2, 117.0, 145.4. The sample for melting point determination was sealed in a capillary tube under argon.

8. The submitters were unsuccessful in repeated attempts to obtain acceptable mass spectral data and combustion analytical data for crystalline (Ipc)$_2$BH (**2**).   Under GCMS (EI) conditions, the only observable peak was due to α-pinene (presumably from retro-hydroboration in the GC). Attempted HRMS analyses (performed at the University of Illinois Mass Spectroscopy Center) under a range of conditions (ESI+, ESI-, CI, MALDI) did not provide any mass fragments that could be attributed to (Ipc)$_2$BH. Finally, attempted combustion analysis (also performed at the University of Illinois; samples sealed under argon) gave acceptable values for hydrogen and boron, but consistently gave low carbon analyses. Calcd for C$_{20}$H$_{35}$B (or C$_{40}$H$_{35}$B$_2$ for the dimer): C, 83.90%; H, 12.32%; B, 3.78%. Found, C, 80.97%; H, 12.19%; B, 3.70%.   The checkers obtained the following results: C, 83.40%; H, 12.28%.

9. (–)-(Diisopinocampheyl)borane ((^{d}Ipc)$_2$BH) has also been synthesized by the submitters using the same procedure starting from (+)-(α)-pinene (98%, ≥91% ee) (Note 2) affording 17.46 g (76%) of white crystals. The enantiomeric purity of the crystalline (–)-(diisopinocampheyl)borane ((^{d}Ipc)$_2$BH) was determined by Mosher ester analysis[5] of (–)-isopinocampheol produced by oxidation with sodium perborate.[6] The ee of the (–)-isopinocampheol thus obtained was >97%.

10. Perborate oxidation of crystalline (+)-(diisopinocampheyl)borane ((^{d}Ipc)$_2$BH) (**2**) was performed as described by Kalbaka.[6] The enantiomeric purity the (+)-isopinocampheol so obtained was determined to be 97% ee by Mosher ester analysis.[5] Thus, a mixture of (+)-isopinocampheol   (0.018   g,   0.117   mmol,   1.0   equiv)   in

dichloromethane (0.80 mL, obtained from Fisher Scientific and dried by passage through activated alumina using a GlassContour solvent purification system (see Note 2)), pyridine (0.038 mL, 0.037 g, 0.47 mmol, 4 equiv; obtained from EMD and distilled from CaH_2 under Ar) and a catalytic amount of dimethylaminopyridine (DMAP; one small crystal; obtained from Sigma-Aldrich and used as obtained) was stirred under Ar at ambient temperature. (R)-(–)-α-Methoxy-α-(trifluoromethyl)phenylacetyl chloride (0.044 mL, 0.059 g, 0.233 mmol, 2 equiv; obtained from Matrix Scientific, used as obtained) was added via syringe. The mixture was stirred at ambient temperature for 18 h, at which point TLC analysis ((9:1 CH_2Cl_2-EtOAc); R_f isopinocampheol = 0.43; R_f for Mosher ester product = 0.98) indicated that the reaction was complete. The mixture was diluted with hexanes (1 mL), filtered to remove the white precipitate, then directly filtered through a short, Pasteur pipette column of silica gel using 30 mL of 9:1 hexanes-EtOAc. The filtrate was collected as a single fraction and concentrated on a rotary evaporator to give the (S)-MTPA ester as an oil. By using the same procedure, the (R)-MPTA ester was prepared (using (S)-(+)-α-methoxy-α-(trifluoromethyl)phenylacetyl chloride, obtained from Alfa Aesar). Key resonances in the ^{19}F and ^1H NMR spectra of the diastereomeric MTPA esters that may be used in making enantiomeric purity determinations are as follows. Partial data for the (S)-MTPA ester of (+)-isopinocampheol: ^{19}F ($CDCl_3$, 376 MHz) δ: –71.46; ^1H (400 MHz, $CDCl_3$) δ: 5.31 (m, 1 H), 2.66 (m, 1 H), 2.35 (m, 1 H), 2.24 (m, 1 H), 1.94 (m, 1 H), 1.85 (m, 1 H), 1.69 (m, 1 H). Partial data for the (R)-MTPA ester of (+)-isopinocampheol: ^{19}F ($CDCl_3$, 376 MHz) δ: –71.56; ^1H (400 MHz, $CDCl_3$) δ: 5.31 (m, 1 H), 2.68 (m, 1 H), 2.37 (m, 1 H), 2.12 (m, 1 H), 1.97 (m, 1 H), 1.82 (m, 2 H).

Working with Hazardous Chemicals

The procedures in *Organic Syntheses* are intended for use only by persons with proper training in experimental organic chemistry. All hazardous materials should be handled using the standard procedures for work with chemicals described in references such as "Prudent Practices in the Laboratory" (The National Academies Press, Washington, D.C., 2011; the full text can be accessed free of charge at http://www.nap.edu/catalog.php?record_id=12654). All chemical waste should be disposed of in accordance with local regulations. For general guidelines for the management of chemical waste, see Chapter 8 of Prudent Practices.

In some articles in *Organic Syntheses*, chemical-specific hazards are highlighted in red "Caution Notes" within a procedure. It is important to recognize that the absence of a caution note does not imply that no significant hazards are associated with the chemicals involved in that procedure. Prior to performing a reaction, a thorough risk assessment should be carried out that includes a review of the potential hazards associated with each chemical and experimental operation on the scale that is planned for the procedure. Guidelines for carrying out a risk assessment and for analyzing the hazards associated with chemicals can be found in Chapter 4 of Prudent Practices.

The procedures described in *Organic Syntheses* are provided as published and are conducted at one's own risk. *Organic Syntheses, Inc.*, its Editors, and its Board of Directors do not warrant or guarantee the safety of individuals using these procedures and hereby disclaim any liability for any injuries or damages claimed to have resulted from or related in any way to the procedures herein.

Discussion

(Diisopinocampheyl)borane ((Ipc)$_2$BH) is a useful chiral organoborane reagent for asymmetric synthesis. (Ipc)$_2$BH can serve as a precursor of a range of reagents, such as (Ipc)$_2$BCl and (Ipc)$_2$BOTf, that have been employed in asymmetric aldol reactions by Paterson.[7] Methanolysis of (Ipc)$_2$BH leads to (Ipc)$_2$BOMe, which is a starting material used for the synthesis of Brown's chiral allylborane[8] and crotylborane[8c,9] reagents. These reagents react with aldehydes to provide homoallylic alcohols with high enantioselectivity.[10] (Ipc)$_2$BH has been widely used in hydroboration reactions with alkenes leading, after oxidation of the resulting B–C, bond to enantioenriched secondary alcohols.[11] Hydroboration reactions of alkynes[12] and allenes[13] with (Ipc)$_2$BH have also been reported. The reaction of (Ipc)$_2$BH with α,β-unsaturated carbonyl derivatives is known to give enolborinates that can be used in aldol reactions.[14] The reductive aldol reaction[15] of 4-acryloylmorpholine with (Ipc)$_2$BH (**2**), as illustrated in the accompanying procedure,[16] leads exclusively to the Z(O)-enolborinate which reacts with a range of aldehydes to give *syn*-aldol adducts with excellent diastereoselectivity and with very high levels of enantioselectivity.[17]

Organic Syntheses has published several procedures in which (diisopinocampheyl)borane ((Ipc)$_2$BH) is generated in situ and used immediately in subsequent transformations.[6,18] Several of these procedures specify the use of >98% ee pinene, which while commercially available is very expensive. A procedure for the synthesis of crystalline (Ipc)$_2$BH, which can be prepared with high enantiomeric purity starting from either (–)-(α)-pinene (**1**) (≥81% ee) or (+)-(α)-pinene (≥91% ee) by using the protocol originally developed by Brown,[3] has not been published in *Organic Syntheses*. This procedure, described in detail above, is the preferred method for making this reagent owing to the bulk availability and very low cost of the two (α)-pinene enantiomers.

Implicit in this procedure is the significant enhancement of enantiomeric purity of (Ipc)$_2$BH, to 97% ee, in comparison to the considerably lower enantiomeric purity of the commercially available, inexpensive (α)-pinene starting materials [(–)-(α)-pinene (**1**) (≥81% ee) and (+)-(α)-pinene (≥91% ee), respectively]. The enhancement of enantiomeric purity is a consequence of the fact that a mixture of *d,d-*, *l,l-*, and *d,l-* isomers of (Ipc)$_2$BH are generated in the hydroboration reaction, and that the major *l,l-* isomer (or *d,d-* isomer, depending on the major (α)-pinene enantiomer in the commercially available starting material) is highly crystalline, whereas the meso (*d,l-*) isomer remains in solution during the crystallization process. It is also instructive to note that as long as the hydroboration reaction gives the statistical mixture of *l,l-*, *d,d-*, and *d,l-*(Ipc)$_2$BH isomers, the minor enantiomer present in the commercial (α)-pinene starting material will be selectively converted into the *d,l-* (meso) isomer of (Ipc)$_2$BH.[19]

By following the procedure described herein, crystalline (Ipc)$_2$BH (**2**) was obtained in 45-52% yield starting from (–)-(α)-pinene (**1**) (98%, ≥81% ee) with an enantiomeric purity of 97% ee as determined by Mosher ester analysis[5] of the (+)-isopinocampheol produced by oxidation of **2** with sodium perborate.[6] The yield of crystalline (Ipc)$_2$BH (76%, >97% ee; Note 8) starting from (+)-(α)-pinene (98%, ≥91% ee) is typically higher than the yield of (Ipc)$_2$BH, owing to the greater enantiomeric purity of the starting material. The colorless crystals of **2,** or its enantiomer, obtained by this procedure can be stored for months at –20 °C in a glovebox, and then weighed out in the exact amount needed for use in any reaction involving

(Ipc)$_2$BH. Alternatively, crystalline (Ipc)$_2$BH (97% ee) can be generated in a pre-tared flask and then used directly in a subsequent reaction without transfer to another reaction vessel, thereby avoiding use of a glovebox. The latter procedure is illustrated in the reductive aldol reaction described in the accompanying procedure.[16]

References

1. E-mail: roush@scripps.edu. This research was supported by the National Institutes of Health (GM038436).
2. (a) Pangborn, A. B.; Giardello, M. A.; Grubbs, R. H.; Rosen, R. K.; Timmers, F. J. *Organometallics* 1996, *15*, 1518–1520. (b) http://www.glasscontour.com/
3. Brown, H. C.; Singaram. B. *J. Org. Chem.* **1984**, *49*, 945–947.
4. Fürstner, A.; Bonnekessel, M.; Blank, J. T.; Radkowski, K.; Seidel, G.; Lacombe, F.; Gabor, B.; Mynott, R. *Chem. Eur. J.* **2007**, *13*, 8762–8783.
5. (a) Dale, J. A.; Dull, D. L.; Mosher, H. S. *J. Org. Chem.* **1969**, *34*, 2543–2549. (b) Dale, J. A.; Mosher, H. S. *J. Am. Chem. Soc.* **1973**, *95*, 512–519.
6. (a) Kabalka, G. W.; Maddox, J. T.; Shoup, T.; Bowers, K. R. *Org. Synth.* **1996**, *73*, 116–119; *Org. Synth.* **1998**, *Coll. Vol. 9*, 522–626. (b) Lane, C. F.; Daniels, J. J. *Org. Synth.* **1972**, *52*, 59–62; *Org. Synth.* **1988**, *Coll. Vol. 6*, 719–721.
7. (a) Paterson, I.; Goodman, J. M.; Lister, M. A.; Schumann, R. C.; McClure, C. K.; Norcross, R. D. *Tetrahedron* **1990**, *46*, 4663–4684. (b) Paterson, I.; Wallace, D. J.; Velázquez, S. M. *Tetrahedron Lett.* **1994**, *35*, 9083–9086.
8. (a) Brown, H. C.; Jadhav, P. K. *J. Am. Chem. Soc.* **1983**, *105*, 2092–2093. (b) Jadhav, P. K.; Bhat, K. S.; Perumal, P. T.; Brown, H. C. *J. Org. Chem.* **1986**, *51*, 432–439. (c) Brown, H. C.; Bhat, K. S.; Randad, R. S. *J. Org. Chem.* **1989**, *54*, 1570–1576.
9. (a) Brown, H. C.; Bhat, K. S. *J. Am. Chem. Soc.* **1986**, *108*, 5919–5923. (b) Brown, H. C.; Jadhav, P. K.; Bhat, K. S. *J. Am. Chem. Soc.* **1988**, *110*, 1535–1538.
10. Selected reviews of allylation and crotylation reactions of aldehydes: (a) Roush, W. R., In *Comprehensive Organic Synthesis*, Trost, B. M., Ed. Pergamon Press: Oxford, **1991**; Vol. 2, p 1. (b) Yamamoto, Y.; Asao, N. *Chem. Rev.* **1993**, *93*, 2207–2293. (c) Denmark, S. E.; Almstead, N. G., In

Modern Carbonyl Chemistry, Otera, J., Ed. Wiley-VCH: Weinheim, **2000**; p 299. (d) Chemler, S. R.; Roush, W. R., In *Modern Carbonyl Chemistry*, Otera, J., Ed. Wiley-VCH: Weinheim, **2000**; p 403. (e) Denmark, S. E.; Fu, J. *Chem. Rev.* **2003**, *103*, 2763–2794. (f) Lachance H.; Hall, D. G. *Org. React.* **2008**, *73*, 1.

11. (a) Brown, H. C.; Zweifel, G. *J. Am. Chem. Soc.* **1960**, *82*, 3222–3223. (b) Brown, H. C.; Zweifel, G. *J. Am. Chem. Soc.* **1961**, *83*, 486–487.

12. (a) Midland, M. M.; Preston, S. B. *J. Am. Chem. Soc.* **1982**, *104*, 2330–2331. (b) Torregrosa, J. L.; Baboulene, M.; Speziale, V.; Lattes, A. *Tetrahedron* **1982**, *38*, 2355–2363. (c) Bhat, N. G.; Aguirre, C. P. *Tetrahedron Lett.* **2000**, *41*, 8027–8032.

13. (a) Brown, H. C.; Narla, G. *J. Org. Chem.* **1995**, *60*, 4686–4687. (b) Flamme, E. M.; Roush, W. R. *J. Am. Chem. Soc.* **2002**, *124*, 13644–13645. (c) Chen, M.; Ess, D. H.; Roush, W. R. *J. Am. Chem. Soc.* **2010**, *132*, 7881–7883. (d) Chen, M.; Roush, W. R. *Org. Lett.* **2011**, *13*, 1992–1995. (e) Chen, M.; Roush, W. R. *J. Am. Chem. Soc.* **2011**, *133*, 5744–5747.

14. (a) Boldrini, G. P.; Mancini, F.; Tagliavini, E.; Trombini, C.; Umani-Ronchi, A. *J. Chem. Soc., Chem. Commun.* **1990**, 1680–1681. (b) Boldrini, G. P.; Bortolotti, M.; Mancini, F.; Tagliavini, E.; Trombini, C.; Umani-Ronchi, A. *J. Org. Chem.* **1991**, *56*, 5820–5826. (c) Allais, C.; Nuhant, P.; Roush, W. R. *Org. Lett.* **2013**, *15*, 3922–3925. (d) Allais, C.; Tsai, A. S. ; Nuhant, P.; Roush, W. R. *Angew. Chem. Int. Ed.* **2013**, *52*, 12888–12891.

15. Selected reviews of reductive aldol reactions: (a) Guo, H.-C.; Ma, J.-A. *Angew. Chem. Int. Ed.* **2006**, *45*, 354–366. (b) Nishiyama, H.; Shiomi, T. *Top. Curr. Chem.* **2007**, *279*, 105–137. (c) Han, S. B.; Hassan, A.; Krische, M. J. *Synthesis* **2008**, *17*, 2669–2679. (d) Garner, S. A.; Han, S. B.; Krische M. J. "Metal Catalyzed Reductive Aldol Coupling," in *Modern Reduction Methods* (Eds. P. Andersson, I. Munslow) Wiley-VCH: Weinheim, **2008**, p 387–408.

16. Abbott, J. R.; Allais, C.; Roush, W. R. *Org. Synth.* **2015**, *92*, 38–57.

17. Nuhant, P.; Allais, C.; Roush, W. R. *Angew. Chem. Int. Ed.* **2013**, *52*, 8703–8707.

18. (a) Partridge, J. J.; Chadha, N. K; Uskokovic, M. R. *Org. Synth.* **1985**, *63*, 44–50; *Org. Synth.* **1990**, *Coll. Vol. 7*, 339–345. (b) Rathke, M. W.; Millard, A. A. *Org. Synth.* **1978**, *58*, 32–35; *Org. Synth.* **1988**, *Coll. Vol. 6*, 943–946.

19. (a) Hoye, T. R.; Suhadolnik, J. C. *J. Am. Chem. Soc.* **1985**, *107*, 5312–5313. (b) Schreiber, S. L.; Schreiber, T. S.; Smith, D. B. *J. Am. Chem. Soc.* **1987**, *109*, 1525–1529. (c) Roush, W. R.; Straub, J. A.; VanNieuwenhze, M. S. *J. Org. Chem.* **1991**, *56*, 1636–1648.

Appendix
Chemical Abstracts Nomenclature (Registry Number)

(+)-(Diisopinocampheyl)borane ((+)-(Ipc)₂BH) or ((Ipc)₂BH: borane, bis[(1S,2R,3S,5S)-2,6,6-trimethylbicyclo[3.1.1]hept-3-yl]; (21947-87-5)
(–)-(Diisopinocampheyl)borane ((–)-(Ipc)₂BH) or ((Ipc)₂BH: borane, bis[(1R,2S,3R,5R)-2,6,6-trimethylbicyclo[3.1.1]hept-3-yl]; (21932-54-7)
Borane-methyl sulfide complex: boron, trihydro[thiobis[methane]]-(T-4)-; (13292-87-0)
(–)-(α)-Pinene: (1S,5S)-2,6,6-trimethylbicyclo[3.1.1]hept-2-ene; (7785-26-4)
(+)-(α)-Pinene: (1R,5R)-2,6,6-trimethylbicyclo[3.1.1]hept-2-ene; (7785-70-8)
diethyl ether; (60-29-7)

William R. Roush is Professor of Chemistry, Executive Director of Medicinal Chemistry, and Associate Dean of the Kellogg School of Science and Technology at the Scripps Research Institute–Florida. His research interests focus on the total synthesis of natural products and the development of new synthetic methodology. Since moving to Scripps Florida in 2005, his research program has expanded into new areas of chemical biology and medicinal chemistry. Dr. Roush was a member of the *Organic Syntheses* Board of Editors from 1993-2002 and was Editor of Volume 78. He currently serves on the *Organic Syntheses* Board of Directors (2003-present).

Jason R. Abbott received his B.S. in Chemistry from Northeastern University in Boston, MA. In 2008, Mr. Abbott enrolled in the Kellogg School of Science and Technology at the Scripps Research Institute–Florida to pursue his Ph. D. in Organic Chemistry. He joined the Roush Group shortly thereafter and defended his Ph. D. in early 2014.

Christophe Allais obtained his Ph. D. in 2010 from Université Paul Cézanne (Marseille, France), under the supervision of Prof. Constantieux and Prof. Rodriguez where he focused on the development of convergent and selective methods to access various heterocycles. In 2011, he joined Prof. Roush's Group as a research associate, expanding his research into the areas of medicinal chemistry, natural product synthesis, and the development of boron-mediated asymmetric methodologies. In March 2014, he joined Pfizer (Groton, CT) as a Senior Scientist.

Simon Breitler, born in Basadingen, Switzerland, studied chemistry at ETH Zurich, which he concluded with a M. Sc. degree in 2011. During his undergraduate education, he carried out research projects in the laboratories of Prof. Erick M. Carreira and Prof. Antonio Togni. After an internship as a research trainee at Syngenta Crop Protection, Stein, Switzerland, he completed his studies with a Master's thesis in the laboratories of Prof. Stephen L. Buchwald at Massachusetts Institute of Technology, Cambridge MA, USA. Currently pursuing a Ph. D. in synthetic organic chemistry with Prof. Erick M. Carreira, his research focuses on natural product synthesis and asymmetric catalysis.

Organic
Syntheses

Simon Krautwald was born in Aachen, Germany, in 1986. He received a M. Sci. degree in Chemistry from Imperial College London in 2010. Simon is currently a Ph. D. candidate in Professor Erick M. Carreira's research laboratory at ETH Zurich, where he is studying iridium-catalyzed enantioselective allylic substitution reactions.

Enantioselective Reductive *Syn*-Aldol Reactions of 4-Acryloylmorpholine: Preparation of (2*R*, 3*S*)-3-Hydroxy-2-methyl-1-morpholino-5-phenylpentan-1-one

Jason R. Abbott, Christophe Allais, and William R. Roush[*1]

Department of Chemistry, The Scripps Research Institute–Florida, 130 Scripps Way #3A2, Jupiter, FL 33458

Checked by Simon Krautwald, Simon Breitler, and Erick M. Carreira

Procedure

A. *(2R, 3S)-3-Hydroxy-2-methyl-1-morpholino-5-phenylpentan-1-one (4) from Crystalline (Diisopinocampheyl)borane (1).* A flame-dried 500-mL, two-necked, round-bottomed flask is equipped with a 5-cm Teflon-coated egg-shaped magnetic stir bar and moved into a glovebox. The flask is charged with crystalline (Ipc)₂BH (**1**) (13.5 g, 47.06 mmol, 1.18 equiv) (Note 1), and capped with rubber septa. The flask is then removed from the glovebox, equipped with an argon line and a thermometer (Note 2), and charged with diethyl ether (190 mL) (Note 3). The resulting white suspension is cooled to 0 °C (Note 4) with an ice/water bath and stirred for 15 min. 4-Acryloylmorpholine (**2**) (Note 3) (6.5 mL, 7.31 g, 51.76 mmol, 1.30 equiv) is added over 5 min using a syringe pump. Upon complete addition, the

mixture is stirred at 0 °C for 2 h. After 15 min the white suspension becomes a clear solution, which gradually gives way to a turbid white suspension (Note 4). The ice/water bath is then replaced with a dry ice/acetone bath and the mixture is cooled to –78 °C and stirred for 15 min at this temperature. Hydrocinnamaldehyde (**3**) (Note 3) (5.3 mL, 5.37 g, 40.00 mmol, 1.00 equiv) is added over 5 min using a syringe pump (Note 5). Upon complete addition, the mixture is stirred at –78 °C for 14 h before a premixed solution of THF/methanol/pH 7 potassium phosphate buffer (Note 3) (1:1:1 v/v/v, 135 mL total) is introduced via syringe. The reaction mixture is allowed to warm to 23 °C and, upon reaching that temperature, stirred vigorously for 6 h. During this period the color of the solution changes from clear to yellow. The mixture is then transferred to a 500-mL separatory funnel with the aid of diethyl ether. The organic layer is removed and the aqueous phase is extracted with two 50-mL portions of diethyl ether. The combined organic layers are washed with two 100-mL

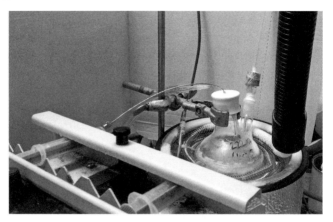

portions of deionized water, one 100-mL portion of brine, and dried over sodium sulfate (Na$_2$SO$_4$) (100 g). The drying agent is removed by vacuum filtration through a 150-mL fritted glass funnel and washed with three 25-mL portions of diethyl ether. The filtrate is concentrated by rotary evaporation (35 °C bath temperature, 460 mmHg initial pressure to 10 mmHg final pressure) to afford 25.0–25.1 g of a clear yellow oil. For purification, the crude product is solubilized in 10-mL of 30% EtOAc-hexanes (70:30 hexanes:EtOAc) (Note 6) and loaded onto a 7.5 cm diameter column containing 200 g of silica gel that is prepacked with 30% EtOAc-hexanes (70:30 hexanes:EtOAc) (Note 7). The flask is washed with three 10-mL portions of 30% EtOAc-hexanes (70:30 hexanes:EtOAc) and the washings are loaded onto the silica gel. Fraction collection (250-mL fractions) is begun and elution proceeds with 3000-mL of 30% EtOAc-hexanes (70:30 hexanes:EtOAc) and then 1000-mL of 40% EtOAc-hexanes (60:40 hexanes:EtOAc). The product is finally eluted from the column using 2000-mL of EtOAc, with fractions 17-24 containing the desired material (Note 8). These fractions are combined and concentrated by rotary evaporation (35 °C bath temperature, 200 mmHg initial pressure to 10 mmHg final pressure) and subsequently dried for 12 h at <5 mmHg (Note 9) to provide 7.8–8.2 g (70–73%) of diastereomerically pure (Notes 10 and 11) (2R, 3S)-3-hydroxy-2-methyl-1-morpholino-5-phenylpentan-1-one (**4**), 97% ee (Note 12), as a white solid, mp 76–78 °C (Note 13).

B. *One-pot Synthesis of (2R, 3S)-3-Hydroxy-2-methyl-1-morpholino-5-phenylpentan-1-one (**4**) from (–)-(α)-pinene (**5**).* A flame-dried 500-mL, two-necked, round-bottomed flask equipped with a 5-cm Teflon-coated egg-shaped magnetic stir bar and rubber septa is purged with argon, and its tare weight is recorded. An argon line is inserted through one of the septa and the other one is replaced with a thermometer (Note 2). The flask is charged with tetrahydrofuran (THF) (80 mL) and borane-methyl sulfide complex (Note 3) (8.2 mL, 6.5 g, 80.1 mmol, 1.75 equiv) is added via syringe. The mixture is cooled to 0 °C (Note 4) with an ice/water bath and (–)-(α)-pinene (**5**) (25.5 mL, 22.3 g, 160.2 mmol, 3.50 equiv) (Note 14) is added over 30 min using a syringe pump. Upon complete addition, the stirring is terminated, the thermometer replaced with a rubber septum, the argon line removed, and the septa are wrapped thoroughly with Parafilm®. The reaction flask is then placed in a 0 °C ice/water bath in a 4 °C cold room for 46 h (Note 15).

After this time, the flask is allowed to warm to room temperature, the Parafilm® is discarded, and the supernatant is removed via cannula. Trituration of the residual chunks of (Ipc)₂BH is performed by introduction of diethyl ether (50 mL) via syringe and subsequent decannulation of the supernatant. The trituration process is repeated two additional times before the cannula is removed and replaced with a needle attached to a vacuum line. The white crystals of (Ipc)₂BH are allowed to dry at <5 mmHg for 3 h. At this time the flask is back-filled with argon, gently shaken to pulverize chunks of solid (Ipc)₂BH with the aid of the magnetic stir bar, and then weighed. This procedure provides 13.0–13.9 g (57–60%, 45.4–48.2 mmol) of (+)-(diisopinocampheyl)borane ((Ipc)₂BH) (**1**) as a white solid (Note 16). The flask is then equipped with an argon line and a thermometer (Note 2), and charged with diethyl ether (215 mL) (Note 3). The resulting white suspension is cooled to 0 °C (Note 17) with an ice/water bath and stirred for

15 min. 4-Acryloylmorpholine (**2**) (Note 3) (6.3 mL, 7.1 g, 50.2 mmol, 1.30 equiv) is added over 5 min using a syringe pump. Upon complete

addition, the mixture is stirred at 0 °C for 2 h. After 15 min the white suspension becomes a clear solution, which gradually converts into a turbid white suspension. The ice/water bath is then replaced with a dry ice/acetone bath and the mixture is cooled to –78 °C and stirred for 15 min at this temperature. Hydrocinnamaldehyde (**3**) (Note 3) (5.1 mL, 5.2 g, 38.6 mmol, 1.00 equiv) is added over 5 min using a syringe pump (Note 5). Upon complete addition, the mixture is stirred at –78 °C for 14 h before a premixed solution of THF/methanol/pH 7 potassium phosphate buffer (Note 3) (1:1:1 v/v/v, 127 mL total) is introduced via syringe. The reaction mixture is allowed to warm to 23 °C and, upon reaching that temperature, stirred vigorously for 6 h. During this period the color of the solution changes from clear to yellow. The mixture is then transferred to a 500-mL separatory funnel with the aid of diethyl ether. The organic layer is removed and the aqueous phase is extracted with two 50-mL portions of diethyl ether. The combined organic layers are washed with two 100-mL portions of deionized water, one 100-mL portion of brine, and dried over sodium sulfate (Na$_2$SO$_4$) (100 g). The drying agent is removed by vacuum filtration through a 150-mL fritted glass funnel and washed with three 25-mL portions of diethyl ether. The filtrate is concentrated by rotary evaporation (35 °C bath temperature, 460 mmHg initial pressure to

10 mmHg final pressure) to afford 22.0 g of a clear yellow oil. For purification, the crude product is solubilized in 10-mL of 30% EtOAc-hexanes (70:30 hexanes:EtOAc) (Note 6) and loaded onto a 7.5 cm diameter column containing 200 g of silica gel that is prepacked with 30% EtOAc-hexanes (70:30 hexanes:EtOAc) (Note 7). The flask is washed with three 10-mL portions of 30% EtOAc-hexanes (70:30 hexanes:EtOAc) and the washings are loaded onto the silica gel. Fraction collection (250-mL fractions) is begun and elution proceeds with 3000-mL of 30% EtOAc-hexanes (70:30 hexanes:EtOAc) and then 1000-mL of 40% EtOAc-hexanes (60:40 hexanes:EtOAc). The product is finally eluted from the column using 2000-mL of EtOAc, with fractions 17-24 containing the desired material (Note 8). These fractions are combined and concentrated by rotary evaporation (35 °C bath temperature, 200 mmHg initial pressure to 10 mmHg final pressure) and subsequently dried for 12 h at <5 mmHg (Note 9) to provide 7.4–8.0 g (69–70%) of diastereomerically pure (Notes 10 and 11) (2*R*, 3*S*)-3-Hydroxy-2-methyl-1-morpholino-5-phenylpentan-1-one (**4**), 97% ee (Note 12), as white solid, mp 76–78 °C (Note 13).

Notes

1. Crystalline (+)-(diisopinocampheyl)borane (**1**) was synthesized and stored in a glovebox as described in the accompanying procedure.[2]
2. The submitters used a single-necked flask and monitored the internal temperature of the reaction mixture using an Oakton Instruments Temp JKT temperature meter with a Teflon-coated thermocouple probe (30.5 cm length, 3.2 mm outer diameter, temperature range –250 to 400 °C).
3. THF (HPLC Grade) and diethyl ether (Certified ACS, stabilized with BHT) were obtained from Fisher Scientific and purified by passage through activated alumina using a GlassContour solvent purification system.[3] Borane-methyl sulfide complex (94%) was obtained from Acros Organics and used as received. (–)-(α)-Pinene (**5**) (98%, ≥81% ee) was obtained from Aldrich Chemical Co., Inc. and used as received. 4-Acryloylmorpholine (**2**) (99%, stabilized with MEHQ) was obtained from TCI, used as received, and stored at −20 °C under argon. Hydrocinnamaldehyde (**3**) (90% technical grade) was obtained from Aldrich Chemical Co., Inc., distilled (13 mmHg, 99-101 °C), and stored

at −20 °C under argon. pH 7.0 Buffer Solution (catalog number SB108-1) was purchased from Fischer and used as received in the reaction workup.

4. The internal temperature of the reaction mixture remained between 0 and 1 °C throughout the course of the reaction.

5. The internal temperature of the reaction mixture remained between –79 °C and –75 °C throughout the course of the aldol reaction.

6. Sonication can be used to help completely solubilize the crude yellow oil.

7. Silica gel (SiliaFlash® F60, 230–400 mesh, 40–63 μm) was obtained from Silicycle. The checkers strongly recommend using this type of Silica gel since a product from Fluka containing calcium oxide led to two mixed fractions. This does not happen when the SiliaFlash gel is used.

8. Individual fractions were analyzed by TLC (Merck Kieselgel 60 F_{254} glass plates precoated with a 0.25 mm thickness of silica gel) using 50% EtOAc-hexanes (50:50 hexanes:EtOAc) and visualized first with a 254-nm UV lamp and then with an aqueous solution of cerium molybdate. In this solvent system, unidentified reaction impurities have Rf values of 0.89, 0.80, and 0.60. The reaction product, **4**, has an Rf value of 0.14 in 50% EtOAc-hexanes (50:50 hexanes:EtOAc) and 0.33 in 25% hexanes-EtOAc (25:75 hexanes:EtOAc).

9. If the product remains as a clear oil after drying for 12 h at <5 mmHg, it may be coevaporated with diethyl ether to induce solidification.

10. The purity of this material was confirmed by spectroscopic and elemental analysis. Syn-**4** exhibits the following properties: white solid; mp 76–78 °C; $[\alpha]_D^{22.8}$ = −10.7 (c = 0.25, CHCl$_3$); ^{1}H NMR (400 MHz, CDCl$_3$) δ: 1.15 (d, J = 7.2 Hz, 3 H), 1.51–1.62 (m, 1 H), 1.92 (dtd, J = 5.4, 9.3, 13.6 Hz, 1 H), 2.52 (dq, J = 2.1, 7.2 Hz, 1 H), 2.68 (ddd, J = 7.1, 9.2, 13.8 Hz, 1 H), 2.88 (ddd, J = 5.2, 9.4, 14.3 Hz, 1 H), 3.43 (br t, J = 4.8 Hz, 2 H), 3.51–3.71 (m, 6 H), 3.93 (ddd, J = 2.1, 3.8, 9.4 Hz, 1 H), 4.39 (s, 1 H), 7.16–7.24 (m, 3 H), 7.26–7.31 (m, 2 H); ^{13}C NMR (100 MHz, CDCl$_3$) δ: 10.2, 32.5, 35.7, 39.0, 41.9, 46.2, 66.8, 66.9, 70.5, 126.0, 128.5 (2C), 128.7 (2C), 142.2, 176.3; IR (neat) 3428, 2921, 2857, 1616, 1454, 1434, 1224, 1114, 1025 cm^{-1}; HRMS (ESI) calcd for C$_{16}$H$_{23}$NNaO$_3$ [M+Na]$^+$ 300.1570. Found 300.1572; Anal. calcd for C$_{16}$H$_{23}$NO$_3$: C, 69.29; H, 8.36; N, 5.05. Found: C, 69.15; H, 8.41; N, 5.07. The diastereomer ratio (syn-**4**/anti-**4**) was determined to be >20:1 from the ratio of resonance integrations at 1.13-1.17 ppm (methyl substituent of syn isomer) and 1.17-1.21 ppm

(methyl substituent of *anti* isomer—see Note 11). Both isomers co-elute by TLC analysis.

11. The *anti*-**4** diastereomer was prepared in low yield (with d.r. ca. 8:1) from *syn*-**4** by Mitsunobu reaction (see discussion) and exhibits the following properties: colorless oil; $[\alpha]_D^{26.4}$ = –9.4 (c = 0.54, CHCl$_3$); ^{1}H NMR (400 MHz, CDCl$_3$) δ: 1.20 (d, J = 7.1 Hz, 3 H), 1.70-1.85 (m, 2 H), 2.59-2.73 (m, 2 H), 2.92 (ddd, J = 5.4, 9.4, 13.7 Hz, 1 H), 3.44-3.49 (m, 2 H), 3.56-3.71 (m, 7 H), 3.94 (d, J = 6.4 Hz, 1 H), 7.16-7.23 (m, 3 H), 7.26-7.31 (m, 2 H); ^{13}C NMR (100 MHz, CDCl$_3$) δ: 15.3, 32.3, 37.2, 40.2, 41.8, 46.1, 66.7, 66.9, 73.4, 125.8, 128.4 (2C), 128.5 (2C), 142.2, 175.1; IR (neat) 3426, 3026, 2922, 2857, 1614, 1496, 1454, 1435, 1361, 1301, 1268, 1220, 1113, 1069, 1026, 934, 846 cm^{-1}.

12. The enantiomeric purity and absolute configuration of *syn*-**4** were determined by Mosher ester analysis.[4] Thus, a mixture of aldol (–)-**4** (0.0064 g, 0.023 mmol, 1.0 equiv) in dichloromethane (0.5 mL, obtained from Fisher and Scientific and dried by passage through activated alumina using a GlassContour solvent purification system (see Note 3)), pyridine (0.0075 mL, 0.007 g, 0.092 mmol, 4 equiv; obtained from EMD and distilled from CaH$_2$ under Ar) and a catalytic amount of dimethylaminopyridine (DMAP; one small crystal; obtained from Sigma-Aldrich and used as obtained) was stirred under Ar at ambient temperature.   (R)-(–)-α-Methoxy-α-(trifluoromethyl)-phenylacetyl chloride (0.0086 mL, 0.012 g, 0.046 mmol, 2 equiv; obtained from Matrix Scientific and used as received) was added via microliter syringe. The mixture was stirred at ambient temperature for 18 h, at which point TLC analysis (1:1, CH$_2$Cl$_2$-EtOAc; R$_f$ **4** = 0.40; R$_f$ for Mosher ester product = 0.79) indicated that the reaction was complete. The mixture was diluted with hexanes (1 mL), filtered to remove the white precipitate, then directly filtered through a short column of silica gel (in a Pasteur pipette) using 15 mL of 4:1 hexanes-EtOAc. The filtrate was collected as a single fraction and concentrated on a rotary evaporator to give the (S)-MTPA ester as an oil. By using the same procedure, the (R)-MPTA ester was prepared (using (S)-(+)-α- methoxy-α-(trifluoromethyl)phenylacetyl chloride, obtained from Alfa Aesar). Key resonances in the ^{19}F and ^{1}H NMR spectra of the diastereomeric MTPA esters that may be used in making enantiomeric purity determinations are as follows. Partial data for the (S)-MTPA ester of (–)-**4**: ^{19}F (CDCl$_3$) δ: –70.88; ^{1}H (400 MHz, CDCl$_3$) δ: 1.14 (d, J = 6.9 Hz, 3H), 1.96 (m, 2H), 2.49 (t, J = 8.0 Hz, 2H), 2.91 (quint, J = 6.8 Hz, 1H), 3.32 (m, 2H), 5.42 (m,

1H). Partial data for the (R)-MTPA ester of (–)-4: ^{19}F (CDCl$_3$) δ: –70.81; ^{1}H (400 MHz, CDCl$_3$) δ: 1.06 (d, J = 6.9 Hz, 3 H), 2.00 (m, 2 H), 2.61 (m, 2 H), 2.89 (quint, J = 6.9 Hz, 1 H), 3.29 (m, 2H), 5.46 (m, 1 H).

13. The melting point was recorded on a Stuart SMP 40 apparatus.

14. Due to the viscosity of (–)-(α)-pinene, it is recommended that a large-gauge (16-18) needle be used.

15. As reported by Brown and Singaram[5] it is imperative that the crystallization be carried out at 0 °C. The submitters observed a significant decrease in yield (from 64–66% at 0 °C to 31% at –18.5 °C) with no discernable increase in reagent purity when the crystallization was carried out at –18.5 °C for 46 h.[2]

16. Crystalline (Ipc)$_2$BH (2) exhibits the following properties: mp 95-98 °C; ^{1}H NMR (500 MHz, d-THF) δ: 0.85 (s), 0.87 (s), 0.89 (s), 0.91 (s), 0.92 (s), 0.93 (s), 0.94 (s), 0.96 (s), 0.97 (s), 0.99 (s), 1.00 (s), 1.02 (s), 1.04 (s), 1.05 (s), 1.06 (s), 1.07 (s), 1.09 (s), 1.12 (s), 1.13 (s), 1.14 (s), 1.15 (s), 1.17 (s), 1.17 (s), 1.19 (s), 1.21 (s), 1.23 (s), 1.24 (s), 1.27 (s), 1.64 (m), 1.65–2.45 (m), 5.18 (m); ^{13}C NMR (125 MHz, d-THF) δ: 21.3, 22.6, 22.9, 23.1, 23.2, 23.28, 23.3, 23.4 (2), 23.7, 25.5, 26.8, 26.9, 27.6, 29.0, 29.2, 29.4, 30.1, 31.9, 32.2, 32.3, 32.6, 34.0, 34.7, 35.1, 35.6, 37.0, 38.9, 39.7, 39.8, 40.2, 40.3, 40.9, 41.6, 41.9, 42.6, 43.1, 43.1, 43.2, 43.3, 48.1, 49.6, 49.7, 49.9, 50.2, 117.0, 145.4. The sample for melting point determination was sealed in a capillary tube under Ar. The NMR sample was prepared under inert atmosphere and it is important to use anhydrous d-THF. See in particular Note 7 of the accompanying procedure.[2]

17. The internal temperature of the reaction mixture remained between 0 and 2 °C throughout the course of the hydroboration.

Working with Hazardous Chemicals

The procedures in *Organic Syntheses* are intended for use only by persons with proper training in experimental organic chemistry. All hazardous materials should be handled using the standard procedures for work with chemicals described in references such as "Prudent Practices in the Laboratory" (The National Academies Press, Washington, D.C., 2011; the full text can be accessed free of charge at http://www.nap.edu/catalog.php?record_id=12654). All chemical waste should be disposed of in accordance with local regulations. For general guidelines for the management of chemical waste, see Chapter 8 of Prudent Practices.

In some articles in *Organic Syntheses*, chemical-specific hazards are highlighted in red "Caution Notes" within a procedure. It is important to recognize that the absence of a caution note does not imply that no significant hazards are associated with the chemicals involved in that procedure. Prior to performing a reaction, a thorough risk assessment should be carried out that includes a review of the potential hazards associated with each chemical and experimental operation on the scale that is planned for the procedure. Guidelines for carrying out a risk assessment and for analyzing the hazards associated with chemicals can be found in Chapter 4 of Prudent Practices.

The procedures described in *Organic Syntheses* are provided as published and are conducted at one's own risk. *Organic Syntheses, Inc.*, its Editors, and its Board of Directors do not warrant or guarantee the safety of individuals using these procedures and hereby disclaim any liability for any injuries or damages claimed to have resulted from or related in any way to the procedures herein.

Discussion

Development of highly diastereo- and enantioselective aldol reactions has captured the attention of numerous research groups for decades.[6,7] Enantioselective reductive aldol reactions[7] are attractive alternatives to conventional aldol, organocatalytic, chiral Lewis acid or chiral Lewis base mediated procedures because bases are not required to generate a reactive enol or metal enolate derivative; instead, the reactive intermediate is generated directly by the 1,4-reduction of the α,β-unsaturated carbonyl substrate. While the vast majority of reductive aldol reactions that have been developed to date use chiral transition metal catalysts, turnover numbers are modest (typically <50) and the reagents (both the metal catalysts as well as the chiral ligands) used in these experiments are expensive or require multi-step syntheses if not commercially available. Cost issues also apply to the vast majority of auxiliary-driven, organocatalytic, and chiral Lewis acid or chiral Lewis base mediated aldol processes.[6] Thus, a significant objective of research in this field increasingly will be on the development of highly cost effective aldol reactions that proceed with exceptional diastereo- and enantioselectivity, that have broad substrate scope (i.e., are not limited to any particular sub-group of substrates such as aromatic aldehydes), and which function with excellent selectivity in the context of aldol reactions with chiral aldehyde substrates (e.g., double asymmetric reactions).[8]

Both enantiomers of α-pinene are widely available in bulk quantities at very low cost. Consequently, a variety of chiral reagents have been developed using α-pinene as the starting material. First among these is (diisopinocampheyl)borane ((Ipc)$_2$BH) which can be generated with excellent enantiomeric purity via the hydroboration of either (–)-(α)-pinene (**5**) (≥81% ee) or (+)-(α)-pinene (≥91% ee) with borane-dimethylsulfide complex followed by crystallization from the reaction mixture.[2,5] The enantiomeric purity enhancement derives from the fact that the minor enantiomer present in the commercial (α)-pinenes is preferentially converted during the hydroboration into the diastereomeric *meso*-(Ipc)$_2$BH which is not crystalline and which is separated during the crystallization step.[2]

(Ipc)$_2$BH is a precursor of a range of chiral reagents such as (Ipc)$_2$BOTf and (Ipc)$_2$BCl that have been employed in asymmetric aldol reactions by Paterson.[9] (Ipc)$_2$BOMe, a starting material used for the synthesis of chiral allylborane[10] and crotylborane[10c,11] reagents, is prepared by methanolysis of (Ipc)$_2$BH. Generation of chiral allylboron reagents via hydroboration of allenes[12] with (Ipc)$_2$BH has also been reported. Enantioenriched secondary alcohols are generated by hydroboration of alkenes with (Ipc)$_2$BH.[13] Finally, enolborinates can be generated by the formal 1,4-reduction of α,β-unsaturated carbonyl compounds with (Ipc)$_2$BH.[14,15]

Hydroboration of 4-acryloylmorpholine (**2**) with (Ipc)$_2$BH (**1**) is performed in diethyl ether at 0 °C for 2 h. The resulting turbid solution contains exclusively the Z(O)-enolborinate (**6**).[15a] The Z(O)-enolborinate solution is cooled to –78 °C and neat hydrocinnamaldehyde (**3**) is added. Mild hydrolytic workup liberates the *syn*-aldol adduct **4**, which is obtained, after purification on silica gel, in 74–76% yield (8.21–8.38 g, Procedure A) or 72% yield (9.08 g, Procedure B), calculated based on aldehyde as the limiting reagent, with complete control of the diastereoselectivity (d.r. >20:1) and excellent enantiomeric purity (97% ee by Mosher ester analysis[4]). The diastereomeric purity of **4** was established by comparison of the spectroscopic data (see Note 11) with those of a sample of the *anti*-diastereomer prepared by Mitsunobu reaction of **4** (*p*-nitrobenzoic acid, diethyl azodicarboxylate, triphenylphosphine) followed by nitrobenzoate ester hydrolysis (potassium carbonate, methanol).

The submitters store crystalline (Ipc)$_2$BH in a glovebox, and transfer this reagent as described in Procedure A. However, recognizing that many investigators may not have access to a glovebox, we developed Procedure B to illustrate that crystalline (Ipc)$_2$BH may be generated in situ by hydroboration of (–)-(α)-pinene and used in a one-pot sequence, with virtually the same efficiency as compared to Procedure A. It should be noted as well, that we intentionally illustrated Procedure B by using the less enantiomerically pure (–)-(α)-pinene ($\geq$81% ee). It stands to reason, that the efficiency of the one-pot procedure (calculated based on pinene) will be greater if (+)-(α)-pinene ($\geq$91 ee) is used.

As summarized in Table 1, the (Ipc)$_2$BH-mediated reductive aldol reactions of 4-acryloylmorpholine (2) with achiral aldehydes furnishes the *syn*-α-methyl-β-hydroxy carboxamides 8 with excellent diastereocontrol (d.r. >20:1) and >96% enantiomeric excess.[15a] The substrate scope spans aliphatic, aromatic and α,β-unsaturated aldehydes.

Table 1. (*I*pc)₂BH-mediated reductive aldol reactions of acryloylmorpholine (2) with achiral aldehydes 7

entry	aldehyde 7	yield[a]	d.r. (syn:anti)[b]	ee[c]
1	(benzaldehyde) CHO	90%	>20:1	97%
2	(furfural) CHO	68%	>20:1	96%
3	(cinnamaldehyde) CHO	68%	>20:1	97%
4	(cyclohexanecarboxaldehyde) CHO	88%	>20:1	97%
5[d]	DMTrO⌒CHO	80%	>20:1	97%

[a] isolated yield; [b] determined by ¹H NMR of the crude reaction mixture; [c] determined by Mosher ester analysis;[4] [d] DMTr = dimethoxytrityl.

Double asymmetric reactions of the chiral Z(O)-enolborinate [derived from **2** and either (Ipc)₂BH (**1**) or (*l*Ipc)₂BH (***ent*-1**)] with a panel of representative chiral, non-racemic aldehydes have also been reported.[15a] Excellent diastereoselectivity (>20:1) is achieved in both the stereochemically matched and mismatched cases for each aldehyde substrate, as shown in Table 2.

Table 2. Double asymmetric reactions of chiral aldehydes with the chiral Z(O)-enolborinates derived from (^{l}Ipc)$_2$BH (1) or (^{d}Ipc)$_2$BH (*ent*-1)

entry	aldehyde 7[a]	(Ipc)$_2$BH	product 8	yield[b] (d.r.)[c]
1	TBDPSO~~~CHO, Me	(^{l}Ipc)$_2$BH (1)	TBDPSO~~~ OH O, Me Me, N O	69% >20:1
2		(^{d}Ipc)$_2$BH (*ent*-1)	TBDPSO~~~ OH O, Me Me, N O	85% >20:1
3	PMBO~~~ TBS TBS, O O, CHO	(^{l}Ipc)$_2$BH (1)	PMBO~~~ TBS TBS O O OH O, Me, N O	82% >20:1
4		(^{d}Ipc)$_2$BH (*ent*-1)	PMBO~~~ TBS TBS O O OH O, Me, N O	78% >20:1
5	DMPMO~~~ TBSO, CHO, Me	(^{l}Ipc)$_2$BH (1)	DMPMO~~~ TBSO OH O, Me Me, N O	71% >20:1
6		(^{d}Ipc)$_2$BH (*ent*-1)	DMPMO~~~ TBSO OH O, Me Me, N O	56% >20:1
7	DMPMO~~~ TBSO, CHO, Me	(^{l}Ipc)$_2$BH (1)	DMPMO~~~ TBSO OH O, Me Me, N O	74% >20:1
8		(^{d}Ipc)$_2$BH (*ent*-1)	DMPMO~~~ TBSO OH O, Me Me, N O	72% >20:1

[a] TBDPS = *tert*-butyldiphenylsilyl; PMB = *p*-methoxybenzyl; TBS = *tert*-butyldimethylsilyl; DMPM = 3,4-dimethoxybenzyl; [b] isolated yields; [c] determined by ^{1}H NMR of the crude reaction mixture. [d] DMTr = dimethoxytrity

In view of the very low cost of all reagents used for the synthesis of either enantiomer of crystalline (Ipc)$_2$BH, the very low cost of 4-acryloylmorpholine (2), and the ease of manipulation of the morpholine amide functionality of the aldol products (which have Weinreb amide-like reactivity),[16] the procedure described here for the synthesis of *syn*-aldols **4** and **8** ranks among the least expensive and most selective *syn*-aldol procedures currently available in the literature.[6,7,14,15a] For laboratory scale experiments, the cost of the raw materials used to generate enolborinate Z(O)-**6** according to this procedure is less than $0.25 per mmol for each aldol reaction.

Extensions of this methodology to the stereocontrolled synthesis of stereochemically defined tetrasubstituted enolborinates (Figure 1),[15b] and to the synthesis of *anti*-aldols from acrylate esters (Figure 2)[15c] have been reported.

Figure 1. Generation of quaternary centers with high enantioselectivity via stereocontrolled generation of tetrasubstituted enolborinate intermediates

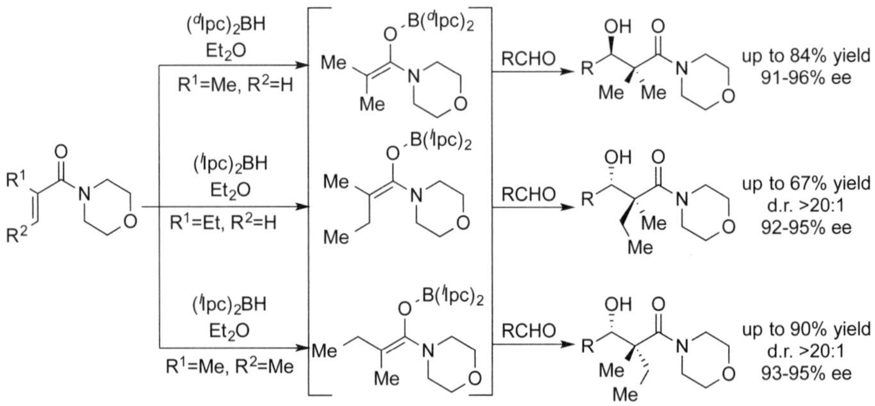

Figure 2. Reductive *anti*-aldol reactions of acrylate esters

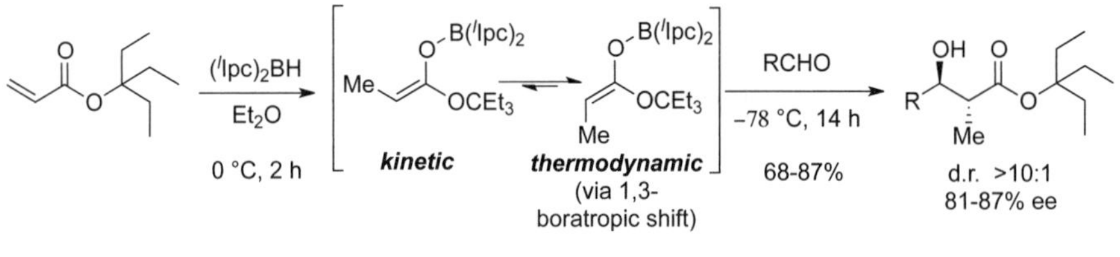

References

1. E-mail: roush@scripps.edu. This research was supported by the National Institutes of Health (GM038436).
2. Abbott, J. R.; Allais, C.; Roush, W. R. *Org. Synth.* **2015**, *92*, 26–37.
3. (a) Pangborn, A. B.; Giardello, M. A.; Grubbs, R. H.; Rosen, R. K.; Timmers, F. J. *Organometallics* **1996**, *15*, 1518–1520. (b) http://www.glasscontour.com/
4. (a) Dale, J. A.; Dull, D. L.; Mosher, H. S. *J. Org. Chem.* **1969**, *34*, 2543–2549. (b) Dale, J. A.; Mosher, H. S. *J. Am. Chem. Soc.* **1973**, *95*, 512–519.
5. Brown, H. C.; Singaram. B. *J. Org. Chem.* **1984**, *49*, 945–947.
6. Selected reviews of enantioselective aldol reactions: (a) Heathcock, C. H., in *Comprehensive Organic Synthesis*, Trost, B. M.; Fleming, I. Eds. Pergamon Press: New York, **1991**, Vol. 2, pp. 181–238. (b) Kim, B. M.; Williams, S. F.; Masamune, S., in *Comprehensive Organic Synthesis*, Trost, B. M.; Fleming, I. Eds. Pergamon Press: New York, **1991**, Vol. 2, pp. 239–275. (c) Cowden, C. J.; Paterson, I. *Org. React.* **1997**, *51*, 1–200. (d) Mahrwald, R., in *Modern Aldol Reactions*, Mahrwald, R. Ed. Wiley-VCH: Weinheim, **2004**, Vol. 2. (e) Denmark, S. E.; Fiujimori, S., in *Modern Aldol Reactions*, Mahrwald, R. Ed. Wiley-VCH: Weinheim, **2004**, Vol. 2, pp. 229–326. (f) Shibasaki, M.; Matsunaga, S.; Kumagai, N., in *Modern Aldol Reactions*; Mahrwald, R. Ed. Wiley-VCH: Weinheim, **2004**, Vol. 2, pp. 197–227. (g) Johnson, J. S.; Nicewicz, D. A., in *Modern Aldol Reactions*, Mahrwald, R. Ed. Wiley-VCH: Weinheim, **2004**, Vol. 2, pp. 69–103. (h) Bisai, V.; Bisai, A.; Singh, V. K. *Tetrahedron* **2012**, *68*, 4541–4580. (i) Matsuo, J.; Murakami, M. *Angew. Chem. Int. Ed.* **2013**, *52*, 9109-9118. (j) Mlynarski, J.; Baś, S. *Chem. Soc. Rev.* **2014**, *43*, 577-587.
7. Selected reviews of reductive aldol reactions: (a) Guo, H.-C.; Ma, J.-A. *Angew. Chem. Int. Ed.* **2006**, *45*, 354–366. (b) Nishiyama, H.; Shiomi, T. *Top. Curr. Chem.* **2007**, *279*, 105–137. (c) Han, S. B.; Hassan, A.; Krische, M. J. *Synthesis* **2008**, *17*, 2669–2679. (d) Garner, S. A.; Han, S. B.; Krische M. J. "Metal Catalyzed Reductive Aldol Coupling," in *Modern Reduction Methods* (Eds. P. Andersson, I. Munslow) Wiley-VCH: Weinheim, **2008**, p 387–408.
8. Masamune, S.; Choy, W.; Petersen, J. S.; Sita, L. R. *Angew. Chem. Int. Ed.* **1985**, *24*, 1–30.
9. (a) Paterson, I.; Goodman, J. M.; Lister, M. A.; Schumann, R. C.; McClure, C. K.; Norcross, R. D. *Tetrahedron* **1990**, *46*, 4663–4684. (b)

Paterson, I.; Wallace, D. J.; Velázquez, S. M. *Tetrahedron Lett.* **1994**, *35*, 9083–9086.

10. (a) Brown, H. C.; Jadhav, P. K. *J. Am. Chem. Soc.* **1983**, *105*, 2092–2093. (b) Jadhav, P. K.; Bhat, K. S.; Perumal, P. T.; Brown, H. C. *J. Org. Chem.* **1986**, *51*, 432–439. (c) Brown, H. C.; Bhat, K. S.; Randad, R. S. *J. Org. Chem.* **1989**, *54*, 1570–1576.

11. (a) Brown, H. C.; Bhat, K. S. *J. Am. Chem. Soc.* **1986**, *108*, 5919–5923. (b) Brown, H. C.; Jadhav, P. K.; Bhat, K. S. *J. Am. Chem. Soc.* **1988**, *110*, 1535–1538.

12. (a) Brown, H. C.; Narla, G. *J. Org. Chem.* **1995**, *60*, 4686–4687. (b) Flamme, E. M.; Roush, W. R. *J. Am. Chem. Soc.* **2002**, *124*, 13644–13645. (c) Chen, M.; Ess, D. H.; Roush, W. R. *J. Am. Chem. Soc.* **2010**, *132*, 7881–7883. (d) Chen, M.; Roush, W. R. *Org. Lett.* **2011**, *13*, 1992–1995. (e) Chen, M.; Roush, W. R. *J. Am. Chem. Soc.* **2011**, *133*, 5744–5747.

13. (a) Brown, H. C.; Zweifel, G. *J. Am. Chem. Soc.* **1960**, *82*, 3222–3223. (b) Brown, H. C.; Zweifel, G. *J. Am. Chem. Soc.* **1961**, *83*, 486–487.

14. (a) Boldrini, G. P.; Mancini, F.; Tagliavini, E.; Trombini, C.; Umani-Ronchi, A. *J. Chem. Soc., Chem. Commun.* **1990**, 1680–1681. (b) Boldrini, G. P.; Bortolotti, M.; Mancini, F.; Tagliavini, E.; Trombini, C.; Umani-Ronchi, A. *J. Org. Chem.* **1991**, *56*, 5820–5826.

15. (a) Nuhant, P.; Allais, C.; Roush, W. R. *Angew. Chem. Int. Ed.* **2013**, *52*, 8703-8707 (b) Allais, C.; Tsai, A. S. ; Nuhant, P.; Roush, W. R. *Angew. Chem. Int. Ed.* **2013**, *52*, 12888-12891. (c) Allais, C.; Nuhant, P.; Roush, W. R. *Org. Lett.* **2013**, *15*, 3922-3925.

16. (a) Concellón, J. M.; Rodríguez-Solla, H.; Méjica, C.; Blanco, E. G. *Org. Lett.* **2007**, *9*, 2981–2984. (b) Dhoro, F.; Kristensen, T. E.; Stockmann, V.; Yap, G. P. A.; Tius, M. A. *J. Am. Chem. Soc.* **2007**, *129*, 7256–7257. (c) Concellón, J. M.; Rodríguez-Solla, H.; Díaz, P. *J. Org. Chem.* **2007**, *72*, 7974–7979. (d) Lin, K.-W.; Tsai, C.-H.; Hsieh, I.-L.; Yan, T.-H. *Org. Lett.* **2008**, *10*, 1927–1930. (e) Concellón, J.; Rodríguez-Solla, H.; del Amo, V.; Díaz, P. *Synthesis* **2009**, 2634–2645. (f) Rye, C. E.; Barker, D. *Synlett* **2009**, 3315–3319.

Appendix
Chemical Abstracts Nomenclature (Registry Number)

(+)-(Diisopinocampheyl)borane ((+)-(Ipc)₂BH) or ((Ipc)₂BH: borane, bis[(1S,2R,3S,5S)-2,6,6-trimethylbicyclo[3.1.1]hept-3-yl]; (21947-87-5)

(–)-(Diisopinocampheyl)borane ((–)-(Ipc)₂BH) or ((Ipc)₂BH: borane, bis[(1R,2S,3R,5R)-2,6,6-trimethylbicyclo[3.1.1]hept-3-yl]; (21932-54-7)

Borane-methyl sulfide complex: boron, trihydro[thiobis[methane]]-(T-4)-; (13292-87-0)

(–)-(α)-Pinene: (1S,5S)-2,6,6-trimethylbicyclo[3.1.1]hept-2-ene; (7785-26-4)

(2R, 3S)-3-Hydroxy-2-methyl-1-morpholino-5-phenylpentan-1-one; (1529772-55-1)

4-Acryloylmorpholine: 2-Propen-1-one, 1-(4-morpholinyl)-; (5117-12-4)

Hydrocinnamaldehyde: 3-Phenylpropionaldehyde; (104-53-0)

Sodium perborate monohydrate; (10332-33-9)

William R. Roush is Professor of Chemistry, Executive Director of Medicinal Chemistry, and Associate Dean of the Kellogg School of Science and Technology at the Scripps Research Institute–Florida. His research interests focus on the total synthesis of natural products and the development of new synthetic methodology. Since moving to Scripps Florida in 2005, his research program has expanded into new areas of chemical biology and medicinal chemistry. Dr. Roush was a member of the *Organic Syntheses* Board of Editors from 1993-2002 and was Editor of Volume 78. He currently serves on the *Organic Syntheses* Board of Directors (2003-present).

Jason R. Abbott received his B.S. in Chemistry from Northeastern University in Boston, MA. In 2008, Mr. Abbott enrolled in the Kellogg School of Science and Technology at the Scripps Research Institute–Florida to pursue his Ph. D. in Organic Chemistry. He joined the Roush Group shortly thereafter and defended his Ph. D. in early 2014.

Christophe Allais obtained his Ph. D. in 2010 from Université Paul Cézanne (Marseille, France), under the supervision of Prof. Constantieux and Prof. Rodriguez where he focused on the development of convergent and selective methods to access various heterocycles. In 2011, he joined Prof. Roush's Group as a research associate, expanding his research into the areas of medicinal chemistry, natural product synthesis, and the development of boron-mediated asymmetric methodologies. In March 2014, he joined Pfizer (Groton, CT) as a Senior Scientist.

Simon Krautwald was born in Aachen, Germany, in 1986. He received a M. Sci. degree in Chemistry from Imperial College London in 2010. Simon is currently a Ph.D. candidate in Professor Erick M. Carreira's research laboratory at ETH Zurich, where he is studying iridium-catalyzed enantioselective allylic substitution reactions.

Simon Breitler, born in Basadingen, Switzerland, studied chemistry at ETH Zurich, which he concluded with a M.Sc. degree in 2011. During his undergraduate education, he carried out research projects in the laboratories of Prof. Erick M. Carreira and Prof. Antonio Togni. After an internship as a research trainee at Syngenta Crop Protection, Stein, Switzerland, he completed his studies with a Master's thesis in the laboratories of Prof. Stephen L. Buchwald at Massachusetts Institute of Technology, Cambridge MA, USA. Currently pursuing a Ph.D. in synthetic organic chemistry with Prof. Erick M. Carreira, his research focuses on natural product synthesis and asymmetric catalysis.

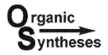

Ligand-Accelerated *ortho*-C–H Olefination of Phenylacetic Acids

Keary M. Engle, Navid Dastbaravardeh, Peter S. Thuy-Boun, Dong-Hui Wang, Aaron C. Sather, and Jin-Quan Yu[1*]

Department of Chemistry, The Scripps Research Institute, 10550 N. Torrey Pines Road, La Jolla, CA 92037, USA

Checked by Sandeep N. Raikar and Huw M. L. Davies

$$
\text{1a} + \text{2a (1.1 equiv)} \xrightarrow[\substack{t\text{-AmylOH (0.5 M)} \\ 90\ °C,\ 1\ atm\ O_2,\ 4\ h}]{\substack{5\ mol\%\ Pd(OAc)_2 \\ 5\ mol\%\ Ac\text{-Ile-OH} \\ 2\ equiv.\ KHCO_3}} \text{3a}
$$

Procedure

Caution! The reaction is run in a sealed vessel at elevated temperature under 1 atm O₂. Though no incidents have been encountered in the submitters' laboratory, it is nonetheless recommended that a blast shield be used as a precaution.

A. *(E)-2-(2-(3-Ethoxy-3-oxoprop-1-enyl)-6-fluorophenyl)acetic acid (3a)*. An oven-dried, single-necked 500-mL round-bottomed flask (Notes 1 and 2) is equipped with a magnetic stir bar (oval shaped, 40 mm length and 16 mm diameter). Palladium(II) acetate (280.6 mg, 1.25 mmol, 0.050 equiv) (Note 3), N-acetyl isoleucine (216.5 mg, 1.25 mmol, 0.050 equiv) (Note 4), 2-fluorophenylacetic acid (**1a**) (3.85 g, 25.0 mmol, 1.00 equiv) (Note 5), and potassium bicarbonate (5.01 g, 50.0 mmol, 2.00 equiv) (Note 6) are added to the round-bottomed flask (Note 7).

t-AmylOH (50.0 mL) (Note 8) and ethyl acrylate (**2a**) (2.93 mL, 27.5 mmol, 1.1 equiv) (Note 9) are added *via* syringe. The reaction flask is tightly fitted with a rubber septum (as shown in the image below). Into the septum is placed an 18G × 1½″ needle attached to a modified 1 mL syringe. (The plunger is removed, and the end opposite from the needle juncture is cut off using scissors). The severed end of the syringe is attached to thick-walled rubber tubing (25 cm length), which is connected to a three-way valve. The second connection of the valve is attached by rubber tubing to a high vacuum line and the third to a balloon filled with O_2. The reaction flask is evacuated briefly (Note 10) under high vacuum (0.1 mmHg, 23 °C) and charged with O_2; this process is repeated three times and then the flask is charged with O_2 for a minute. The needle attached to the rubber septum is removed. The

reaction mixture is stirred (Note 11) at room temperature for 5 min, during which time it is yellow to orange in color. The reaction flask is placed in an oil bath and then heated to 90 °C and stirred for 4 h (Note 12). While heating, a blast shield is placed in front of the reaction set-up as a safety precaution. During the course of the reaction, the solution becomes green/yellow to dark brown in color. The reaction flask is removed from the oil bath and allowed to cool to room temperature and then placed in an ice bath for 5 min. The stopper is then removed, and 2.0 N aqueous HCl solution (50 mL) is added dropwise over 20 min with vigorous stirring (Note 13). The mixture is transferred to a 500 mL separatory funnel and extracted with EtOAc (3 × 100 mL) (Notes 14 and 15). The organic layers are combined in a 500 mL Erlenmeyer flask, and anhydrous sodium sulfate (Note 16) is added. The solution is allowed to stand for 5 min. Sodium sulfate is removed using a Büchner filter funnel, and the filtrate is concentrated *in vacuo* (30 mmHg, 40 °C) to obtain the crude product as a viscous red/brown oil (Note 17). The resulting residue is purified by silica gel flash column chromatography (Notes 18 and 19) using 66:33:1 hexanes:EtOAc:HOAc as the eluent, giving product **3a** as an off-white solid (5.17 g, 82%) (Notes 20 and 21). The product is further purified by recrystallization from hexanes/EtOAc (Notes 22–26) to provide 4.91 g (78%) of a white solid.

Notes

1. In the submitters' experiment, a 350-mL Schlenk flask is used. The reaction vessel contains a high-vacuum valve with PTFE O-ring at the tip and PTFE wiper to protect the O-ring on the shaft.
2. Prior to use, the reaction vessel and stir bar are cleaned in the following way. They are first allowed to soak in concentrated nitric acid for two or more hours to remove trace transition metal precipitates. The nitric acid is poured off, and the items are rinsed five times with distilled water, then three times with acetone, and allowed to dry in an oven for two hours.
3. Palladium acetate trimer (>99.99% trace metals basis) is purchased from Pressure Chemical Co. (product 1730) and used as received.
4. *N*-Acetyl-isoleucine (Ac-Ile-OH) (>99%) is purchased from Bachem (product E-1080.0005) and used as received (white crystalline solid).
5. 2-Fluorophenylacetic acid (**1a**) (98%) is purchased from Oakwood (product 001292 (25 g)) and used as received (white crystalline solid).

6. Potassium bicarbonate (>99.5%) is purchased from Fisher (product P235-500) and used as received (white crystalline solid).

7. The reagents are weighted out on individual pieces of weighing paper, folded diagonally, and then added to the reaction flask in the aforementioned order. To avoid spilling the solid reagents when adding them to the flask, an additional piece of weighing paper is gently rolled (corner to opposite corner) and fitted inside the neck of the flask prior to transferring in the solid reagents, then removed before addition of liquid reagents. None of the reagents are air- or moisture-sensitive, so they can be measured out without special precautions.

8. 2-Methyl-2-butanol (*t*-AmylOH) (99%) is purchased from Sigma Aldrich (product 152463-1L) and used as received (colorless liquid). In the submitters' experience, newly opened bottles give the most consistent results. Solvent obtained from older bottles can give inconsistent results, unless the solvent is distilled before use.

9. Ethyl acrylate (**2a**) (99%) containing 10–20 ppm hydroquinone monomethyl ether as an inhibitor is purchased from Sigma Aldrich (product E9706-100ML) and used as received (colorless liquid).

10. Both *t*-AmylOH and ethyl acrylate are volatile under reduced pressure. Thus, each evacuation should be done carefully. The three-way valve is adjusted and connected to the vacuum until gentle bubbling of the solution is observed, and the exposure to high vacuum should last no longer than 5 seconds. Slowly swirling the reaction vessel, while the solution is under vacuum helps modulate bubbling.

11. The solution was stirred at 500–600 rpm, which is the highest controlled stirring rate that could be maintained.

12. The oil bath temperature is 90 °C. Complete conversion of **1a** to **3a** is normally observed after 90–120 min. In this procedure, the extended reaction time is to ensure complete conversion.

13. A 2.0 N hydrochloric acid solution is prepared by adding 12.1 N "concentrated" HCl (EMD Chemicals, product HX0603-3) to deionized water (17:83, HCl,H$_2$O, vol:vol).

14. Prior to purification, it is recommended that the conversion be assayed by collecting a ^{1}H NMR spectrum of the crude reaction mixture. To do this, a small aliquot of the organic phase (the top layer in the separatory funnel after quenching with HCl and adding EtOAc), is concentrated *in vacuo* and dissolved in CDCl$_3$. The resulting solution is added to an NMR tube and analyzed. The conversion is determined by integration of the benzylic methylene protons, which appear as singlets

(approximately 3.72 ppm for **1a** and 3.87 ppm for **3a**, as referenced from the residual chloroform peak at 7.26 ppm).

15. All bulk organic solvents are purchased from Fisher Scientific and used as received.

16. Anhydrous sodium sulfate was purchased from Fischer Scientific (product S421-10) and used as received.

17. ^{1}H NMR analysis of the crude material after extraction reveals only the desired product, trace quantities of Ac-Ile-OH, and residual *t*-amylOH. The pure product is a white crystalline solid. Thus, the yellow/orange/red color of the crude material is possibly due to unidentifiable oligomeric/polymeric byproducts or organometallic species. The purification procedures outlined herein are designed to remove these byproducts.

18. Silica gel (32–63 μm, 60 Å) was purchased from Dynamic Adsorbents, Inc. (product 02826-25) and used as received.

19. The column is 60 mm wide in diameter, and the height of the silica is 20 cm inside the column. The eluent was collected in 100 mL fractions.

20. Column chromatography gives the product as an off-white solid. Though it appears >95% pure by ^{1}H NMR analysis, the pale yellow color is evidence of trace impurities, which are challenging to remove.

21. A reaction on half-scale provided 2.61 g (83%) of an off-white solid before recrystallization.

22. The column-purified product is dissolved in 4 mL of hot EtOAc (70 °C), then allowed to cool to room temperature and diluted with 1 mL of hexanes. The solution is cooled to –5 °C and kept overnight. The resulting crystals are collected by vacuum filtration on a Büchner funnel and washed with 2 mL of ice-cold hexanes. The mother liquor is concentrated *in vacuo* (30 mmHg, 40 °C), and the crystallization procedure is repeated with 1 mL EtOAc and 1 mL hexanes. The combined crystals were transferred to a 50-mL, round-bottomed flask and dried overnight at 0.37 mmHg to provide 4.91 g (78%) of a white solid.

23. Analytically pure material suitable for elemental analysis is obtained by taking the column-purified product and recrystallizing it from EtOAc/hexanes twice using the procedure in Note 22.

24. The product is indefinitely stable to air and moisture.

25. Single crystals suitable for X-ray diffraction are obtained by preparing a concentrated solution of **3a** in Et$_2$O in a 2 mL vial, placing this smaller vial in a 20 mL scintillation vial containing 5 mL of hexanes, sealing the

larger vial, and allowing the system to stand unperturbed overnight (approximately 8 hours).

26. Analytical data for product **3a**: TLC (hexanes:EtOAc:HOAc, 66:33:1) R_f = 0.29; M.p = 97–99 °C; ^{1}H NMR (600 MHz, CDCl$_3$) δ: 1.33 (t, J = 7.2 Hz, 3 H), 3.87 (d, J = 1.8 Hz, 2 H), 4.27 (q, J = 6.6 Hz, 2 H), 6.38 (d, J = 15.6 Hz, 1 H), 7.10 (t, J = 9.6 Hz, 1 H), 7.29 (td, J_1 = 8.4, J_2 = 6.0 Hz, 1 H), 7.38 (d, J = 7.8 Hz, 1 H), 7.84 (d, J = 15.6 Hz, 1 H); ^{13}C NMR (150 MHz, CDCl$_3$) δ: 14.2, 30.5 (d, J_{C-F} = 5.1 Hz), 60.9, 116.5 (d, J_{C-F} = 22.7 Hz), 120.5 (d, J_{C-F} = 16.1 Hz), 122.1, 122.5 (d, J_{C-F} = 3.2 Hz), 129.1 (d, J_{C-F} = 9.0 Hz), 136.3 (d, J_{C-F} = 3.5 Hz), 140.3 (d, J_{C-F} = 3.4 Hz), 161.4 (d, J_{C-F} = 245.6 Hz), 166.5, 175.9; Decoupled ^{19}F NMR (375 MHz, CDCl$_3$) δ: –115.11 (s, 1 F); IR (film): 2984, 1735, 1705, 1638, 1574, 1462, 1369, 1318, 1264, 1241, 1183, 1156, 1001, 971, 866, 802 cm^{-1}; HRMS (ESI-TOF) m/z calcd for C$_{13}$H$_{14}$FO$_4$ [M+H]$^+$: 253.0876, found: 253.0871; Anal. calcd. for C$_{13}$H$_{13}$FO$_4$: C 61.90, H 5.20, F 7.53, found: C 62.07, H 5.36, F 7.42.

Working with Hazardous Chemicals

The procedures in *Organic Syntheses* are intended for use only by persons with proper training in experimental organic chemistry. All hazardous materials should be handled using the standard procedures for work with chemicals described in references such as "Prudent Practices in the Laboratory" (The National Academies Press, Washington, D.C., 2011; the full text can be accessed free of charge at http://www.nap.edu/catalog.php?record_id=12654). All chemical waste should be disposed of in accordance with local regulations. For general guidelines for the management of chemical waste, see Chapter 8 of Prudent Practices.

In some articles in *Organic Syntheses*, chemical-specific hazards are highlighted in red "Caution Notes" within a procedure. It is important to recognize that the absence of a caution note does not imply that no significant hazards are associated with the chemicals involved in that procedure. Prior to performing a reaction, a thorough risk assessment should be carried out that includes a review of the potential hazards associated with each chemical and experimental operation on the scale that is planned for the procedure. Guidelines for carrying out a risk assessment and for analyzing the hazards associated with chemicals can be found in Chapter 4 of Prudent Practices.

The procedures described in *Organic Syntheses* are provided as published and are conducted at one's own risk. *Organic Syntheses, Inc.*, its Editors, and its Board of Directors do not warrant or guarantee the safety of individuals using these procedures and hereby disclaim any liability for any injuries or damages claimed to have resulted from or related in any way to the procedures herein.

Discussion

Transition metal–catalyzed reactions that form C(sp^2)–C(sp^2) linkages are an invaluable component of modern organic synthesis.[2] Prominent among this family of synthetic transformations is the Pd(0)-catalyzed Mizoroki–Heck reaction,[2,3] which couples aryl halides and olefins. The first step in the catalytic cycle for the Mizoroki–Heck reaction is the oxidative addition of [Pd(0)L$_n$] to the aryl halide (ArX) to form a [Pd(II)(Ar)(X)L$_n$] species, which goes on to react *via* olefin coordination, 1,2-migratory insertion, and β-hydride elimination. Reductive elimination from the resulting [Pd(II)(H)(X)] species then regenerates the active [Pd(0)L$_n$] species. From the vantage point of synthetic efficiency, an attractive alternative would be to take advantage of a Pd(II)-mediated C(aryl)–H cleavage event[4,5] (which, mechanistically, is redox neutral with concomitant loss of HX), to generate an analogous [Pd(II)(Ar)L$_n$] species, which could then proceed through the above sequence of elementary steps. Reoxidation of Pd(0) at the end of the catalytic cycle with an external oxidant would regenerate the active catalyst. This approach is potentially advantageous because it obviates the need for the regioselective installation of a halide onto the arene, which can take several steps or be infeasible altogether in intricately functionalized or sensitive organic molecules.

Since the seminal discoveries of Fujiwara and Moritani in the late 1960s,[6] Pd(II)-mediated C(aryl)–H olefination has steadily progressed to improved levels of catalytic efficiency and broadened substrate scope.[7,8] However, the practical utility of these reactions in organic synthesis has remained limited for a several reasons. Firstly, there is a lack of positional selectivity with substituted arenes (*e.g.*, toluene). Secondly, large stoichiometric excess of the arene must typically be used to overcome the low affinity between the arene and the metal center. Thirdly, reactivity with electron-poor arenes is typically low because the mechanism of C–H cleavage is thought to involve a Friedel–Crafts-type electrophilic palladation event. Solutions to these problems include the use of electron-rich heteroarenes and/or substrates containing a coordinating functional group which can direct *ortho*-C–H cleavage.[9–12] Indeed, with electron-rich nitrogen heterocycles, the power of Pd(II)-mediated C(aryl)–H olefination for the expedient synthesis of complex carbon skeletons has been demonstrated in the total synthesis of several alkaloid natural products.[13]

During the past 10 years, our research group has worked extensively to develop synthetically enabling C–H functionalization reactions.[5] As part of this effort, we have focused on the design, discovery, and optimization of ligand scaffolds to control and accelerate C–H activation with Pd(II) catalysts.[5g] In 2010, we found that mono-*N*-protected amino acid (MPAA) ligands, which we originally used to achieve chiral induction in enantioselective C–H activation,[14] also enhanced kinetic reactivity in Pd(II)-catalyzed C–H olefination of phenylacetic acid substrates.[15] Mechanistic studies were consistent with the observed rate increases stemming from acceleration of the C–H cleavage step.[15c,d] Since this series of initial reports,[15] MPAA ligands have found use in promoting a broad range Pd(II)-catalyzed C–H functionalization reactions.[16–18]

As previously demonstrated, the MPAA-accelerated *ortho*-C–H olefination reaction tolerates a range of electron-donating and -withdrawing substituents on the aryl group of the phenylacetic acid (Table 1).[15c] Additionally, hydrocinnamic acids are also competent substrates.[15a,b] Substrates bearing mono-substitution at the α-position are compatible with the MPAA-accelerated conditions, including commercial non-steroidal anti-inflammatory drugs, such as ketoprofen and naproxen.[15a,c] A similar acceleration effect using MPAAs with α,α-disubstituted substrates has not been observed in our studies.[15a,c] In the absence of a blocking group at the *ortho* or *meta* position, formation of the corresponding 2,6-diolefinated product is observed, and a separate procedure has been optimized to facilitate that transformation (Scheme 1).[15b] Various functionalized olefin coupling partners are tolerated; acrylates, vinyl ketones, and styrenes are all reactive (Table 2).[15c] Simple terminal alkenes also participate in the reaction but give the formal C–H allylation product.[15c]

Table 1. Pd(II)-catalyzed *ortho*-C–H olefination with representative phenylacetic acids[a]

$$R^1 \text{—} \underset{1}{\overset{R^2}{\bigcirc}} CO_2H + \underset{(2 \text{ equiv.})}{\overset{R^3}{\diagup}} \xrightarrow[\substack{t\text{-AmylOH (0.2 M)} \\ 90\ ^\circ C,\ 1\ atm\ O_2,\ 2\ h}]{\substack{5\ mol\%\ Pd(OAc)_2 \\ 10\ mol\%\ Ac\text{-Ile-OH} \\ 2\ equiv.\ KHCO_3}} R_1 \text{—} \underset{3}{\overset{R_2}{\bigcirc}} \overset{CO_2H}{\underset{R^3}{}}$$

Entry	Phenylacetic Acid	Olefin	Product	Yield (%)
1	**1b** (CF₃, CH₂CO₂H)	**2a** (CO₂Et)	**3b**	96
2	**1c** (Me, CH₂CO₂H)	**2a** (CO₂Et)	**3c**	97
3	**1d** (F₃C, CH₂CO₂H)	**2a** (CO₂Et)	**3d**	98
4	**1e** (Me, CH₂CO₂H)	**2a** (CO₂Et)	**3e**	83[b]
5	**1f** (OMe, MeO, CH₂CO₂H)	**2a** (CO₂Et)	**3f**	95
6	**1g** (NO₂, CH₂CO₂H)	**2a** (CO₂Et)	**3g**	70[c]
7	**1h** (Ph-C(O), Me, CHCO₂H)	**2a** (CO₂Et)	**3h**	98
8	**1i** (MeO-naphthyl, Me, CHCO₂H)	**2a** (CO₂Et)	**3i**	94

[a] Isolated yields. [b] An additional 11% of the 2,6-diolefinated product was formed. [c] An additional 6% of the decarboxylated product was formed.

Table 2. Pd(II)-catalyzed *ortho*-C–H olefination with representative olefin coupling partners[a,b]

Entry	Phenylacetic Acid	Olefin	Product	Yield (%)
1	1b	2b	3j	99
2	1b	2c	3k	90
3	1b	2d	3l	99
4	1b	2e	3m	65[c]
5	1b	2f	3n E:Z = 4:1	62[c]
6	1b	2g	3o	94
7	1b	2h	3p	97[c]

[a] The reaction conditions are identical to those in Table 1. [b] Isolated yields. [c] 10 h.

Scheme 1. Pd(II)-catalyzed 2,6-diolefination of a representative phenylacetic acid

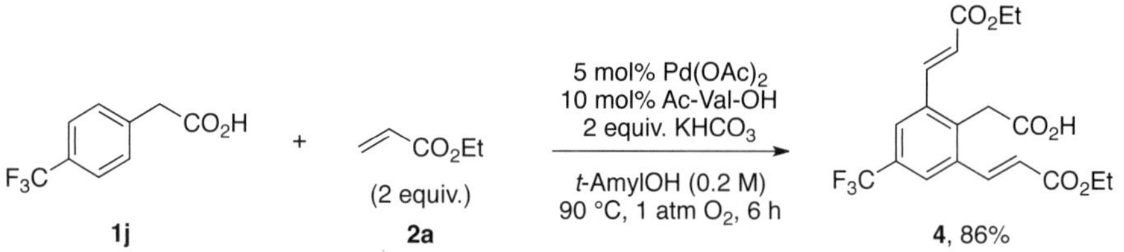

F3C — (phenylacetic acid) —CO2H (**1j**) + (CO2Et) (**2a**, 2 equiv.)

5 mol% Pd(OAc)2
10 mol% Ac-Val-OH
2 equiv. KHCO3
————————————
t-AmylOH (0.2 M)
90 °C, 1 atm O2, 6 h

→ **4, 86%** (diolefinated product with CO2Et groups, F3C substituted)

Overall, the method is high-yielding, operationally simple, and tolerant towards a variety of functional groups (see Table 1). The reaction is atom-economical in the sense that every atom from both starting materials (discounting the formal loss of "H2") is incorporated into the product,[19] and the only byproduct that is produced is H2O (or H2O2). The synthetic utility of this C–H olefination reaction has recently been demonstrated in the total synthesis of (+)-lithospermic acid and related analogs.[20]

In this manuscript, we describe an optimized version of our MPAA-accelerated *ortho*-C–H olefination reaction that is more amenable to scale-up. Specifically, compared to our earlier report,[15c] we have reduced the olefin and MPAA ligand equivalents. Furthermore, we demonstrate that with a representative phenylacetic acid substrate, the reaction can be run at a higher concentration (0.5 M) and can be performed on over a 5-gram scale. The connectivity of the product and the olefin stereochemistry have been confirmed *via* single-crystal X-ray diffraction (Figure 1). This work demonstrates the essential role that the ligand plays in developing practical Pd(II)-catalyzed C–H functionalization reactions that are reliable in preparative-scale synthesis.

Figure 1. X-ray crystal structure of 3a

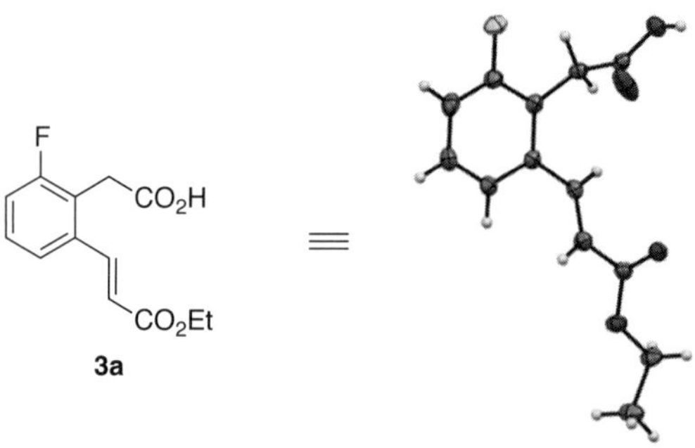

(Structure **3a**: fluorinated phenyl with CO2H and CH=CH–CO2Et groups) ≡ (X-ray crystal structure)

3a

References

1. Department of Chemistry, The Scripps Research Institute (TSRI), 10550 N. Torrey Pines Road, La Jolla, CA 92037 (USA). E-mail: yu200@scripps.edu. This research was supported by the NIH (NIGMS, 1 R01 GM084019-02), the NSF (NSF CHE-1011898), Amgen, and Eli Lilly. Individual awards and fellowships were granted by the NSF GRFP, the NDSEG Fellowship Program, TSRI, and the Skaggs-Oxford Scholarship program (K.M.E.); the Austrian Science Fund (N.D.); the EPA STAR Graduate Fellowship Program (Agreement No. FP917296-01-0) (P.S.T.-B.); the China Scholarship Council (D.-H.W.); and the ARCS and Donald and Delia Baxter Foundations (A.C.S.). TSRI Manuscript No. 21472.

2. Nicolaou, K. C.; Bulger, P. G.; Sarlah, D. *Angew. Chem. Int. Ed.* **2005**, *44*, 4442–4489.

3. (a) Mizoroki, T.; Mori, K.; Ozaki, A. *Bull. Chem. Soc. Jpn.* **1971**, *44*, 581. (b) Heck, R. F.; Nolley, J. P. *J. Org. Chem.* **1972**, *37*, 2320–2322.

4. Selected reviews on Pd-catalyzed C–H functionalization from other groups, see: (a) Jia, C.; Kitamura, T.; Fujiwara, Y. *Acc. Chem. Res.* **2001**, *34*, 633–639. (b) Campeau, L.-C.; Stuart, D. R.; Fagnou, K. *Aldrichim. Acta* **2007**, *40*, 35–41. (c) Satoh, T.; Miura, M. *Chem. Lett.* **2007**, *36*, 200–205. (d) Seregin, I. V.; Gevorgyan, V. *Chem. Soc. Rev.* **2007**, *36*, 1173–1193. (e) Ackermann, L.; Vicente, R.; Kapdi, A. R. *Angew. Chem. Int. Ed.* **2009**, *48*, 9792–9826. (f) Daugulis, O.; Do, H.-Q.; Shabashov, D. *Acc. Chem. Res.* **2009**, *42*, 1074–1086. (g) Lyons, T. W.; Sanford, M. S. *Chem. Rev.* **2010**, *110*, 1147–1169. (h) Yeung, C. S.; Dong, V. M. *Chem. Rev.* **2011**, *111*, 1215–1292. (i) Wencel-Delord, J.; Dröge, T.; Liu, F.; Glorius, F. *Chem. Soc. Rev.* **2011**, *40*, 4740–4761.

5. For reviews from our group, see: (a) Yu, J.-Q.; Giri, R.; Chen, X. *Org. Biomol. Chem.* **2006**, *4*, 4041–4047. (b) Chen, X.; Engle, K. M.; Wang, D.-H.; Yu, J.-Q. *Angew. Chem. Int. Ed.* **2009**, *48*, 5094–5115. (c) Giri, R.; Shi, B.-F.; Engle, K. M.; Maugel, N.; Yu, J.-Q. *Chem. Soc. Rev.* **2009**, *38*, 3242–3272. (d) Wasa, M.; Engle, K. M.; Yu, J.-Q. *Isr. J. Chem.* **2010**, *50*, 605–616. (e) Engle, K. M.; Mei, T.-S.; Wasa, M.; Yu, J.-Q. *Acc. Chem. Res.* **2012**, *45*, 788–802. (f) Mei, T.-S.; Kou, L.; Ma, S.; Engle, K. M.; Yu, J.-Q. *Synthesis* **2012**, *44*, 1778–1791. (g) Engle, K. M.; Yu, J.-Q. *J. Org. Chem.* **2013**, *78*, 8927–8955. (h) Farmer, M. E.; Laforteza, B. N.; Yu, J.-Q. *Biorg. Med. Chem.* **2014**, *22*, 4445-4452.

6. Initial reports of C(sp^2)–H olefination: (a) Moritani, I.; Fujiwara, Y. *Tetrahedron Lett.* **1967**, *8*, 1119–1122. (b) Fujiwara, Y.; Moritani, I.; Matsuda, M.; Teranishi, S. *Tetrahedron Lett.* **1968**, *9*, 633–636.

7. For selected examples of subsequent work to improve the catalytic turnover and expand the substrate scope of Pd(II)-mediated C–H olefination, see: (a) Fujiwara, Y.; Moritani, I.; Danno, S.; Asano, R.; Teranishi, S. *J. Am. Chem. Soc.* **1969**, *91*, 7166–7169. (b) Asano, R.; Moritani, I.; Fujiwara, Y.; Teranishi, S. *J. Chem. Soc. D: Chem. Commun.* **1970**, 1293. (c) Shue, R. S. *J. Chem. Soc. D: Chem. Commun.* **1971**, 1510–1511. (d) Fujiwara, Y.; Asano, R.; Moritani, I.; Teranishi, S. *J. Org. Chem.* **1976**, *41*, 1681–1683. (e) Tsuji, J.; Nagashima, H. *Tetrahedron* **1984**, *40*, 2699–2702. (f) Jia, C.; Lu, W.; Kitamura, T.; Fujiwara, Y. *Org. Lett.* **1999**, *1*, 2097–2100. (g) Jia, C.; Piao, D.; Oyamada, J.; Lu, W.; Kitamura, T.; Fujiwara, Y. *Science* **2000**, *287*, 1992–1995. (h) Yokota, T.; Tani, M.; Sakaguchi, S.; Ishii, Y. *J. Am. Chem. Soc.* **2003**, *125*, 1476–1477. (i) Dams, M.; DeVos, D. E.; Celen, S.; Jacobs, P. A. *Angew. Chem. Int. Ed.* **2003**, *42*, 3512–3515. (j) Kubota, A.; Emmert, M. H.; Sanford, M. S. *Org. Lett.* **2012**, *14*, 1760–1763.

8. For a review of Pd(II)-mediated C–H olefination, see: Le Bras, J.; Muzart, J. *Chem. Rev.* **2011**, *111*, 1170–1214.

9. Examples of non-directed Pd(II)-catalyzed C–H olefination with electron-rich heterocycles: (a) Fujiwara, Y.; Maruyama, O.; Yoshidomi, M.; Taniguchi, H. *J. Org. Chem.* **1981**, *46*, 851–855. (b) Itahara, T.; Ikeda, M.; Sakakibara, T. *J. Chem. Soc., Perkin Trans. 1* **1983**, 1361–1363. (c) Ferreira, E. M.; Stoltz, B. M. *J. Am. Chem. Soc.* **2003**, *125*, 9578–9579. (d) Beccalli, E. M.; Broggini, G. *Tetrahedron Lett.* **2003**, *44*, 1919–1921. (e) Ma, S.; Yu, S. *Tetrahedron Lett.* **2004**, *45*, 8419–8422. (f) Liu, C.; Widenhoefer, R. A. *J. Am. Chem. Soc.* **2004**, *126*, 10250–10251. (g) Grimster, N. P.; Gauntlett, C.; Godfrey, C. R. A.; Gaunt, M. J. *Angew. Chem. Int. Ed.* **2005**, *44*, 3125–3129.

10. Examples of Pd(II)-catalyzed C–H olefination of substrates containing directing groups (a) Diamond, S. E.; Szalkiewicz, A.; Mares, F. *J. Am. Chem. Soc.* **1979**, *101*, 490–491. (b) Satoh, T.; Itaya, T.; Miura, M.; Nomura, M. *Chem. Lett.* **1996**, *25*, 823–824. (c) Miura, M.; Tsuda, T.; Satoh, T.; Pivsa-Art, S.; Nomura, M. *J. Org. Chem.* **1998**, *63*, 5211–5215. (d) Boele, M. D. K.; van Strijdonck, G. P. F.; de Vries, A. H. M.; Kamer, P. C. J.; de Vries, J. G.; van Leeuwen, P. W. N. M. *J. Am. Chem. Soc.* **2002**, *124*, 1586–1587. (e) Cai, G.; Fu, Y.; Li, Y.; Wan, X.; Shi, Z. *J. Am. Chem. Soc.* **2007**, *129*, 7666–7673. (f) Houlden, C. E.; Bailey, C. D.; Ford, J. G.;

Gagné, M. R.; Lloyd-Jones, G. C.; Booker-Milburn, K. I. *J. Am. Chem. Soc.* **2008**, *130*, 10066–10067. (g) Li, J.-J.; Mei, T.-S.; Yu, J.-Q. *Angew. Chem. Int. Ed.* **2008**, *47*, 6452–6455. (h) Nishikata, T.; Lipshutz, B. H. *Org. Lett.* **2010**, *12*, 1972–1975. (i) Zhu, C.; Falck, J. R. *Org. Lett.* **2011**, *13*, 1214–1217. (j) Wrigglesworth, J. W.; Cox, B.; Lloyd-Jones, G. C.; Booker-Milburn, K. I. *Org. Lett.* **2011**, *13*, 5326–5329. (k) García-Rubia, A.; Urones, B.; Gómez Arrayás, R.; Carretero, J. C. *Angew. Chem. Int. Ed.* **2011**, *50*, 10927–10931. (l) Gandeepan, P.; Cheng, C.-H. *J. Am. Chem. Soc.* **2012**, *134*, 5738–5741. (m) Yu, M.; Xie, Y.; Xie, C.; Zhang, Y. *Org. Lett.* **2012**, *14*, 2164–2167.

11. (a) Capito, E.; Brown, J. M.; Ricci, A. *Chem. Commun.* **2005**, 1854–1856. (b) Maehara, A.; Tsurugi, H.; Satoh, T.; Miura, M. *Org. Lett.* **2008**, *10*, 1159–1162. (c) García-Rubia, A.; Arrayás, R. G.; Carretero, J. *Angew. Chem. Int. Ed.* **2009**, *48*, 6511–6515.

12. Cho, S. H.; Hwang, S. J.; Chang, S. *J. Am. Chem. Soc.* **2008**, *130*, 9254–9256.

13. Examples in which Pd(II)-mediated C–H olefination of an electron-rich heteroarene has been used in total synthesis: (a) Trost, B. M.; Godleski, S. A.; Genêt, J. P. *J. Am. Chem. Soc.* **1978**, *100*, 3930–3931. (b) Cushing, T. D.; Sanz-Cervera, J. F.; Williams, R. M. *J. Am. Chem. Soc.* **1993**, *115*, 9323–9324. (c) Baran, P. S.; Corey, E. J. *J. Am. Chem. Soc.* **2002**, *124*, 7904–7905. (d) Garg, N. K.; Caspi, D. D.; Stoltz, B. M. *J. Am. Chem. Soc.* **2004**, *126*, 9552–9553. (e) Beck, E. M.; Hatley, R.; Gaunt, M. J. *Angew. Chem. Int. Ed.* **2008**, *47*, 3004–3007. (f) Bowie, A. L., Jr.; Trauner, D. *J. Org. Chem.* **2009**, *74*, 1581–1586.

14. Amino acid ligands were identified as effective ligand scaffolds for Pd(II)-catalyzed C–H functionalization reactions during our studies of enantioselective C–H functionalization: (a) Shi, B.-F.; Maugel, N.; Zhang, Y.-H.; Yu, J.-Q. *Angew. Chem. Int. Ed.* **2008**, *47*, 4882–4886. (b) Shi, B.-F.; Zhang, Y.-H.; Lam, J. K.; Wang, D.-H.; Yu, J.-Q. *J. Am. Chem. Soc.* **2010**, *132*, 460–461. For an early report describing stoichiometric asymmetric cyclopalladation of (dimethylamino)methylferrocene in the presence of Ac-Val-OH, see: (c) Sokolov, V. I.; Troitskaya, L. L.; Reutov, O. A. *J. Organomet. Chem.* **1979**, *182*, 537–546.

15. (a) Wang, D.-H.; Engle, K. M.; Shi, B.-F.; Yu, J.-Q. *Science* **2010**, *327*, 315–319. (b) Engle, K. M.; Wang, D.-H.; Yu, J.-Q. *Angew. Chem. Int. Ed.* **2010**, *49*, 6169–6173. (c) Engle, K. M.; Wang, D.-H.; Yu, J.-Q. *J. Am. Chem. Soc.* **2010**, *132*, 14137–14151. (d) Baxter, R. D.; Sale, D.; Engle, K. M.; Yu, J.-Q.; Blackmond, D. G. *J. Am. Chem. Soc.* **2012**, *134*, 4600–4606.

16. Non-stereoselective, MPAA-promoted reactions: (a) Lu, Y.; Wang, D.-H.; Engle, K. M.; Yu, J.-Q. *J. Am. Chem. Soc.* **2010**, *132*, 5916–5921. (b) Dai, H.-X.; Stepan, A. F.; Plummer, M. S.; Zhang, Y.-H.; Yu, J.-Q. *J. Am. Chem. Soc.* **2011**, *133*, 7222–7228. (c) Huang, C.; Chattopadhyay, B.; Gevorgyan, V. *J. Am. Chem. Soc.* **2011**, *133*, 12406–12409. (d) Lu, Y.; Leow, D.; Wang, X.; Engle, K. M.; Yu, J.-Q. *Chem. Sci.* **2011**, *2*, 967–971. (e) Engle, K. M.; Thuy-Boun, P. S.; Dang, M.; Yu, J.-Q. *J. Am. Chem. Soc.* **2011**, *133*, 18183–18193. (f) Novák, P.; Correa, A.; Gallardo-Donaire, J.; Martin, R. *Angew. Chem. Int. Ed.* **2011**, *50*, 12236–12239. (g) Leow, D.; Li, G.; Mei, T.-S.; Yu, J.-Q. *Nature* **2012**, *486*, 518–522. (h) Wang, Y.-N.; Guo, X.-Q.; Zhu, X.-H.; Zhong, R.; Cai, L.-H.; Hou, X.-F. *Chem. Commun.* **2012**, *48*, 10437–10439. (i) Cong, X.; You, J.; Gao, G.; Lan, J. *Chem. Commun.* **2013**, *49*, 662–664. (j) Li, G.; Leow, D.; Wan, L.; Yu, J.-Q. *Angew. Chem. Int. Ed.* **2013**, *52*, 1245–1247. (k) Presset, M.; Oehlrich, D.; Rombouts, F.; Molander, G. A. *Org. Lett.* **2013**, *15*, 1528–1531. (l) Meng, X.; Kim, S. *Org. Lett.* **2013**, *15*, 1910–1913. (m) Dai, H.-X.; Li, G.; Zhang, X.-G.; Stepan, A. F.; Yu, J.-Q. *J. Am. Chem. Soc.* **2013**, *135*, 7567–7571. (n) Li, Y.; Ding, Y.-J.; Wang, J.-Y.; Su, Y.-M.; Wang, X.-S. *Org. Lett.* **2013**, *15*, 2574–2577. (o) Rosen, B. R.; Simke, L. R.; Thuy-Boun, P. S.; Dixon, D. D.; Yu, J.-Q.; Baran, P. S. *Angew. Chem. Int. Ed.* **2013**, *52*, 7317–7320. (p) Wen, Z.-K.; Xu, Y.-H.; Loh, T.-P. *Chem. Sci.* **2013**, *4*, 4520–4524. (q) Meng, X.; Kim, S. *J. Org. Chem.* **2013**, *78*, 11247–11254. (r) Wang, H.-L.; Hu, R.-B.; Zhang, H.; Zhou, A.-X.; Yang, S.-D. *Org. Lett.* **2013**, *15*, 5302–5305. (s) Cong, X.; Tang, H.; Wu, C.; Zeng, X. *Organometallics* **2013**, *32*, 6565–6575. (t) Thuy-Boun, P. S.; Villa, G.; Dang, D.; Richardson, P.; Su, S.; Yu, J.-Q. *J. Am. Chem. Soc.* **2013**, *135*, 17508–17513. (u) Wan, L.; Dastbaravardeh, N.; Li, G.; Yu, J.-Q. *J. Am. Chem. Soc.* **2013**, *135*, 18056–18059. (v) Chan, K. S. L.; Wasa, M.; Chu, L.; Laforteza, B. N.; Miura, M.; Yu, J.-Q. *Nat. Chem.* **2014**, *6*, 146–150.

17. Stereoselective, MPAA-promoted reactions: (a) Wasa, M.; Engle, K. M.; Lin, D. W.; Yoo, E. J.; Yu, J.-Q. *J. Am. Chem. Soc.* **2011**, *133*, 19598–19601. (b) Cheng, X.-F.; Li, Y.; Su, Y.-M.; Yin, F.; Wang, J.-Y.; Sheng, J.; Vora, H. U.; Wang, X.-S.; Yu, J.-Q. *J. Am. Chem. Soc.* **2013**, *135*, 1236–1239. (c) Gao, D.-W.; Shi, Y.-C.; Gu, Q.; Zhao, Z.-L.; You, S.-L. *J. Am. Chem. Soc.* **2013**, *135*, 86–89. (d) Pi, C.; Li, Y.; Cui, X.; Zhang, H.; Han, Y.; Wu, Y. *Chem. Sci.* **2013**, *4*, 2675–2679. (e) Chu, L.; Wang, X.-C.; Moore, C. E.; Rheingold, A. L.; Yu, J.-Q. *J. Am. Chem. Soc.* **2013**, *135*, 16344–16347. (f) Xiao, K.-J.; Lin, D. W.; Miura, M.; Zhu, R.-Y.; Gong, W.; Wasa, M.; Yu, J.-Q. *J. Am. Chem. Soc.* **2014**, *136*, 8138–8142.

18. For computational studies of MPAA ligands, see Ref. 16s and the following: (a) Musaev, D. G.; Kaledin, A.; Shi, B.-F.; Yu, J.-Q. *J. Am. Chem. Soc.* **2012**, *134*, 1690–1698. (b) Cheng, G.-J.; Yang, Y.-F.; Liu, P.; Chen, P.; Sun, T.-Y.; Li, G.; Zhang, X.; Houk, K. N.; Yu, J.-Q.; Wu, Y.-D. *J. Am. Chem. Soc.* **2013**, *136*, 894–897.

19. Trost, B. M. *Science* **1991**, *254*, 1471–1477.

20. (a) Wang, D.-H.; Yu, J.-Q. *J. Am. Chem. Soc.* **2011**, *133*, 5767–5769. (b) Ghosh, A. K.; Cheng, X.; Zhou, B. *Org. Lett.* **2012**, *14*, 5046–5049. (c) Wang, H.; Li, G.; Engle, K. M.; Yu, J.-Q.; Davies, H. M. L. *J. Am. Chem. Soc.* **2013**, *135*, 6774–6777.

Appendix
Chemical Abstracts Nomenclature (Registry Number)

N-Acetyl isoleucine (Ac-Ile-OH); (2*S*, 3*S*)-2-acetamido-3-methylpentanoic acid (3077-46-1)

Potassium bicarbonate; (298-14-6)

Palladium(II) acetate; (3375-31-3)

2-Fluorophenylacetic acid; 2-(2-fluorophenyl)acetic acid (451-82-1)

tert-Amyl alcohol (*t*-AmylOH); 2-methyl-2-butanol (75-85-4)

Ethyl acrylate; Ethyl propenoate (140-88-5)

Keary Mark Engle graduated *Phi Beta Kappa* and *summa cum laude* from the University of Michigan, where he worked with Prof. Adam Matzger studying self-assembled monolayers. A Fulbright Scholar, he spent the 2007–2008 academic year studying under the tutelage of Prof. Manfred Reetz at the Max-Planck-Institut für Kohlenforschung (Germany). He completed his graduate work jointly at The Scripps Research Institute and the University of Oxford, under the supervision of Prof. Jin-Quan Yu and Prof. Véronique Gouverneur, respectively, earning a Ph.D. in Chemistry and a DPhil in Biochemistry. During graduate school, his honors included NSF and NDSEG Predoctoral Fellowships. Presently, Keary is an NIH Postdoctoral Fellow with Prof. Robert H. Grubbs at Caltech.

Navid Dastbaravardeh studied chemistry at the Ludwig Maximilian University of Munich (LMU, Germany) and received his diploma in 2008. He then moved to the Vienna University of Technology (VUT, Austria), where he completed his Ph.D. in Organic Chemistry under the supervision of Profs. Michael Schnürch and Marko D. Mihovilovic. Supported by an Erwin-Schrödinger Research Fellowship, he is currently pursuing postdoctoral research with Prof. Jin-Quan Yu at The Scripps Research Institute, focusing on palladium-catalyzed C–H bond-functionalization reactions.

Peter S. Thuy-Boun carried out undergraduate research with Prof. Lijuan Li and Prof. Paula Diaconescu, graduating with a BS in Chemistry from UCLA. Since 2010, he has been a graduate student at The Scripps Research Institute working on Pd-catalyzed C–H alkylation methodology under the supervision of Prof. Jin-Quan Yu and on proteomic characterization of the human gut microbiome under the guidance of Prof. Dennis W. Wolan. His graduate work is supported by an EPA STAR Predoctoral Fellowship.

Dong-Hui Wang completed his BSc at Lanzhou University in 2000 then carried out research at the Shanghai Institute of Organic Chemistry under Prof. Zhaoguo Zhang. In 2004, he began graduate studies under Prof. Jin-Quan Yu. He earned his MSc from Brandeis University before relocating to The Scripps Research Institute, where he completed his Ph.D. in 2010. His thesis research was recognized with the Chinese Government Award for Outstanding Self-Financed Students Abroad. Dong-Hui worked as a postdoctoral research fellow in the laboratory of Prof. Stephen Buchwald at MIT (2010–2012), and as a medicinal chemist at Abide Therapeutics (2012–2014). Presently, he is an Assistant Professor at the Shanghai Institute of Organic Chemistry (China).

Organic
Syntheses

Aaron C. Sather earned his BSc at the University of Oregon, working with Prof. Darren W. Johnson on arsenic remediation and anion recognition. He graduated *cum laude* with the distinction of departmental honors. He then moved to The Scripps Research Institute to study molecular recognition under the guidance of Prof. Julius Rebek, Jr., where he was both a Baxter Fellow and an ARCS Fellow. Presently, Aaron is a NIH Postdoctoral Fellow at MIT in the research laboratory of Prof. Stephen Buchwald.

Jin-Quan Yu received his BSc in Chemistry from East China Normal University and his MSc from the Guangzhou Institute of Chemistry. In 2000, he obtained his PhD at the University of Cambridge with Prof. J. B. Spencer. Following time as a junior research fellow at Cambridge, he joined the laboratory of Prof. E. J. Corey at Harvard University as a postdoctoral fellow. He then began his independent career at Cambridge (2003–2004), before moving to Brandeis University (2004–2007), and finally to The Scripps Research Institute, where he is currently Frank and Bertha Hupp Professor of Chemistry. His group studies transition metal–catalyzed C–H activation.

Dr. Sandeep N. Raikar did his BSc and MSc from Sri Satya Sai Institute of Higher Learning-India. He then moved to University of Kansas where he worked with Prof. Helena C. Malinakova and obtained his Ph.D. degree in 2013. Currently he is a postdoctoral associate in Prof. Huw M. L. Davies laboratory at Emory University.

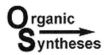

Palladium-catalyzed β-Selective C(sp³)–H Arylation of N-Boc-Piperidines

Anthony Millet and Olivier Baudoin[*1]

Université Claude Bernard Lyon 1, CNRS UMR 5246 – Institut de Chimie et Biochimie Moléculaires et Supramoléculaires, CPE Lyon, 43 Boulevard du 11 Novembre 1918, 69622 Villeurbanne, France

Checked by Louis C. Morrill and Richmond Sarpong

A.

1

n-BuLi / TMEDA
then i-Pr$_2$PCl

n-Hexane
20 °C to reflux, 4 h

2

B.

3

s-BuLi / TMEDA, Et$_2$O –78 °C, 3 h
then ZnCl$_2$, –78 to 20 °C, 1.5 h
then Pd$_2$dba$_3$, **2**, 3-bromoanisole

Toluene
60 °C, 17 h

4

Procedure

Caution! sec-Butylithium is very pyrophoric and must not be allowed to come into contact with the atmosphere. This reagent should only be handled by individuals trained in its proper and safe use. It is recommended that transfers be carried out by using a 20-mL or smaller glass syringe filled to no more than 2/3 capacity or by cannula.

A. *2-(Diisoproylphosphanyl)-1-phenylpyrrole (2)*. A 250-mL, three-necked, round-bottomed flask equipped with a 3.7-cm Teflon-coated magnetic stirbar, a condenser capped by a rubber septum and a thermometer is placed in a 19 °C water bath and charged with 1-phenylpyrrole (**1**) (4.30 g, 30 mmol, 1.0 equiv) (Note 1). The third neck of the flask is closed by a rubber septum and connected to a combined nitrogen/vacuum line and evacuated/back-filled with nitrogen twice. *n*-Hexane (45 mL) (Note 2) is added by syringe, followed by TMEDA, also by syringe, (6.75 mL, 5.23 g, 45 mmol, 1.5 equiv) (Note 3), and the mixture is stirred (200 rpm) at 19 °C until complete dissolution. *n*-Butyllithium (13.3 mL, 30 mmol, 1.0 equiv) (2.26 M in hexanes) (Note 4) is added dropwise over 10 min using a syringe pump (Note 5). The rubber septum on the 250-mL, three-necked flask is replaced by a glass stopper and the water bath is replaced by an oil bath. The reaction mixture is heated to reflux for 3 h, giving a brown solution. Chlorodiisopropylphosphine (4.77 mL, 4.58 g, 30 mmol, 1.0 equiv) (Note 6) in 10 mL of hexanes (Note 2) was added dropwise at reflux through the condenser over 5 min using a syringe pump to give a beige precipitate, and the mixture is refluxed for an additional hour. The reaction is cooled to between 0-5 °C using an ice/water bath before slow addition of 50 mL of degassed water (Note 7). The quenched reaction is stirred (200 rpm) for 10 min at 0–5 °C to give a clear solution. The solution is transferred to a 250-mL separatory funnel, the layers are separated, and the aqueous layer is extracted with hexanes (2 x 75 mL) (Note 2). The organic layer are combined and washed with brine (75 mL) (Note 8), dried over 5 g of magnesium sulfate (Note 9), filtered through a 150-mL medium porosity sintered glass funnel into a 500-mL round-bottomed flask. The residue is washed with hexanes (25 mL) (Note 2), and the resulting solution is concentrated by rotary evaporation (40 °C, 20 mmHg). The oil is transferred to a 100-mL round-bottomed flask, washing with 5 mL of hexanes, and concentrated by rotary evaporation (40 °C, 20 mmHg), to afford 7.87 g of a brown oil. The crude mixture is dried under high-vacuum (20 °C, 0.5 mmHg) and under stirring using a 1-cm Teflon-coated magnetic stirbar until crystallization of a brown solid. The flask is sealed with a rubber septum, connected to a combined nitrogen/vacuum line and evacuated/back-filled with nitrogen twice. The solid is dissolved in 10 mL of degassed methanol (Note 10) upon heating to 50 °C. The solution is allowed to cool to room temperature, and then cooled at –18 °C using a low temperature freezer for 15 h. The resulting crystals are collected by filtration through a 25-mL medium porosity sintered glass funnel, washed with 5 mL of ice-cold methanol twice, and then transferred to a 50-mL round-bottomed flask and dried for 4 h at

0.05 mmHg to provide 3.61–3.73 g (46–48% yield) of compound **2** as a white powder (Note 11).

Reaction Apparatus

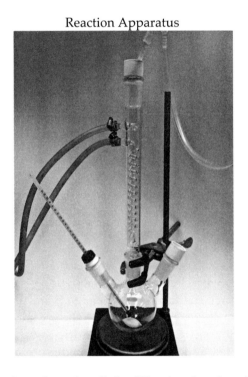

B. *tert-Butyl 3-(3-methoxyphenyl)piperidine-1-carboxylate* (**4**). A 500-mL, three-necked, round-bottomed flask equipped with a 4.7-cm Teflon-coated magnetic stirbar, a thermometer and a 250-mL pressure equalizing dropping funnel capped by a rubber septum is charged with *tert*-butyl piperidine-1-carboxylate (**3**) (7.98 mL, 7.69 g, 41.5 mmol, 1.0 equiv) (Note 12). The third neck of the flask is connected to a combined nitrogen/vacuum line using a glass adapter and evacuated/back-filled with nitrogen twice. Diethyl ether (65 mL) (Note 13) is added to the stirred (200 rpm) reaction mixture followed by the addition of TMEDA (7.47 mL, 5.79 g, 49.8 mmol, 1.2 equiv) by syringe (Note 3). The solution is subsequently cooled to –78 °C using a dry-ice/acetone bath. *s*-Butyllithium (48.8 mL, 49.8 mmol, 1.2 equiv) (1.02 M in cyclohexane) (Note 14) is added dropwise over 30 min *via* the pressure equalizing dropping funnel (Note 15), to give a yellow cloudy solution which is stirred for 3 h at –78 °C. Zinc chloride (99.6 mL, 49.8 mmol, 1.2 equiv) (0.5 M in THF) (Note 16) is added dropwise over 45 min *via* the pressure equalizing dropping funnel (Note 17). The resulting mixture is stirred for 30 min at –78 °C, allowed to warm to 20 °C by removing the dry-ice/acetone bath, and stirred for 30 min. The reaction mixture is rapidly transferred to a 500-mL round-bottomed flask, which has been previously evacuated/back-filled with nitrogen twice. The clear orange solution is concentrated by rotary evaporation (44 °C, 80 mmHg). A 4.7-cm Teflon-coated magnetic stirbar is added to the flask before it is closed with a rubber septum and connected to a combined nitrogen/vacuum line. The resulting white cloudy solution is concentrated for 15 min under vacuum (0.05 mmHg) and then back-filled with nitrogen. Meanwhile, a 100-mL round-bottomed flask containing a 2.5-cm Teflon-coated magnetic stirbar is charged with tris(dibenzylideneacetone)dipalladium(0) (950 mg, 1.04 mmol, 0.025 equiv) (Note 18) and 2-(diisopropylphosphanyl)-1-phenylpyrrole (**2**) (538 mg, 2.07 mmol, 0.05 equiv), closed with a rubber septum, connected to a combined nitrogen/vacuum line and evacuated/back-filled with nitrogen twice. Toluene (60 mL) (Note 19) is added by syringe , and the solution is stirred (200 rpm) for 20 min at 20 °C. The resulting catalyst solution is added to the above piperidinylzinc reagent *via* syringe, and the 100 mL flask is washed with toluene (35 mL) (Note 19), before addition of 3-bromoanisole (3.65 mL, 29.1 mmol, 0.7 equiv) (Note 20). The resulting red-brown solution is heated by oil bath (60 °C) and stirred (200 rpm) for 17 h. After cooling to room temperature, a saturated aqueous solution of ammonium chloride (150 mL) (Note 21) is added, followed by ethyl acetate (75 mL) (Note 22). The bi-phasic solution is transferred to a 500-mL separatory funnel and layers are separated. The aqueous layer is extracted with ethyl acetate (2 x 75 mL) (Note 22). The combined organic layers are washed with brine (150 mL) (Note 8), dried over magnesium sulfate (10 g) (Note 9), filtered through a 150-mL medium porosity sintered glass funnel, which is washed with ethyl acetate (25 mL) (Note 22). The resulting solution is concentrated by rotary evaporation (45 °C, 40 mmHg) to afford 22.4 g of an orange oil containing a precipitate (Note 23). This crude mixture is dissolved in dichloromethane (100 mL) (Note 24), charged with silica gel (30 g) (Note 25) then concentrated by rotary evaporation (20 °C, 35 mmHg), followed by 10 min under 0.05 mmHg at 20 °C. The silica-adsorbed reaction mixture is charged on a column (9 cm width, 10 cm height) containing 275 g of silica gel and eluted with 4 L of 5% ethyl acetate-*n*-hexanes collecting 50 mL fractions. A first fraction containing a mixture of α- and β-arylated products is obtained in fractions 43-50, which are concentrated by rotary evaporation (45 °C, 50 35 mmHg) to afford 1.36 g of a mixture of compounds. The second fraction containing the desired product (4.13–4.21 g) is obtained in fractions 51-80, which are concentrated by rotary evaporation (45 °C, 5 mmHg). The first fraction is dissolved in dichloromethane (50 mL) and silica gel (6 g) is added before being concentrated by rotary evaporation (20 °C, 35 mmHg), followed by 10 min under 0.05 mmHg at 20 °C. The silica-adsorbed reaction mixture is charged on a column (4 cm width, 12 cm height) containing 70 g of silica gel and eluted with 1.5 L of 5% ethyl acetate/n-hexanes, collecting 20 mL fractions. The desired product (0.82–0.90 g) is obtained in fractions 30-70, which are concentrated by rotary evaporation (45 °C, 5 mmHg). The combined product containing fractions are dried for 2 h under 0.05 mmHg at 20 °C to afford 4.95–5.11 g (58–60% yield) of compound **4** as a yellow oil (Notes 26 and 27).

Reaction Apparatus

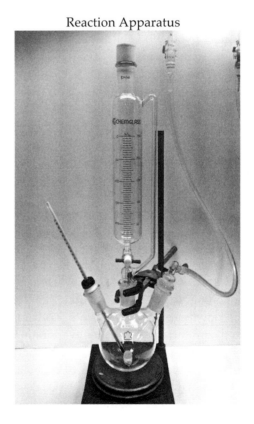

1. 1-Phenylpyrrole (99%) was purchased from Sigma-Aldrich and used as received.
2. *n*-Hexane (puriss., pa, ACS Reagent, ≥99% (GC)) were purchased from Sigma-Aldrich, and distilled under nitrogen from sodium benzophenone ketyl. The solvent is withdrawn from the receiver flask with a syringe. Hexanes (mixture of isomers, Sigma-Aldrich) were used for extractions and chromatography.
3. *N*,*N*,*N'*,*N'*-Tetramethylethylenediamine (99%) was purchased from Sigma-Aldrich and distilled under nitrogen from calcium hydride before use.
4. A 2.5 M solution of *n*-butyllithium in hexanes under Sure/Seal was purchased from Sigma-Aldrich, and the concentration was determined by titration with biphenyl-4-methanol to be 2.26 M prior to use according to the reported method: Juaristi, E.; Martinez-Richa, A.; Garcia-Rivera, A.; Cruz-Sánchez, J. S. *J. Org. Chem.* **1983**, *48*, 2603-2606.
5. The temperature increases to 31 °C during the addition, and a yellow solution is obtained.
6. Chlorodiisopropylphosphine (96%) was purchased from Acros-Organics, and used as received.
7. Water was degassed by argon-bubbling for 1 h.
8. Sodium chloride (ACS reagent, ≥99%) was purchased from Sigma-Aldrich and added to water until saturation.
9. Magnesium sulfate (Sec Pur) was purchased from Fisher scientific and used as received.
10. Methanol (puriss, pa, ACS reagent, ≥99.9% (GC)) was purchased from Avantor Performance Materials and degassed by argon-bubbling for 1 h before being passed through an activated alumina column using a GlassContour solvent system. The solvent is withdrawn from the receiver flask with a syringe.
11. Analytical data for 2-(diisopropylphosphanyl)-1-phenylpyrrole (**2**): R_f = 0.22 (100% hexanes, TLC: silica gel 60 Å porosity SiliaplateTM glass backed TLC plates/250 μm thickness, F-254 indicator obtained from Silicycle, visualized with 254 nm UV lamp) (1-phenylpyrrole (**3**) R_f = 0.35, 100% hexanes); ^1H NMR (300 MHz, CDCl$_3$) δ: 0.90 (d, *J* = 7.0 Hz, 3 H), 0.93–0.99 (m, 6 H), 1.01 (d, *J* = 7.0 Hz, 3 H), 1.98 (hept, *J* = 7.0 Hz, 2 H), 6.37 (dd, *J* = 3.6, 2.9 Hz, 1 H), 6.54 (dd, *J* = 3.6, 1.6 Hz, 1 H),

7.00 (td, J = 2.9, 1.6 Hz, 1 H), 7.30–7.46 (m, 5 H); ^{13}C NMR (101 MHz, CDCl$_3$) δ: 19.1 (d, J = 8.3 Hz), 20.1 (d, J = 18.2 Hz), 24.6 (d, J = 8.3 Hz), 109.1, 117.2 (d, J = 4.3 Hz), 126.2, 127.4, 127.9 (d, J = 4.4 Hz), 128.1, 128.6, 141.1; ^{31}P{^{1}H} NMR (162 MHz, CDCl$_3$) δ: –18.7; ATR-FTIR (cm^{-1}): v 959, 1023, 1128, 1187, 1315, 1420, 1453, 1493, 1510, 1595, 2861, 2898, 2922, 2944, 2962, 3040, 3061, 3089; HRMS (EI) calculated for C$_{16}$H$_{22}$NP ([M$^{+\bullet}$]): 259.1490, found: 259.1488; mp = 55–56 °C (MeOH); Anal calcd for C$_{16}$H$_{22}$NP: C, 74.10; H, 8.55; N, 5.40, Found: C, 74.02; H, 8.49; N, 5.60; compound **4** was stored under nitrogen before use.

12. *N*-Boc piperidine (98%) was purchased from Sigma-Aldrich and used as received.

13. Diethyl ether (puriss., pa, ACS certified, BHT stabilized, ≥99% (GC)) was purchased from Fisher Scientific and degassed by argon-bubbling for 1 h before being passed through an activated alumina column using a GlassContour solvent system. The solvent is withdrawn from the receiver flask with a syringe.

14. A 1.4 M solution of *s*-butyllithium in cyclohexane under Sure/Seal was purchased from Sigma-Aldrich, and the concentration was determined by titration with biphenyl-4-methanol to be 1.02 M prior to use according to the reported method: Juaristi, E.; Martinez-Richa, A.; Garcia-Rivera, A.; Cruz-Sánchez, J. S. *J. Org. Chem.* **1983**, *48*, 2603-2606.

15. The temperature increases to –73 °C during the addition.

16. Zinc Chloride (0.5 M solution in THF) was purchased from Acros Organics under AcroSeal, and used as received.

17. The internal temperature increases to –70 °C, and a white cloudy solution is obtained.

18. Tris(dibenzylideneacetone)dipalladium(0) (97%) was purchased from Sigma-Aldrich, and used as received.

19. Toluene (puriss., pa, ACS certified, HPLC grade, ≥99.8% (GC)) was purchased from Fisher Scientific and degassed by argon-bubbling for 1 h before being passed through an activated alumina column using a GlassContour solvent system. The solvent is withdrawn from the receiver flask with a syringe.

20. 3-Bromoanisole (98%+) was purchased from Alfa Aesar and used as received.

21. Ammonium chloride (99.5%) was purchased from Acros Organics, and added to water until saturation.

22. Ethyl acetate (puriss., pa, ACS certified, ≥99.9% (GC)) was purchased from Fisher Scientific and used as received.

23. TLC of crude mixture (10% EtOAc/hexanes, visualized with KMnO$_4$ stain): *tert*-butyl piperidine-1-carboxylate (**2**) (R$_f$ = 0.53), *tert*-butyl 2-(3-methoxyphenyl)piperidine-1-carboxylate (R$_f$ = 0.46), *tert*-butyl 3-(3-methoxyphenyl)piperidine-1-carboxylate (R$_f$ = 0.41).

24. Dichloromethane (puriss., pa, ACS certified, ≥99.9% (GC)) was purchased from Fisher Scientific and used as received.

25. Silica 60 Å (0.04-0.063 mm) was purchased from Fisher Scientific.

26. Analytical data for *tert*-butyl 3-(3-methoxyphenyl)piperidine-1-carboxylate (**4**): R$_f$ = 0.41 (10% EtOAc/hexanes, TLC: silica gel 60 Å porosity SiliaplateTM glass backed TLC plates/250 μm thickness, F-254 indicator obtained from Silicycle, visualized with 254 nm UV lamp and subsequently using KMnO$_4$ stain (see Note 28)); ^{1}H NMR (400 MHz, CDCl$_3$) δ: 1.47 (s, 9 H), 1.52–1.68 (m, 2 H), 1.71–1.79 (m, 1 H), 1.96–2.06 (m, 1 H), 2.52–2.91 (br m, 3 H), 3.80 (s, 3 H), 4.00–4.35 (br m, 2 H), 6.74–6.80 (m, 2 H), 6.83 (d, J = 7.7 Hz, 1 H), 7.20–7.27 (m, 1 H); ^{13}C NMR (101 MHz, CDCl$_3$) δ: 25.5, 28.5, 31.8, 42.6, 44.2 (br), 50.8 (br), 55.2, 79.4, 111.7, 113.2, 119.5, 129.5, 145.3, 154.9, 159.7; ATR-FTIR (cm^{-1}): ν 699, 786, 853, 1040, 1132, 1147, 1155, 1231, 1263, 1364, 1390, 1582, 1601, 1688, 2837, 2856, 2933, 2974, 3003; HRMS (EI) calculated for C$_{17}$H$_{25}$NO$_3$ ([M$^{+\bullet}$]): 291.1834, found: 291.1840; Anal calcd for C$_{17}$H$_{25}$NO$_3$: C, 70.07; H, 8.65; N, 4.81, Found: C, 69.96; H, 8.71; N, 4.83;

27. Analytical gas chromatography with mass spectroscopy (GC/MS) analysis was carried out on a Shimadzu QP2010 GCMS apparatus under electronic impact (EI) and equipped with a SLB-5ms column (15.0 m x 0.10 mm) containing a 0.10 μm film thickness with a flow rate of 0.58 mL per min with 474 kPa He. The temperature profile is the following one: 1 min at 90 °C, temperature increase to 220 °C at a rate of 8 °C per min, then temperature increase to 300 °C at a rate of 40 °C per min, then 300 °C for 4.5 min. Retention times: *tert*-butyl piperidine-1-carboxylate (**3**): 6.53 min; 1-phenylpyrrole (**1**): 6.98 min; 2-(diisopropylphosphanyl)-1-phenylpyrrole (**2**): 14.46 min; *tert*-butyl 2-(3-methoxyphenyl)piperidine-1-carboxylate (α-arylated product): 17.47 min; *tert*-butyl 3-(3-methoxyphenyl)piperidine-1-carboxylate (**4**): 18.32 min.

28. The KMnO$_4$ stain was prepared using 1.5 g of KMnO$_4$ and 10 g of K$_2$CO$_3$ dissolved in 200 mL of water and 1.25 mL of 10% weight NaOH solution.

Working with Hazardous Chemicals

The procedures in *Organic Syntheses* are intended for use only by persons with proper training in experimental organic chemistry. All hazardous materials should be handled using the standard procedures for work with chemicals described in references such as "Prudent Practices in the Laboratory" (The National Academies Press, Washington, D.C., 2011; the full text can be accessed free of charge at http://www.nap.edu/catalog.php?record_id=12654). All chemical waste should be disposed of in accordance with local regulations. For general guidelines for the management of chemical waste, see Chapter 8 of Prudent Practices.

In some articles in *Organic Syntheses*, chemical-specific hazards are highlighted in red "Caution Notes" within a procedure. It is important to recognize that the absence of a caution note does not imply that no significant hazards are associated with the chemicals involved in that procedure. Prior to performing a reaction, a thorough risk assessment should be carried out that includes a review of the potential hazards associated with each chemical and experimental operation on the scale that is planned for the procedure. Guidelines for carrying out a risk assessment and for analyzing the hazards associated with chemicals can be found in Chapter 4 of Prudent Practices.

The procedures described in *Organic Syntheses* are provided as published and are conducted at one's own risk. *Organic Syntheses, Inc.*, its Editors, and its Board of Directors do not warrant or guarantee the safety of individuals using these procedures and hereby disclaim any liability for any injuries or damages claimed to have resulted from or related in any way to the procedures herein.

Discussion

3-Arylpiperidines are important building blocks in pharmaceutical research.[2] They are usually synthesized through the construction of the piperidine ring or the reduction of a 3-arylpyridine precursor, because the inert character of the C–H bonds in β-position to the nitrogen atom is commonly thought to preclude direct β-functionalization.[3] From the end of the 1990's, the direct arylation in position 2 of *N*-Boc-piperidine **3** (Boc =

tert-butoxycarbonyl) has been developed, via a sequence of Boc-directed-lithiation α to the nitrogen atom, followed by transmetalation to zinc and Negishi cross-coupling (Scheme 1).[4] In 2011, Knochel and co-workers reported an unexpected case of diastereoselective arylation of *N*-Boc-2-methylpiperidine occurring in β-position, in contrast to other *N*-Boc-substituted-piperidines undergoing α-arylation.[4e] Inspired by this observation and in extension to our previous work on the migrative β-arylation of ester enolates,[5] we turned to the development of a general palladium-catalyzed migrative β-arylation of *N*-Boc-piperidines involving ligand-controlled selectivity (Schemes 1-2).[6] This methodology allows a rapid access to a large variety of 3-arylpiperidines from commercially available *N*-Boc-piperidine and aryl bromides, with good β/α arylation selectivities (up to 97:3),[7] and moderate-to-good yields (up to 71%).

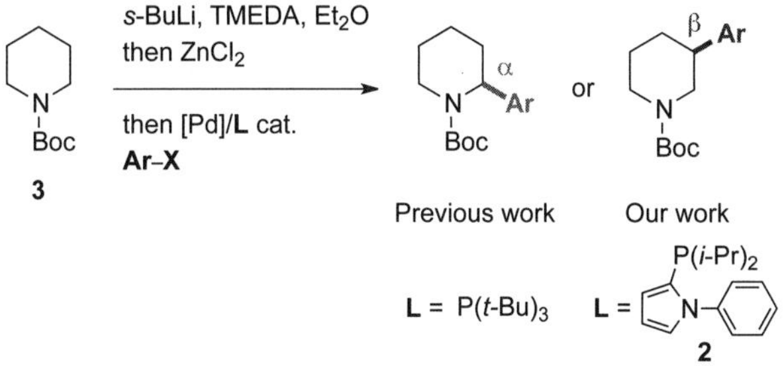

Scheme 1. α vs. β-Arylation

DFT calculations showed that the Pd migration occurs from a Pd–CH agostic complex in which the piperidine ring adopts the twist-boat conformation, and proceeds through a sequence of β-H-elimination/rotation/insertion. Calculations also indicated that α and β-reductive eliminations are the two selectivity-determining steps. In addition, it was shown that the ligand flexibility is a key element in the control of the arylation selectivity, with flexible biarylphosphines such as **2** favoring the β-arylation product, whereas more rigid phosphines such as P(*t*-Bu)₃ favor the α-arylated product.

In our initial work, typical reactions were performed from 0.5 mmol (92.6 mg) of *N*-Boc-piperidine **3** ([**3**] = 0.33 M) and 0.35 mmol (44 µL) of 3-bromoanisole using 2.5 mol% of Pd₂dba₃ and 5 mol% of ligand **2**. Under optimized conditions on a larger scale (7.69 g of *N*-Boc-piperidine), the reaction concentration could be increased to 0.43 M, whereas a decrease of the catalyst loading led to incomplete conversion of 3-bromoanisole. The β-arylation of N-Boc-piperidine was shown to be effective with a large variety of aryl and heteroaryl electrophiles, including those bearing sensitive substituents, and containing halide or triflate leaving groups (Scheme 2, **4a-e**). The reaction of aryl bromides containing an *ortho* electron-withdrawing group failed in presence of the standard phenylpyrrole-based ligand **2** due to the almost exclusive formation of ene-carbamate **5**, but a good reactivity was recovered using DavePhos as the ligand (Scheme 2, **4f-g**). The reactions in Scheme 2 and 3 have been performed using the original conditions.[6]

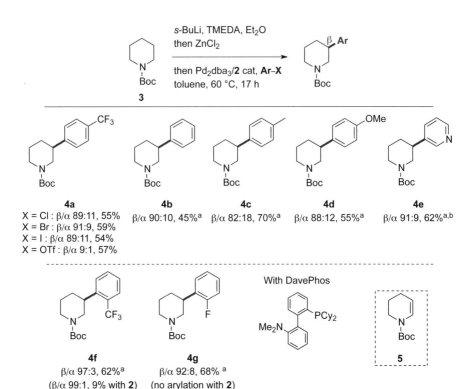

β/α arylation selectivity was determined on crude mixture by GCMS or ^{1}H NMR analysis.
[a] X = Br; [b] 80 °C.

Scheme 2. β-Arylation of N-Boc-Piperidine

The study of substituted piperidines led to interesting observations. The reaction of *N*-Boc-4-methylpiperidine was found to be diastereoselective (r.d. > 95:5), but occurred in α position (Scheme 3, **7a**). In contrast, 2,4-disubstituted *N*-Boc-piperidines and *N*-Boc-decahydroquinoline furnished trisubstituted piperidines in a highly β-selective and *trans*-diastereoselective manner (Scheme 3, **8a-10c**), due to favorable conformational effects.

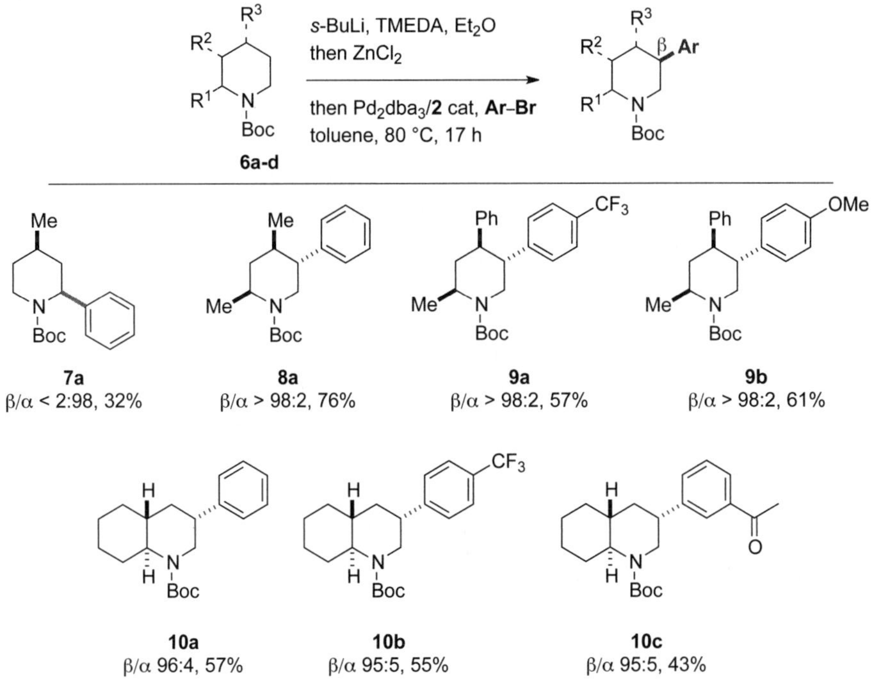

s-BuLi, TMEDA, Et2O
then ZnCl2
then Pd2dba3/2 cat, Ar–Br
toluene, 80 °C, 17 h
6a-d
7a
β/α < 2:98, 32%
8a
β/α > 98:2, 76%
9a
β/α > 98:2, 57%
9b
β/α > 98:2, 61%
10a
β/α 96:4, 57%
10b
β/α 95:5, 55%
10c
β/α 95:5, 43%

β/α selectivity was determined on the crude mixture by GCMS or ^{1}H NMR analysis.

Scheme 3. β-Arylation of Substituted *N*-Boc-Piperidines

In summary, a general and practical palladium-catalyzed migrative Negishi coupling was developed to directly access 3-aryl-*N*-Boc-piperidines in good to excellent selectivity, and with yields between 43 and 76%. The design of a new flexible phenylpyrrole-based phosphine ligand proved to be a key element to induce both efficiency and β-selectivity.

References

1. Université Claude Bernard Lyon 1, CNRS UMR 5246 – Institut de Chimie et Biochimie Moléculaires et Supramoléculaires, CPE Lyon, 43 Boulevard du 11 Novembre 1918, 69622 Villeurbanne, France. E-mail: olivier.baudoin@univ-lyon1.fr. This work was financially supported by Agence Nationale de la Recherche (Programme Blanc 2011 "EnolFun") and Institut Universitaire de France.

2. (a) Wilkström, H.; Sanchez, D.; Lindberg, P.; Hacksell, L.-E.; Johansson, A. M.; Thorberg, S.-O.; Nilsson J. L. G.; Svensson, U.; Hjorth, S.; Clarck, D.; Carlsson, A. *J. Med. Chem.* **1984**, *27*, 1030–1036. (b) Amat, B.; Cantó, M.; Llor, N.; Escolano, C.; Molins, E.; Espinosa, E.; Bosch, J. *J. Org. Chem.* **2002**, *67*, 5343–5351. (c) Wallace, D. J.; Brands, K. J. M.; Bremeyer, N.; Brewer, S. E.; Desmond, R.; Emerson, K. M.; Foley, J.; Fernandez, P.; Hu, W.; Keen, S. P.; Mullens, P.; Muzzio, D.; Sajonz, P.; Tan, L.; Wilson, R. D.; Zhou, G.; Zhou, G. *Org. Process Res. Dev.* **2011**, *15*, 831–840.

3. Buffat, M. G. T. *Tetrahedron*, **2004**, *60*, 1701–1729.

4. (a) Dieter, R. K.; Li, S. *J. Org. Chem.* **1997**, *62*, 7726–7735. (b) Campos, K. R.; Klapars, A.; Waldman, J. H.; Dormer, P. G.; Chen, C.-y. *J. Am. Chem. Soc.* **2006**, *128*, 3538–3539. (c) Coldham. I.; Leonori, D. *Org. Lett.* **2008**, *10*, 3923–3925. (d) Beng T. K.; Gawley, R. E. *Org. Lett.* **2011**, *13*, 394–397. (e) Seel, S.; Thaler, T.; Takatsu, K.; Zhang, C.; Zipse, H.; Straub, B. F.; Mayer, P.; Knochel, P. *J. Am. Chem. Soc.* **2011**, *133*, 4774–4777.

5. (a) Renaudat, A.; Jean-Gérard, L.; Jazzar, R.; Kefalidis C. E.; Clot, E.; Baudoin, O. *Angew. Chem. Int. Ed.* **2010**, *49*, 7261–7265. (b) Larini, P.; Kefalidis, C. E.; Jazzar, R.; Renaudat, A.; Clot, E.; Baudoin, O. *Chem.– Eur. J.* **2012**, *18*, 1932–1944. (c) Aspin, S. ; Goutierre, A.-S.; Larini, P.; Jazzar, R.; Baudoin, O. *Angew. Chem. Int. Ed.* **2012**, *51*, 10808–10811. (d) Aspin, S.; López-Suárez, L.; Larini, P.; Goutierre, A.-S.; Jazzar, R.; Baudoin, O. *Org. Lett.* **2013**, *15*, 5056–5059.

6. Millet, A.; Larini, P.; Clot, E.; Baudoin, O. *Chem. Sci.* **2013**, *4*, 2241–2247.

7. The α- and β-arylated products can be separated by silica gel chromatography.

Appendix
Chemical Abstracts Nomenclature (Registry Number)

1-Phenylpyrrole (635-90-5)

N,N,N',N'-Tetramethylethylenediamine (110-18-9)

n-Butyllithium (109-72-8)

Chlorodiisopropylphosphine (40244-90-4)

s-Butyllithium (598-30-1)

Zinc chloride (7646-85-7)

Tris(dibenzylideneacetone)dipalladium (0) (51364-51-3)

3-Bromoanisole (2398-37-0)

Anthony Millet was born in France in 1986. After an internship at Chimie Paristech in the group of Dr. Véronique Michelet, he received his M.Sc. in organic chemistry from University Pierre et Marie Curie (Paris VI). He is currently a Ph.D. student in the group of Prof. Olivier Baudoin at the University of Lyon (France). His research interests include the intermolecular functionalization of unactivated C(sp³)–H bonds.

Olivier Baudoin studied chemistry at Ecole Nationale Supérieure de Chimie de Paris (1995). He completed his Ph.D. in 1998 in the group of Prof. Jean-Marie Lehn in Paris. He then worked as a post-doctoral fellow with Prof. K. C. Nicolaou in the Scripps Research Institute (La Jolla). He joined Institut de Chimie des Substances Naturelles (Gif-sur-Yvette) in 1999 as a CNRS researcher and obtained his Habilitation diploma in 2004. In 2006, he was appointed as a Professor at the University Claude Bernard Lyon 1, and was promoted to First Class Professor in 2011. He was the recipient of the CNRS Bronze Medal in 2005, the Scholar Award of the French Chemical Society, Organic Division in 2010, and was appointed as a junior member of Institut Universitaire de France in 2009-14.

Organic
Syntheses

Louis C. Morrill studied chemistry at the University of St Andrews, obtaining his MChem (2010) and Ph.D. (2014) degrees under the direction of Prof. Andrew Smith, investigating Lewis base organocatalysis. He is currently a Postdoctoral Research Fellow with Prof. Richmond Sarpong at UC Berkeley, exploring the total synthesis of complex diterpenoid alkaloid natural products. Commencing June 2015, he will take up an independent position at Cardiff University as a University Research Fellow (URF) in synthetic organic chemistry.

Preparation of Mono-Cbz Protected Guanidines

Kihyeok Kwon, Travis J. Haussener and Ryan E. Looper[*1]

Department of Chemistry, University of Utah, 315 S 1400 E, Salt Lake City, Utah 84112, United States

Checked by Joshua N. Payette and Mohammad Movassaghi

Procedure

A. *Carbonylbenzyloxycyanamide (1)*. A one-necked 500-mL round-bottomed flask open to the atmosphere, equipped with a magnetic stirring bar (Note 1) is charged with cyanamide (50 weight % solution in H₂O) (12.6 g, 11.7 mL, 0.15 mol) (Notes 2 and 3) and distilled water (100 mL). Sodium hydroxide pellets (6.16 g, 0.154 mol, 2.05 equiv) are then added in portions (~3 x 2 g) over a 15 min period. The mixture is then stirred for 30 min at room temperature and then cooled to 0 °C (Note 4). The flask is fitted with a 100 mL addition funnel and the addition funnel charged with

Published on the Web 3/25/2015
© 2015 Organic Syntheses, Inc.

benzyl chloroformate (12.8 g, 10.7 mL, 0.075 mol, 1.00 equiv) (Note 4). The benzyl chloroformate solution is then added dropwise (~ 1 drop/second) over a span of 15 min. After addition of the benzyl chloroformate the reaction is stirred for an additional 3 h at room temperature. The mixture is transferred to a 250 mL separatory funnel and washed with diethyl ether (1 x 50 mL). The aqueous layer is then transferred to a 1-L Erlenmeyer flask equipped with a magnetic stirring bar (1.7 mm x 3.8 mm) and acidified to pH = 2 (Note 6) with conc. HCl (approx. 7 mL). Dichloromethane (100 mL) is then added to dissolve the solids and the mixture transferred to a 250 mL separatory funnel. After separation of the layers, the aqueous fraction is extracted with dichloromethane (2 x 50 mL) and the combined organics dried over anhydrous Na_2SO_4. The organics are filtered through a sintered glass funnel and the resultant sodium sulfate is washed with dichloromethane (2 x 25 mL). The solvent is removed on a rotary evaporator, and then the flask transferred to a high vacuum line (3.0 mmHg) for 3 h. The title compound is obtained as a viscous light-yellow oil. This material is used in the next step without further purification (Note 7).

B. *Potassium benzyloxycarbonylcyanamide (2).* A 250-mL round-bottomed flask open to the atmosphere is equipped with a magnetic stirring bar (1.0 mm x 2.5 mm) and charged with MeOH (100 mL) (Note 8). Potassium hydroxide flakes (4.20 g, 0.075 mol, 1.00 equiv) are then added in portions (~4 x 1 g) and the mixture stirred until all the KOH is dissolved. The flask is then cooled by placement in an ice bath, the internal temperature measured to be ~0 °C, and the flask fitted with a 125 mL addition funnel. The crude material from Step A is dissolved in MeOH (25 mL) and added dropwise via the addition funnel over 30 min, at which point the solution turns a milky white. The flask is stoppered and allowed to stand in a –20 °C freezer overnight. The crude solid is collected on a Büchner funnel and washed with cold MeOH (2 x 25 mL) to give the first crop of the title compound (Note 9) as a fine white powder (9.46 g, 59%). The filtrate is further concentrated to half the original volume and stored in the freezer to yield a second crop (1.31 g, 8%) (Note 10). The combined solids are dried under air overnight (Note 11).

C. *N-Benzyl, N'-Cbz-guanidine (3).* A flame-dried 250-mL single-necked round-bottomed flask equipped with a magnetic stirring bar (1.0 mm x 2.5 mm), and a rubber septum is placed under a nitrogen atmosphere. The flask is then charged with potassium carbobenzyloxycyanamide (2) (6.42 g, 30.0 mmol). Acetonitrile (100 mL) is then added and the mixture stirred vigorously for 15 min. Trimethylsilyl chloride (3.58 g, 4.20 mL, 33.0 mmol, 1.10 equiv) is then added dropwise via syringe over a period of 10 min. The mixture is stirred at room temperature for 30 min and the solution becomes milky. Benzylamine (3.54 g, 3.60 mL, 33.0 mmol, 1.10 equiv) is then added in a single portion, and the solution immediately becomes more opaque. The mixture is then stirred for 1 h. The reaction flask is then transferred to a rotary evaporator and concentrated to dryness under reduced pressure (25 mmHg). The resulting solid is slurried in dichloromethane (350 mL) and transferred to a 500 mL separatory funnel. The organic phase is washed with 1M Na_2CO_3 (100 mL), sat. NaCl (100 mL) and dried over anhydrous Na_2SO_4. The organics are decanted, and the sodium sulfate is washed with dichloromethane (3 x 50 mL). The combined organics are then concentrated under reduced pressure (25 mmHg) in a 1 L round-bottomed flask to give an off-white solid. The solid is slurried in ethyl acetate (100 mL) and transferred to a 500 mL Erlenmeyer flask. The evaporation flask is rinsed with an additional portion of ethyl acetate (40 mL). A stir bar (1.7 mm x 3.8 mm) is then added to the Erlenmeyer flask and the mixture heated to reflux in an oil bath. Methanol is added in portions (20 mL; 120 mL total) until all the solids dissolve. The solution is then passed through filter paper and the filtrate stored in the freezer (–20 °C) for 4 h. The resulting microcrystals were collected on a sintered glass funnel, washed with diethyl ether (40 mL) and then dried under vacuum to give 7.34 g (86%) of a white solid. The mother liquor is then concentrated until solid began to appear (~ 1/3 volume), and the mixture is then allowed to stand in the freezer (–20 °C) for 4 h. The resulting microcrystals were collected on a sintered glass funnel, washed with diethyl ether (40 mL) and then dried under vacuum to give 0.74 g (9%) of a white solid, which was combined with the original crop of crystals to provide 8.08 g (95%) of the product (Notes 12 and 13).

Notes

1. Mixtures were stirred between 600 and 1000 rpm any more or less would cause the stirring to cease when solid precipitates formed. The stir bar used was egg shaped and 3.3 cm in length.
2. All chemicals were purchased from Sigma Aldrich Chemical Co. and were used without further purification. Benzyl chloroformate was

purchased as a technical grade (≥98%) reagent and cyanamide was purchased as a 50% weight solution in water.

3. An excess (2 equiv) of sodium cyanamide are required as the second equivalent deprotonates the acidic acylcyanamide that is formed.

4. Temperature was monitored by a thermocouple placed directly in the reaction mixture.

5. Benzyl chloroformate is moisture sensitive and can produce toxic and corrosive fumes. When used it was open to the atmosphere; exercise caution as accidental contact with water can have severe consequences. Store and handle this chemical in a fume hood with adequate ventilation.

6. pH was determined by EMD Millipore colorpHast® pH Test Strips.

7. This material should be carried to the next step as soon as possible as it is subject to numerous decomposition pathways at room temperature but is reasonably stable in the freezer up to 5 days. The crude material displays the following spectroscopic properties: ^{1}H NMR (CDCl$_3$, 400 MHz) δ: 5.22 (s, 2 H), 7.34–7.36 (m, 5 H).

8. The crude material from step 1 is very viscous and requires extensive hand stirring (5-20 min) in MeOH before it will completely dissolve. Once completely dissolved in the required amount of MeOH it is much easier to work with.

9. Potassium benzyloxycarbonylcyanamide has been previously shown to be an inhibitor of alcohol dehydrogenase; care should be exercised when handling this compound and it should be handled in a fume hood with adequate ventilation.

10. In a second experiment the checkers received 8.9 g (55%) in the first crop and 3.2 g (20%) in the second crop of crystals.

11. This compound exhibits the following physiochemical properties: R$_f$ = 0.71 (ethyl acetate). mp 219–221 °C. ^{1}H NMR (DMSO-d_6, 400 MHz) δ: 4.86 (s, 2 H), 7.24–7.35 (m, 5 H); ^{13}C NMR (DMSO-d_6, 100 MHz) δ: 65.2, 122.1, 127.1, 127.3, 128.1, 138.7, 162.5; IR (powder) ν 3060, 3030 (both w), 2189 (s) 2143 (m), 1622 (s), 1398, 1342, 1306, 1178, 1147 (all m), 779, 732, 697 (all s) cm^{-1}. HRMS (ESI-) Calculated for C$_9$H$_7$N$_2$O$_2$ [M-] m/z 175.0513, obsd 175.0510. Anal calcd for C$_9$H$_7$KN$_2$O$_2$: C, 50.45; H, 3.29; N, 13.07. Found C, 50.56; H, 3.43; N, 13.22.

12. This compound exhibits the following physiochemical properties: $R_f =$ 0.47 (ethyl acetate). mp 160–162 °C. ^1H NMR (DMSO-d_6, 400 MHz) δ: 4.37 (d, J = 4.7 Hz, 2 H), 4.96 (s, 2 H), 7.23–7.36 (m, 10 H). ^{13}C NMR (DMSO-d_6, 100 MHz) δ: 43.4, 65.2, 126.9, 127.1, 127.4, 127.6, 128.2, 128.4, 138.0, 161.5, 163.1. IR (powder) ν 3475, 3283 (both m), 1640, 1617, 1588, 1559 (all s), 1426 (m), 1275, 1131 (both s) cm^{-1}. HRMS (ESI+) Calculated for $C_{16}H_{18}N_3O_2$ m/z (M+H) 284.1394, obsd. 284.1385. Anal. calcd for $C_{16}H_{17}N_3O_2$: C, 67.83; H, 6.05; N, 14.83. Found C, 67.91; H, 6.16; N, 14.83.

13. The checkers obtained 4.17 g (77%) of the product when the reaction was performed on approximately half-scale.

Working with Hazardous Chemicals

The procedures in *Organic Syntheses* are intended for use only by persons with proper training in experimental organic chemistry. All hazardous materials should be handled using the standard procedures for work with chemicals described in references such as "Prudent Practices in the Laboratory" (The National Academies Press, Washington, D.C., 2011; the full text can be accessed free of charge at http://www.nap.edu/catalog.php?record_id=12654). All chemical waste should be disposed of in accordance with local regulations. For general guidelines for the management of chemical waste, see Chapter 8 of Prudent Practices.

In some articles in *Organic Syntheses*, chemical-specific hazards are highlighted in red "Caution Notes" within a procedure. It is important to recognize that the absence of a caution note does not imply that no significant hazards are associated with the chemicals involved in that procedure. Prior to performing a reaction, a thorough risk assessment should be carried out that includes a review of the potential hazards associated with each chemical and experimental operation on the scale that is planned for the procedure. Guidelines for carrying out a risk assessment and for analyzing the hazards associated with chemicals can be found in Chapter 4 of Prudent Practices.

The procedures described in *Organic Syntheses* are provided as published and are conducted at one's own risk. *Organic Syntheses, Inc.*, its Editors, and its Board of Directors do not warrant or guarantee the safety of individuals using these procedures and hereby disclaim any liability for any injuries or damages claimed to have resulted from or related in any way to the procedures herein.

Discussion

The guanidinium ion has proven itself as a valuable motif for molecular recognition, through both electrostatic and hydrogen bond donor-acceptor patterns.[2] These properties have allowed this functional group to be deployed in a vast range of chemical fields from catalysis to medicinal chemistry.[3,4,5,6] It is thus not surprising that its utility has prompted the development of numerous reagents and synthetic methods to introduce the guanidine unit (Figure 1). To avoid problems associated with the high basicity and potential nucleophilicity of the guanidine, most methods typically install a diacylated (or di-protected) guanidine. The most commonly utilized reagent for guanylation is di-Boc-*S*-Me-pseudothiourea

(1), which generates a highly reactive carbodiimide intermediate upon reaction with a thiophilic metal salt such as Hg(II), Cu(I) or Ag(I).[7] The parent thiourea 2 can also be used in conjunction with an activating agent (typically Mukaiyama's reagent (3)[8] or other suitable peptide coupling reagents such as EDC[9]), although this reagent is very expensive. Goodman's reagent (4)[10] is very reactive and is capable of guanylating weakly nucleophilic amines. The pyrazole transfer reagents 5 and 6 have also been developed to obviate the use of toxic metal salts, but these reagents remain prohibitively expensive for use on a preparative scale.[11]

Figure 1. Common reagents for guanylation

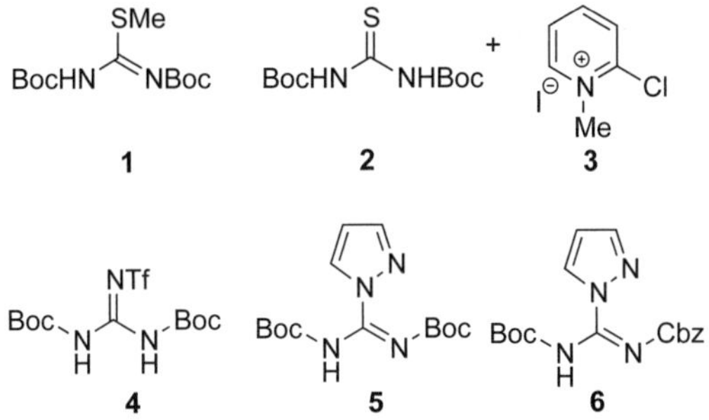

These reagents (1-6) are typically used to install the guanidine unit as an N,N-disubstituted guanidine. For more highly substituted substrates, however, it becomes useful to generate peripheral C-N bonds around the intact guanidine unit, in a selective manner.[12] We became interested in exploiting mono-N-acylguanidines in these methodologies, anticipating the ability to exploit a resultant open nitrogen valence for further derivitization. The synthesis of mono-N-acylguanidines, via acylation with acid chlorides or anhydrides is often complicated by over acylation due to the increased acidity of the initially formed mono-N-acylguanidine.[13] Alternatively they can be accessed by the controlled hydrolysis of diacyl-protected guanidines.[14] Several methods have also been developed for their synthesis from N-acyl-thioureas[15] or N-acyl-pseudothioureas.[15c] The use of the mono-protected versions of the reagent 5 have also been reported, but are unreactive or poorly reactive toward amines because of their decreased electrophilicity.[11,12i] The differentially protected pyrazole 6 can be employed followed by removal of the Boc or Cbz protecting group.[16] All in all, these methods are difficult to employ in a preparative setting as they utilize costly reagents, provide poor yields, and require multiple synthetic manipulations.

While it is known that anilines undergo addition to acylcyanamides, these reactions typically require harsh reaction conditions which can significantly decompose the cyanamide.[17] These addition reactions are limited to anilines because acylcyanamides are quite acidic with pK_a's ~ 2-4.[18] If this reaction could be extended to a variety of amines it would provide an important cheap and practical method for both the installation of a guanidine unit, but also to a selectively protected unit. We considered that silylation might be capable of temporarily activating the

Table 1. TMSCl-mediated Guanylation of Amines with Potassium Benzyloxycarbonylcyanamide

cyanamide and masking the acidic proton. Indeed, treatment of potassium carbobenzyloxycyanamide with TMSCl is capable of generating a silylcarbodiimide intermediate which is highly reactive and capable of guanylating a variety of amines except very electron deficient aniline derivatives (Table 1).

References

1. Department of Chemistry, University of Utah, Salt Lake City, UT, 84112. r.looper@utah.edu; REL is grateful to the National Institutes of Health, General Medical Sciences (R01-GM090082) for financial support of this research.
2. For recent reviews see: a) Schug, K. A.; Lindner, W. *Chem. Rev.* **2005**, *105*, 67. b) Blondeau, P.; Segura, M.; de Mendoza, J.; Pérez-Fernández, R. *Chem. Soc. Rev.* **2007**, *36*, 198. c) Sullivan, J. D.; Giles, R. L.; Looper, R. E. *Curr. Bioact. Compd.* **2009**, *5*, 39.
3. Berlink, R. G. S.; Burtoloso, A. C. B.; Kossuga, M. H. *Nat. Prod. Rep.* **2008**, *25*, 919.
4. a) Tsukamoto, S.; Yamashita, T.; Matsunaga S.; Fusetani, N. *Tetrahedron Lett.* **1999**, *40*, 737. b) Tsukamoto, S.; Yamashita, T.; Matsunaga, S.; Fusetani, N. *J. Org. Chem.* **1999**, *64*, 3794.
5. Orner, B. P.; Hamilton, A. D. *J. Inclus. Phenom. Mol.* **2001**, *41*, 141.
6. Corey, E. J.; Grogan, M. J. *Org. Lett.* **1999**, *1*, 157.
7. For recent reviews of guanylating reagents and their uses see: a) Manimala, J. C.; Anslyn, E. V. *Eur. J. Org. Chem.* **2002**, 3909. b) Katritzky, A. R.; Rogovoy, B.V. *Arkivoc* **2005**, 49.
8. Yong, Y. F.; Kowalski, J. A.; Lipton, M. A. *J. Org. Chem.* **1997**, *62*, 1540.
9. Manimala, J. C.; Anslyn, E. V. *Tetrahedron Lett.* **2002**, *43*, 565.
10. Feichtinger, K.; Zapf, C.; Sings, H. L.; Goodman, M. *J. Org. Chem.* **1998**, *63*, 3804.
11. Bernatowicz, M. S.; Wu, Y.; Matsueda, G. R. *Tetrahedron Lett.* **1993**, *34*, 3389.
12. a) Overman, L. E.; Rabinowitz, M. H. *J. Org. Chem.* **1993**, *58*, 3235. b) Cohen, F.; Overman, L. E.; Sakata, S. K. L. *Org. Lett.* **1999**, *1*, 2169. c) Kim, M.; Vulcahy, J. V.; Espino, C. G.; Du Bois, J. *Org. Lett.* **2006**, *8*, 1073. d) Snider, B. B.; Xie, C. Y. *Tetrahedron Lett.* **1998**, *39*, 7021. e) Arnold, M. A.; Day, K. A.; Durón, S. G.; Gin, D. Y. *J. Am. Chem. Soc*

2006, *128*, 13255. f) Hövelmann, C. H. ; Streuff, J.; Brelot, L.; Muñiz, K. *Chem. Commun.* **2008**, 2334. g) Allingham, M. T.; Howard-Jones, A.; Murphy, P. J.; Thomas D. A.; Caulkett, P.W.R. *Tetrahedron Lett.* **2003**, *44*, 8677. h) Albrecht, C.; Barnes, S.; Böckemeier, H.; Davies, D.; Dennis, M.; Evans, D. M.; Fletcher, M. D.; Jones, I.; Leitmann, V.; Murphy, P. J.; Rowles, R.; Nash, R.; Stephenson, R. A.; Horton, P. N.; Hursthouse, M. B. *Tetrahedron Lett.* **2008**, *49*, 185. i) Davies, D.; Fletcher, M. D.; Franken, H.; Hollinshead, J.; Kaehm, K.; Murphy, P. J.; Nash, R.; Potter, D. M. *Tetrahedron Lett.* **2010**, *51*, 6825. j) Giles, R. L.; Sullivan, J. D.; Steiner, A. M.;. Looper, R. E. *Angew. Chem. Int. Ed.* **2009**, *48*, 3116. k) Gainer, M. J.; Bennett, N. R.; Takahashi, Y.; Looper, R. E. *Angew. Chem. Int. Ed.* **2011**, *50*, 684.

13. Takikawa, H.; Nozawa, D.; Mori, K. *J. Chem. Soc., Perkins Trans. 1* **2001**, 657.

14. a) Dodd, D. S.; Zhao, Y. *Tetrahedron Lett.* **2001**, *42*, 1259. b) Ghosh, A. K.; Hol, W. G. J.; Fan,E. *J. Org. Chem.* **2001**, *66*, 2161. c) Poss, M. A.; Iwanowicz, E.; Reid, J. A.-I.; Lin, R.; Gu, Z. *Tetrahedron Lett.* **1992**, *33*, 5933.

15. a) Shinada, T.; Umezawa, T.; Ando, T.; Kozuma, H.; Ohfune, Y. *Tetrahedron Lett.* **2006**, *47*, 1945-1947. b) Lee, S.; Yi, K. Y.; Hwang, S. K.; Lee, B. H.; Yoo, S.; Lee, K. *J. Med. Chem.* **2005**, *48*, 2882. c) Padmanabhan, S.; Lavin, R. C.; Thakker, P. M.; Durant, G. J. *Synth. Commun.* **2001**, *31*, 2491. d) Padmanabhan, S.; Lavin, R. C.; Thakker, P. M.; Guo, J.; Zhang, L.; Moore, D.; Perlman, M. E.; Kirk, C.; Daly, D.; Burke- Howie, K. J.; Wolcott, T.; Chari, S.; Berlove, D.; Fischer, J. B.; Holt, W. F.; Durant, G. J.; McBurney, R. N. *Bioorg. Med. Chem. Lett.* **2001**, *11*, 3151.

16. Sawayama, Y.; Nishikawa, T. *Synlett*, **2011**, 5, 651.

17. Eschalier, A.; Dureng, G.; Duchene-Marullaz, P.; Berecoechea, J.; Anatol, J. *Eur. J. Med. Chem.* **1983**, *18*, 139-45.

18. a) Bader, R. Z. *Phys. Chem.* **1890**, *6*, 304. b) Howard, J. C. *J. Org. Chem.* **1964**, *29*, 761.

Appendix
Chemical Abstracts Nomenclature (Registry Number)

Cyanamide; (420-04-2)
Benzyl chloroformate: Benzyloxycarbonyl chloride; (501-53-1)

Organic
Syntheses

Carbonylbenzyloxycyanamide: Carbamic acid, *N*-cyano-, phenylmethyl ester; (86554-53-2)
Potassium benzyloxycarbonylcyanamide: Carbamic acid, *N*-cyano-, phenylmethyl ester, potassium salt; (50909-46-1)
N-Benzyl, *N'*-Cbz-guanidine: Carbamic acid, *N*-[imino[(phenylmethyl)amino]methyl]-, phenylmethyl ester; (22102-72-3)

Ki Hyeok Kwon was born in 1976 in Seoul, Korea. He studied chemical engineering at the University of Seoul, where he received his B.S. degree in 2002 and M.S. degree in 2005 under the supervision of Prof. Do. W. Lee. After received his degree, he moved to Marquette University where he earned his Ph. D. degree in 2011 for the catalytic coupling reaction involving C-H bond activation under the guidance Dr. Chae S. Yi. He is now pursuing post-doctoral research in the group of Prof. Ryan Looper at the University of Utah. His research focuses on the synthesis and methodology development of complex molecules with biological activity.

Travis Haussener was born in Elmira, NY in 1987. He attended The Pennsylvania State University where he earned a B.S. degree in chemistry in 2009. He is currently working on his Ph. D. at The University of Utah under the direction of Ryan Looper. Travis is interested in the total synthesis of Pactamycin. In his spare time he enjoys climbing, backcountry skiing, and running ultramarathons.

Ryan Looper was born in Banbury, England in 1976. He returned to the U.S. to attend Western Washington University where he earned a B.S. degree in Chemistry in 1998 and an M.S. degree in 1999 under the guidance of J. R. Vyvyan. He obtained his Ph.D. degree in 2004 at Colorado State University in the laboratories of R. M. Williams. After an NIH post-doctoral fellowship at Harvard University with S. L. Schreiber, he began his independent career at the University of Utah in 2007. He is primarily interested in the synthesis of complex natural products with biomedical significance.

Joshua Payette was born in Memphis, Tennessee and graduated from Wheaton College, Illinois in 2005 with a B.S. in Chemistry. In 2005 he joined the research group of Professor Hisashi Yamamoto at The University of Chicago. In 2010 he earned a PhD for his studies in chiral oxazaborolidinium catalyzed cycloaddition reactions. As a postdoctoral research associate in Professor Mohammad Movassaghi 's group at MIT, he is currently pursuing the total synthesis of alkaloid natural products.

Preparation of *N*-(Boc)-Allylglycine Methyl Ester Using a Zinc-mediated, Palladium-catalyzed Cross-coupling Reaction

N. D. Prasad Atmuri and William D. Lubell*

Département de Chimie, Université de Montréal, P.O. Box 6128, Station Centre-ville, Montréal, QC H3C 3J7, Canada

Checked by Jean-Nicolas Desrosiers, Nizar Haddad, and Chris H. Senanayake

A.

BocHN—CHOH—CO₂Me
1

Iodine (1.3 equiv), Ph₃P (1.3 equiv),
Imidazole (1.3 equiv)
DCM, 0 °C - rt
———————————→
82%

BocHN—CHI—CO₂Me
2

B.

BocHN—CHI—CO₂Me
2

1. Zn (6 equiv), DMF

Br⌒⌒Br (0.6 equiv), 60 °C
TMSCl (0.2 equiv), rt - 35 °C
———————————→
2. ⌒Br (1 M in THF)
Pd₂(dba)₃ (0.02 equiv)
P(*o*-tol)₃ (0.1 equiv)
−78 °C - rt, 12 h
65%

BocHN—CH=CH₂—CO₂Me
3

Procedure

A. *tert-Butyl (R)-1-(methoxycarbonyl)-2-iodoethylcarbamate* (**2**). An oven-dried 1000-mL, three-necked, round-bottomed flask containing an egg-shaped Teflon®-coated magnetic stir bar (7 cm long) is equipped with a rubber septum with a thermometer, a 125 mL addition funnel and an argon inlet adaptor. The apparatus is purged with argon (Note 1). Keeping a positive flow of argon, the septum is removed temporarily and the flask is charged with triphenylphosphine (Note 2) (32.66 g, 124.5 mmol, 1.3 equiv) and 400 mL of dichloromethane (Note 3). The solution is stirred at room temperature and imidazole (8.47 g, 124.5 mmol, 1.3 equiv) (Note 2) is added in one portion. The resulting mixture is stirred at room temperature for 10 min until full dissolution of imidazole is observed. The solution is cooled in an ice bath to 0 °C and maintained at that temperature during the addition of iodine (31.60 g, 124.5 mmol, 1.3 equiv) (Note 2) in four portions over a period of 20 min. The reaction mixture is stirred in the dark. The solution is warmed to room temperature, stirred for 10 min, and cooled to 0 °C. The septum is replaced with a dropping funnel (see photo), which is charged with *tert*-butyl (S)-1-(methoxycarbonyl)-2-hydroxyethylcarbamate (**1**, 21.00 g, 95.8 mmol) (Note 4) in 100 mL of dichloromethane. To the

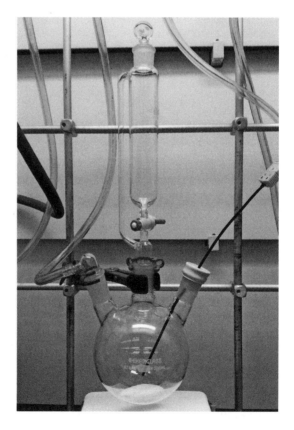

reaction mixture at 0 °C, the solution of alcohol **1** in dichloromethane is added drop-wise over 60 min. The resulting slurry is stirred at 0 °C for 1 h, allowed to warm up to room temperature over 1 h, and stirred at that temperature for 1.5 h (Note 5). The reaction mixture is filtered through 150 g of silica gel dry-packed in a 9 cm column using 50:50 ether:hexanes (~700 mL) as eluent and concentrated under reduced pressure to give 32.5 g of brown oil, which is purified by column chromatography (Note 6). Evaporation of the collected fractions provides a colorless oil (26.1 g, 82%) (Notes 7 and 8), which converts to a white solid in the freezer at –20 °C.

B. *tert-Butyl (S)-1-(methoxycarbonyl)but-3-enylcarbamate* (**3**). An oven dried 250-mL three-necked, round-bottomed flask containing an egg-shaped Teflon®-coated magnetic stir bar (4 cm long) is equipped with a reflux condenser fitted with an argon inlet adaptor, a thermometer and a rubber septum (Note 1) (see photo). The apparatus is purged with argon. Keeping a positive flow of argon, the septum is removed temporarily and the flask is charged with zinc dust (11.92 g, 182.3 mmol, 6 equiv) (Note 9).

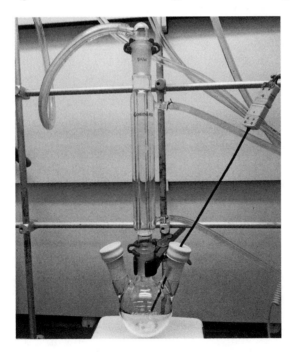

Dry DMF (20 mL) (Note 10) is then added to the flask via a syringe. 1,2-Dibromoethane (1.57 mL, 3.42 g, 18.2 mmol, 0.6 equiv) (Note 11) is added next to the stirred suspension via a syringe. The mixture is stirred and heated to 60 °C and stirred at 60 °C for 45 min (Note 12). The mixture is cooled to room temperature. Chlorotrimethyl silane (TMS-Cl; 0.77 mL, 6.0 mmol) (Note 13) is added via a syringe to the slurry, which is stirred for 40 min at room temperature (Note 14). A solution of *tert*-butyl (*R*)-1-(methoxycarbonyl)-2-iodoethylcarbamate (**2**, 10 g, 30.39 mmol) (Note 15) in dry DMF (20 mL) is added via a syringe (Note 16) to the room temperature mixture of activated zinc, which is then heated in a 35 °C oil bath and stirred for 60 min. The zinc insertion was judged complete by TLC analysis (Note 17). After complete zinc insertion, the reaction mixture is cooled to room temperature, and charged with $Pd_2(dba)_3$ (779 mg, 0.85 mmol, 0.028 equiv) (Note 18) and tri(*o*-tolyl)phosphine (925 mg, 3.03 mmol, 0.1 equiv) (Note 19). The resulting mixture is cooled to –78 °C. A solution of vinyl bromide (1 M in THF, 42.5 mL, 42.5 mmol, 1.4 equiv, Note 20) is added drop-wise via a cannula to the stirred –78 °C suspension. After the addition of the vinyl bromide is complete, the cold bath is removed and the reaction mixture is allowed to warm to room temperature with stirring for 12 h (Note 21). The reaction mixture is diluted with ethyl acetate (200 mL) and transferred to a 1 L Erlenmeyer. Water (200 mL) is added and the resulting mixture is filtered through a pad of 35 g of Celite™ in a 6 cm diameter filter funnel. The pad is washed with ethyl acetate (300 mL). The filtrate and washings are combined and transferred to a separating funnel. The organic layer is separated. The aqueous layer is extracted with ethyl acetate (2 x 200 mL). The combined organic layers are washed with brine (400 mL), dried over anhydrous sodium sulfate, filtered, and concentrated under reduced pressure to give 8.2 g of brown oil, which is purified by column chromatography (Note 22). Evaporation of the collected fractions yields 4.50 g (65%) of a brown oil (Notes 23 and 24).

Notes

1. The submitters conducted this procedure under an atmosphere of argon.
2. Triphenylphosphine (99 %), imidazole (purity ≥99 %), and iodine (purity ≥99.9%) were obtained from Aldrich and used as provided. An exotherm of Δ 7 °C per added portion of iodine was observed by the checkers.

3. The submitters used dry dichloromethane from a solvent filtration system (Glass Contour, Irvine, CA). The checkers used anhydrous dichloromethane under inert atmosphere with a sure seal bottle from Aldrich chemical company, Inc.

4. The checkers purchased N-(*tert*-butoxycarbonyl)-L-serine methyl ester (**1**) from Sigma-Aldrich. The material was prepared by the submitters according to the procedure reported by Trost, et al.[2.]

5. TLC analysis was performed on Merck Aluminum silica gel plates 60 F_{254}. Reaction conversion was ascertained using the following procedure. The reaction mixture was spotted directly on the TLC plate, which was eluted with (4:1) hexanes/ethyl acetate, and visualized with 254 nm UV light and $KMnO_4$ stain after heating: starting material **1** (R_f = 0.075, a UV inactive and $KMnO_4$ active spot); iodide **2** (R_f = 0.55, a UV active and $KMnO_4$ active spot).

6. Iodoalanine **2** is purified on a silica column in a dark place. The crude residue is absorbed onto 70 g of silica gel, and added to a column (18.4 cm diameter x 20 cm length), which is packed with a slurry of 360 g of silica (high purity grade Silica gel particle size 230-400 mesh ASTM, Merck Ltd.) in hexanes (700 mL). An eluent of 5% diethyl ether in hexanes (~1000 mL) is first flushed through the column to remove the less polar spot *tert*-butyl 1-(methoxycarbonyl)vinylcarbamate. The eluent is switched to 10% diethyl ether/hexanes (~2500 mL) to elute the iodide (R_f = 0.55, 4:1 hexanes/ethyl acetate, visualized with 254 nm UV light and KMnO4 stain on heating).

7. The reaction performed at half scale provided product in 79% yield.

8. Iodide **2**: [1H] NMR (400 MHz, $CDCl_3$) δ: 1.44 (s, 9 H), 3.51–3.59 (m, 2 H), 3.78 (s, 3 H), 4.48–4.53 (m, 1 H), 5.32–5.38 (broad d, 1H); [13C] NMR (100 MHz, $CDCl_3$) δ: 7.9, 28.4, 53.10, 53.8, 80.6, 154.9, 170.1; IR (film): 3346.6, 2981.6, 1731.6, 1688.9, 1521.4, 1438.3,1312.9, 1273.2, 1248.4, 1155.9, 1140.3, 1032.4, 1009.5, 862.9 cm^{-1}. HRMS calcd for $C_9H_{16}NIO_4$ [M+1]: 330.0124. Found: 330.0197. $[\alpha]_D$ +42.0 (c 1.0, $CHCl_3$). Calcd for $C_9H_{16}NIO_4$: C, 32.84; H, 4.90; N, 38.56. Found: C, 33.06; H, 4.82; N, 4.28. The enantiomeric purity of compound **2** was >98 % by SFC (eluent 10 % MeOH, pressure 150 bar, flow rate 3 mL/min, injection volume 25 μL into a 20 μL loop, column AD-H, 25 cm x 5 μm, column temp. 35 °C, t_r = 1.68 min); injection of a sample containing an incremental addition of 0.1 mg of the *R*-isomer (t_r = 2.25 min) into 10 mg of *S*-**2** established the limits of detection to be at least 1:99.

9. Zinc dust (particle size <10 μm, >98 %) was purchased and used as received from Aldrich chemical company.

10. The submitters used dry DMF from a solvent filtration system (Glass Contour, Irvine, CA). The checkers used anhydrous DMF under inert atmosphere with a sure seal bottle from Aldrich chemical company.

11. 1,2-Dibromoethane was purchased from Baker Chemicals and used as received. The checkers purchased 1,2-dibromoethane from Aldrich Chemical Company, Inc. An exotherm of Δ 35 °C and gas evolution were observed by the checkers after addition was completed.

12. The submitters used an oil bath kept at 60 °C external temperature once the exotherm stopped.

13. TMS-Cl (purity ≥97%) was purchased from Aldrich chemical company and used as received.

14. Evolution of gas was observed after the addition of TMS-Cl.

15. During addition of iodide **2**, the round-bottomed flask was covered with aluminum foil and stirred in the dark until the reaction was complete, because iodide **2** is light sensitive.

16. The time for reagent addition varied with reaction scale. In the reported experiment, the addition was completed over 25 min in order to maintain the internal temperature below 35 °C.

17. TLC analysis was performed on Merck Aluminum silica gel plates 60 F_{254}. Reaction conversion was monitored by spotting the reaction mixture on a TLC plate and eluting with 2:1 hexanes/ethyl acetate. Using 254 nm UV light, consumption of starting material (R_f = 0.7) and another spot slightly above the base line were visualized.

18. $Pd_2(dba)_3$ (purity 97%) was purchased from Aldrich chemical company and used as received.

19. Tri(*o*-tolyl)phosphine (purity ≥97%) was purchased from Aldrich chemical company and used as received.

20. Anhydrous THF (50 mL) in a flame-dried measuring cylinder fitted with a septum was cooled to –78 °C. On cooling, the volume of the THF contracted to 45 mL. Vinyl bromide gas was bubbled into the THF until the total volume rose from 45 mL to 48.1 mL yielding a 1M solution. Employment of vinyl bromide solutions of >1 molar augmented formation of methyl *N*-(Boc)alaninate and diminished iodide yield. Checkers used commercially available 1 M vinyl bromide in THF from Aldrich chemical company. Methyl *N*-(Boc)alaninate is removed by chromatography: TLC R_f = 0.21 (9:1 hexanes/ethyl acetate), visualized as a UV inactive and $KMnO_4$ active spot. The checkers stopped the

positive flow of argon after the addition was completed in order to avoid any loss of vinyl bromide. Excess pressure is released through a bubbler connected to the argon inlet adapter. Submitters used a balloon connected to a three-way stopcock to contain the excess pressure of vinyl bromide.

21. Reaction conversion is ascertained on an aliquot of the reaction mixture (100 μL), which was partitioned between ethyl acetate (200 μL) and water (500 μL). The ethyl acetate layer was analyzed by TLC using 9:1 hexanes/ethyl acetate as eluant, and the plate was visualized with 254 nm UV light as well as with KMnO$_4$ stain after heating: methyl N-(Boc)alaninate (R$_f$ = 0.21, a UV inactive and KMnO$_4$ active spot); iodide **2** (R$_f$ = 0.29, a UV and KMnO$_4$ active spot); olefin **3** (R$_f$ = 0.29, a UV inactive and KMnO$_4$ active spot).

22. Olefin **3** is purified on a silica column. The residue is absorbed onto 30 g of silica gel. The column (18.4 cm diameter) is packed with slurry of 250 g of silica (high purity grade Silica gel particle size 230-400 mesh ASTM, Merck Ltd.) in hexanes (1000 mL). Elution with 2% ethyl acetate/hexanes removes first all non-polar spots. Switching to 7% ethyl acetate in hexanes (~1500 mL) elutes the product (TLC R$_f$ = 0.29, 9:1 hexanes/ethyl acetate), which is visualized as a UV inactive and KMnO4 active spot on heating.

23. The reaction performed at half scale provided product in 64% yield.

24. N-(Boc)Allylglycine methyl ester (**3**): [α]$_D$ +20.2 (c 1.5, CHCl$_3$); lit.[3] [α]$_D$ +18.8 (c 1.0, CHCl$_3$); ^{1}H NMR (500 MHz, CDCl$_3$) δ: 1.40 (s, 9 H), 2.41–2.53 (m, 2 H), 3.70 (s, 3 H), 4.32–4.38 (m, 1 H), 5.04 (br s, 1 H), 5.08 (s, 1 H), 5.10–5.12 (m, 1 H), 5.62–5.71 (m, 1 H); ^{13}C NMR (125 MHz, CDCl$_3$) δ: 28.37, 36.86, 52.29, 53.0, 79.94, 119.12, 132.42, 155.27, 172.62; IR (film): 3359.0, 2978.3, 1745.9, 1715.7, 1505.7, 1437.9, 1366.3, 1249.9, 1162.6, 1051.2, 1022.7, 920.6 cm^{-1}; HRMS calcd for C$_{11}$H$_{19}$NO$_4$ [M+1]: 230.1314; Found: 230.13867. In lieu of combustion analysis the checkers determined the purity of N-(Boc)allylglycine methyl ester (98.1%) by quantitative ^{1}H NMR assay using dimethyl fumarate as a standard. Attempts to ascertain the enantiomeric purity of olefin **3**, as well as the hydrochloride obtained on treating **3** with HCl gas in dichloromethane, both were unsuccessful using SFC on a chiral column. To assess enantiomeric purity, diastereomeric amides were synthesized using respectively Boc-L-Ala (1.5 equiv), i-Pr$_2$NEt (2 equiv), HATU (1.5 equiv) (Note 25). The residue was examined by ^{1}H NMR spectroscopy in CD$_3$CN at 700 MHz and 400 MHz. Incremental addition of (S,S)-**6** into

(S,R)-**9** (Note 25) and observation of the methyl ester singlets at 3.691 and 3.697 ppm demonstrated the diastereomers were of >99:1 dr. Hence, olefin **3** is assumed to be of >98% enantiomeric purity.

25. Synthetic procedure for making diastereomers (S,S)- **6** and (S,R)-**9**:

(S)-Methyl 2-aminopent-4-enoate hydrochloride (**4**). Dry HCl gas was bubbled into a stirred solution of *tert*-butyl (S)-1-(methoxycarbonyl)but-3-enylcarbamate (**3**, 70 mg, 0.30 mmol) in dry dichloromethane at room temperature. Consumption of SM was observed after 3h (TLC). The resulting solution was concentrated under reduced pressure to give (S)-**4** as a brown solid: ^{1}H NMR (400 MHz, CD$_3$OD) δ: 2.68-2.75 (m, 2 H), 3.86 (s, 3 H), 4.16-4.19 (m, 1 H), 5.27-5.33 (m, 2 H), 5.77-5.81 (m, 1 H). (R)-Methyl 2-aminopent-4-enoate hydrochloride R-**8** was made by an analogous method as that used to prepare S-**4**.

(S)-Methyl 2-((S)-2-((*tert*-butoxycarbonyl)amino)propanamido)pent-4-enoate (S,S-**6**). A stirred solution of Boc-L-Ala (**5**, 85.1 mg, 0.45 mmol, 1.5 equiv) in DCM (5 mL) was treated with amine hydrochloride (S)-**4** (50 mg, 0.30 mmol, 1 equiv), DIEA (77.5 mg, 0.6 mmol, 2 equiv), and HATU (171.1 mg, 0.45 mmol, 1.5 equiv), stirred at room temperature for 16 h, diluted with DCM (~10 mL) and washed with saturated aqueous NaHCO$_3$. The layers were separated. The aqueous layer was extracted with DCM (~10 mL). The combined organic layers were washed with brine, dried over anhydrous sodium sulfate, filtered, and concentrated under reduced pressure to give (S,S)-**6** as brown oil, which was analyzed without further purification. This following spectral data was provided by the submitters. (S,S)-**6**: ^{1}H NMR (400 MHz, CD$_3$CN) δ: 1.27–1.28 (d, J = 7.2, 3 H), 1.44 (s, 9 H), 2.43–2.51 (m, 1 H), 2.53–2.60 (m,

1 H), 3.70 (s, 3 H), 4.06–4.10 (m, 1 H), 4.44–4.49 (m, 1 H), 5.10–5.19 (m, 2 H), 5.63 (br s, 1 H), 5.72–5.82 (m, 1 H), 6.91 (br s, 1 H); (*S,R*)-**9** was made by the analogous method used to prepare (*S,S*)-**6** using *R*-**8**: (*S,R*)-**9**: ^{1}H NMR (400 MHz, CD$_3$CN) δ: 1.26–1.28 (d, *J* = 7.2, 3 H), 1.44 (s, 9 H), 2.42–2.50 (m, 1 H), 2.53–2.60 (m, 1 H), 3.70 (s, 3 H), 4.05–4.09 (m, 1 H), 4.44–4.49 (m, 1 H), 5.10–5.18 (m, 2 H), 5.62 (br s, 1 H), 5.68–5.81 (m, 1 H), 6.90 (br s, 1 H).

Working with Hazardous Chemicals

The procedures in *Organic Syntheses* are intended for use only by persons with proper training in experimental organic chemistry. All hazardous materials should be handled using the standard procedures for work with chemicals described in references such as "Prudent Practices in the Laboratory" (The National Academies Press, Washington, D.C., 2011; the full text can be accessed free of charge at http://www.nap.edu/catalog.php?record_id=12654). All chemical waste should be disposed of in accordance with local regulations. For general guidelines for the management of chemical waste, see Chapter 8 of Prudent Practices.

In some articles in *Organic Syntheses*, chemical-specific hazards are highlighted in red "Caution Notes" within a procedure. It is important to recognize that the absence of a caution note does not imply that no significant hazards are associated with the chemicals involved in that procedure. Prior to performing a reaction, a thorough risk assessment should be carried out that includes a review of the potential hazards associated with each chemical and experimental operation on the scale that is planned for the procedure. Guidelines for carrying out a risk assessment and for analyzing the hazards associated with chemicals can be found in Chapter 4 of Prudent Practices.

The procedures described in *Organic Syntheses* are provided as published and are conducted at one's own risk. *Organic Syntheses, Inc.*, its Editors, and its Board of Directors do not warrant or guarantee the safety of individuals using these procedures and hereby disclaim any liability for any injuries or damages claimed to have resulted from or related in any way to the procedures herein.

Discussion

In the field of peptide chemistry, unsaturated amino acids and their esters are useful starting materials for the synthesis of amino acids, constrained peptides and peptide mimics. Allylglycine esters have been used as versatile building blocks for the synthesis of macrocyclic dipeptide □-turn mimics,[4] azabicyclo[X.Y.0]alkanone amino acids,[5] the key intermediate of diaminopimelate metabolism L-tetrahydrodipicolinic acid,[6] potent macrocyclic HCV NS3 protease inhibitors,[7] bicyclic amino acid substrates for intramolecular Pauson-Khand cyclizations,[8] and anti-bacterial cyclic peptides.[9] The double bond of the unsaturated amino acid has been functionalized by various chemical processes, including Diels–Alder reactions,[10] and cycloadditions,[11] cross-metathesis,[12] as well as Heck[13] and Suziki-Miyaura cross coupling reactions.[14]

Although a variety of methods have provided allylglycinates in enantiomerically enriched forms,[3,15-25] they require often longer reaction sequences. Diastereoselective syntheses of allylglycinate have been achived using chiral auxiliaries which may be removed or destroyed,[23] such as ephedrine-derived imidazolidinone glycinimides,[22] menthone-derived nitrones,[18] and camphor-derived glycine derivatives.[17,20,24] Enantioselective approaches to allylglycinate have featured allylation of ketoester oximes employing chiral bis(oxazoline) ligands,[21] and allylation of *tert*-butyl glycinate using tartrate-derived and Cinchona alkaloid-derived quaternary ammonium phase-transfer catalysts.[19,28] In addition, allylglycinates have been prepared from amino acids as chiral educts. For example, glutamate served as starting material for the syntheisis of allyl 2-(Boc)amino-4-triphenylphosphonium butanoate, which reacted with various aldehydes and paraformaldehyde in Wittig-Horner-Wadsworth-Emmons reactions to yield unsaturated amino acids.[16] Similarly, ylide from methyltriphenylphosphonium bromide reacted with α-*tert*-butyl *N*-(PhF) aspartate β-aldehyde to provide protected allylglycinate.[4] In the context of our research in peptide mimicry,[4,5] we required an efficient, atom economical route to enantiomerically pure allylglycine analogues. Building on the established precedent of zinc-mediated, palladium-catalyzed cross-coupling reactions of commercially available and inexpensive *tert*-butyl (*R*)-1-(methoxycarbonyl)-2-iodoethylcarbamate,[2,15,26,27,29] this extension employs vinyl bromide to give effective access to allylglycine in enantiomerically pure form on multi-gram scale.

The zinc insertion reaction of the alininyl iodide **2** are typically performed in DMF to thwart the chelation of the ester and carbamate functions with zinc, which has been suggested to promote β-elimination to the corresponding amino acrylate, particulary when performed in THF.[26] The organozinc intermediate is relatively stable towards air and moisture. Attempts to perform the related Kumuda coupling using vinylmagnesium bromide were unsuccessful and resulted in β-elimination affording amino acrylate. Negishi cross coupling of the alininyl zinc intermediate with vinyl bromide is effectively mediated by $Pd_2(dba)_3$ and tri-(o-tolyl)phosphine and has been examined at lower temperature to give the N-(Boc)-allylglycine methyl ester **3** with improved yield.

References

1. N. D. Prasad Atmuri, William D. Lubell, Department of Chemistry, Université de Montréal, P.O. Box 6128, Succursal Centre-ville, Montréal, QC H3C 3J7, Canada. E-mail: william.lubell@umontreal.ca We thank the Natural Sciences and Engineering Research Council of Canada for financial support. Shastri Indo-Canadian institute is thanked for a Quebec Tuition Fee Exemption grant to N.D. Prasad Atmuri.

2. Trost, B. M.; Rudd, M. T. *Org. Lett.* **2003**, 5, 4599.

3. Collier, P. N.; Campbell, A. D.; Patel, I.; Taylor, R. J. *Tetrahedron* **2002**, 58, 6117.

4. Kaul, R.; Surprenant, S.; Lubell, W. D. *J. Org. Chem.* **2005**, 70, 3838.

5. Surprenant, S.; Lubell, W. D. *Org. Lett.* **2006**, 8, 2851.

6. Caplan, J. F.; Sutherland, A.; Vederas, J. C. *J. Chem. Soc., Perkin Trans. 1* **2001**, 2217.

7. Nair, L. G.; Bogen, S.; Bennett, F.; Chen, K.; Vibulbhan, B.; Huang, Y.; Yang, W.; Doll, R. J.; Shih, N.-Y.; Njoroge, F. G. *Tetrahedron Lett.* **2010**, 51, 3057.

8. Bolton, G. L.; Hodges, J. C.; Ronald Rubin, J. *Tetrahedron* **1997**, 53, 6611.

9. Boyle, T. P.; Bremner, J. B.; Coates, J.; Deadman, J.; Keller, P. A.; Pyne, S. G.; Rhodes, D. I. *Tetrahedron* **2008**, 64, 11270.

10. Kotha, S. *Acc. Chem. Res.* **2003**, 36, 342.

11. Avenoza, A.; Busto, J. H.; Canal, N.; Peregrina, J. M.; Pérez-Fernández, M. *Org. Lett.* **2005,** 7, 3597.

12. Wang, Z. J.; Jackson, W. R.; Robinson, A. J. *Org. Lett.* **2013**, 15, 3006.

13. Collier, P. N.; Patel, I.; Taylor, R. J. *Tetrahedron Lett.* **2002**, 43, 3401.

14. Krebs, A.; Ludwig, V.; Pfizer, J.; Dürner, G.; Göbel, M. W. *Chem-Eur. J.* **2004**, 10, 544.

15. A synthesis of methyl *N*-(Cbz)allyglycinate without experimental procedure is reported in: Olsen, J. A.; Severinsen, R.; Rasmussen, T. B.; Hentzer, M.; Givskov, M.; Nielsen, J. *Bioorg. Med. Chem. Lett.* **2002**, 12, 325.

16. Rémond, E.; Bayardon, J.; Ondel-Eymin, M.-J. l.; Jugé, S. *J. Org. Chem.* **2012**, 77, 7579.

17. Workman, J. A.; Garrido, N. P.; Sançon, J.; Roberts, E.; Wessel, H. P.; Sweeney, J. B. *J. Am. Chem. Soc.* **2005,** 127, 1066.

18. Katagiri, N.; Okada, M.; Morishita, Y.; Kaneko, C. *Tetrahedron* **1997,** 53, 5725.

19. Ohshima, T.; Shibuguchi, T.; Fukuta, Y.; Shibasaki, M. *Tetrahedron* **2004,** 60, 7743.

20. Yeh, T.-L.; Liao, C.-C.; Uang, B.-J. *Tetrahedron* **1997**, 53, 11141.

21. Hanessian, S.; Yang, R.-Y. *Tetrahedron Lett.* **1996,** 37, 8997.

22. Guillena, G.; Nájera, C. *J. Org. Chem.* **2000**, 65, 7310.

23. Myers, A. G.; Gleason, J. L. *Org. Synth.* **1999**, 76, 57.

24. Xu, P.-F.; Lu, T.-J. *J. Org. Chem.* **2003,** 68, 658.

25. Kitagawa, O.; Hanano, T.; Kikuchi, N.; Taguchi, T. *Tetrahedron Lett.* **1993**, 34, 2165.

26. Jackson, R. F.; Moore, R. J.; Dexter, C. S.; Elliott, J.; Mowbray, C. E. *J. Org. Chem.* **1998**, 63, 7875.

27. Carrillo-Marquez, T.; Caggiano, L.; Jackson, R. F.; Grabowska, U.; Rae, A.; Tozer, M. *J. Org. Biomol. Chem.* **2005**, 3, 4117.

28. Bojan, V.; Maja, G. –P.; Radomir, M.; Radomir, N. S. *Org. Lett.* **2014**, 16, 34.

29. Jackson, R. F.; Perez - Gonzalez, M. *Org. Synth.* **2005**, 77–88.

Appendix
Chemical Abstracts Nomenclature (Registry Number)

1,2-Dibromoethane; (106-93-4)
Chlorotrimethylsilane; (75-77-4)

Vinyl bromide; (593-60-2)
Tris(dibenzylideneacetone)dipalladium(0); (51364-51-3)
Tri(o-tolyl)phosphine; (6163-58-2)
Zinc dust (particle size <10 μm); (7440-66-6)
Triphenylphosphine; (603-35-0)
Imidazole; (288-32)
Iodine; (7553-56-2)

N. D. Prasad Atmuri was born in India, in 1984. After receiving his M.Sc. degree in Organic Chemistry from Nagarjuna University in 2006, he worked 4.6 years as an Associate Scientist for Syngene International Ltd, India. In 2013, he began graduate studies at the Université de Montréal under the direction of Professor William D. Lubell. His current research focuses on the development of methods for synthesizing azabicyclo[X.Y.0]alkanones by way of ring closing metathesis and transannular cyclization with specific interest on novel indolizidinone modulators of the prostaglandin F2☐ receptor to develop improved therapeutics for preventing preterm labor.

Dr. William D. Lubell obtained his Ph.D. in 1989 from the University of California, Berkeley under the supervision of Professor Henry Rapoport, and studied as a postdoctoral fellow with Professor Ryoji Noyori at Nagoya University, Japan. In 1991, he joined the Chemistry Department of the Université de Montréal, where he is Full Professor. Advancing applications of peptides in drug discovery, Lubell has innovated methods for constraining amino acids and peptides to study structure-activity relationships and evolve peptidomimetic drug candidates with enhanced pharmacokinetic properties, including submonomer azapeptide synthesis, aminolactam scanning and azabicycloalkane amino acid libraries.

Jean-Nicolas (Nick) Desrosiers received his bachelor degree University of Montreal in 2003. He obtained his Ph.D. in 2008 under the supervision of Prof. André Charette, with whom he performed studies on enantioselective processes catalyzed by chiral bisphosphine monoxide copper complexes. In 2008, he joined the research group of Prof. Eric N. Jacobsen at Harvard University as a NSERC postdoctoral fellow. His postdoctoral research mainly focused on silyl ketene acetals acylation through anion-binding catalysis. He is currently working in the Chemical Development department of Boehringer-Ingelheim, Ridgefield, CT.

Dr. Nizar Haddad was born and raised in Israel. He received his B.A. degree (1984) and D.Sc. degree in chemistry (1988, Professor D. Becker) from the Technion, Israel Institute of Technology. After his postdoctoral research at the University of Chicago with Professor J. D. Winkler and additional postdoctoral work at Harvard University with Professor Y. Kishi, he joined the faculty in the Chemistry Department at the Technion, Israel in 1991. Following one-year sabbatical leave with Professor K. C. Nicolaou at the Scripps Research Institute, he joined Boehringer Ingelheim Pharm. Inc. in 1998 where he gained wide experience in process research, process development and scale up operations. Currently, as Senior Principal Scientist, he is heading the Chemical Catalysis group since 2007. His research interests include the development and application of catalytic asymmetric reactions and the application of lab automation in the development of new and economical processes.

Copper(II) Triflate as Additive in Low Loading Au(I)-Catalyzed Hydroalkylation of Unactivated Alkenes

Weizhen Fang, Marc Presset, Amandine Guérinot, Christophe Bour, Sophie Bezzenine-Lafollée,* and Vincent Gandon[1]*

ICMMO (UMR CNRS 8182), Université Paris-Sud, 91405 Orsay (France)

Checked by Pavel K. Elkin and Viresh H. Rawal

Procedure

A. *N-Allyl-N-benzylamine (1)*. A 500-mL one-necked, round-bottomed flask equipped with a magnetic stirring bar (5 x 2 cm Teflon-coated, ovoid-shaped) is charged with potassium carbonate (27.7 g, 0.2 mol, 1.2 equiv)

Org. Synth. **2015**, *92*, 117-130
DOI: 10.15227/orgsyn.092.0117

Published on the Web 5/5/2015
© 2015 Organic Syntheses, Inc.

(Note 1) and then sealed with a rubber septum, into which is inserted a needle connected to a nitrogen/vacuum inlet line. The flask is evacuated and filled with nitrogen. Allylamine (150 mL, 114.5 g, 2.0 mol, 11.9 equiv) (Notes 1 and 2) is added via cannula into the flask. The septum is replaced with a 100 mL pressure-equalizing addition funnel, and the funnel is charged with benzyl bromide (20.0 mL, 28.6 g, 0.17 mol, 1.0 equiv) (Note 1), then sealed with the septum containing the nitrogen inlet needle. Benzyl bromide is then added dropwise over 30 min under a slight positive pressure of nitrogen. The reaction mixture is stirred at 25 °C for 3 h (Notes 3 and 4), then diluted with 50 mL of CH_2Cl_2, added via the addition funnel. The resulting mixture is filtered through a fritted funnel (porosity 3, 10 cm diameter). The solid residue is rinsed with CH_2Cl_2 (2 x 100 mL), and the filtrate is concentrated by rotary evaporation (40 °C, 10 mmHg) to afford a pale yellow liquid. The crude product is purified by vacuum distillation (69 °C, 1 mmHg) to afford 21.0–21.7 g (85–88%) of *N*-allyl-*N*-benzyl-amine **1** as a colorless oil (Notes 5, 6, 7, and 8).

B. *N-Allyl-N-benzyl-2-oxocyclohexanecarboxamide (2)*. A 250-mL one-necked, round-bottomed flask equipped with a magnetic stirring bar (3 x 1.5 cm Teflon-coated, ovoid-shaped) is sealed with a rubber septum, into which inserted a needle connected to a nitrogen/vacuum inlet line. The flask is evacuated and filled with nitrogen, then charged with *N*-allyl-*N*-benzylamine **1** (15.0 g, 101.9 mmol, 2.0 equiv) using a syringe. The septum is removed and 4-dimethylaminopyridine (1.9 g, 15.6 mmol, 0.3 equiv), toluene (50 mL), and ethyl 2-oxocyclohexanecarboxylate (8.6 mL, 9.1 g, 53.6 mmol, 1.0 equiv) are added (Note 9). The neck is equipped with a 30 cm Graham-type water-cooled reflux condenser, the top of which is sealed with the septum containing a nitrogen inlet needle, and positive pressure of nitrogen is maintained. The resulting mixture is stirred at reflux in a pre-heated oil bath at 130 °C (external temperature) for 2 days (Notes 3 and 10). The reaction mixture is then allowed to cool down to room temperature (25 °C) and transferred to a 1 L separatory funnel. The flask is rinsed with EtOAc (50 mL) and the rinsate is added to the funnel. The solution is washed successively with 150 mL of aqueous 1 M HCl solution and 100 mL of a saturated aqueous NaCl solution, dried over 20 g of $MgSO_4$, filtered through a fritted funnel (porosity 3, 5 cm diameter), and concentrated by rotary evaporation (40 °C, 10 mmHg) to afford a brown oil. The crude product is purified by flash chromatography on a silica gel column (8 x 40 cm, 250 g of silica gel) using EtOAc:hexanes (15:85) (Note 11). Four 250 mL fractions are collected, then the solvent is changed to EtOAc:hexanes

(25:75) and another six fractions (250 mL each) are collected. The desired product is obtained in fractions 6-10, which are concentrated by rotary evaporation (40 °C, 10 mmHg) and dried under vacuum (1.0 mmHg) for 2 h to give 10.7–13.0 g (73–90%) of *N*-allyl-*N*-benzyl-2-oxocyclohexane-carboxamide **2** as a yellow oil (Notes 12 and 13).

C. *N-Benzyl-4-methyl-2-azaspiro[4.5]decane-1,6-dione (3)*. An oven-dried 16 x 4 cm (150 mL) Schlenk tube equipped with a Teflon-coated magnetic stir bar (1.3 x 2.5 cm) is evacuated and filled with nitrogen. This procedure is repeated 3 times. While maintaining a stream of nitrogen, the tube is charged with copper (II) triflate (673 mg, 1.86 mmol, 0.1 equiv) and toluene (50 mL), followed by a solution of *N*-allyl-*N*-benzyl-2-oxocyclohexanecarboxamide **2** (5.0 g, 18.4 mmol, 1.0 equiv) in toluene (20 mL) and a solution of JohnPhosAuCl (5 mL of a 2.0 g/L in toluene, 0.02 mmol, 0.1 mol%), both added by syringe (Notes 14 and 15). The tube is sealed with a glass stopper and the stirred mixture is immersed in a preheated oil bath at 110 °C (external temperature) for 2 h (see photo) (Notes 3, 16, and 17). The reaction mixture is allowed to cool down to room temperature (25 °C), then filtered through a fritted funnel (porosity 3, 5 cm diameter) into a 250-mL round-bottomed flask. The reaction vessel was rinsed with 50 mL of EtOAc, and the rinsate was passed through the fritted funnel. The filter cake was washed with EtOAc (2 x 15 mL), and the filtrate was concentrated by rotary evaporation (40 °C, 10 mmHg) to afford a brown oil (Note 18). The crude product is dissolved in 100 mL of CH_2Cl_2 and treated with 25 g of silica gel. The resulting mixture is placed on a rotary evaporator (25 °C, 100 mmHg) until the silica appears dry. The silica gel-adsorbed crude product is placed on a silica gel column (6 x 30 cm, 250 g of silica gel) prepared using EtOAc:hexanes (15:85). The column is eluted using the same solvent system and 100 mL fractions are collected. After 15 fractions are collected, the solvent is changed to EtOAc:hexanes (25:75) and another 20 fractions are collected. The minor diastereomer is found in fractions 4-16 (determined by TLC), which are combined, concentrated, and placed under vacuum (25 °C, 1 mmHg) to afford 1.52–1.53 g (31%) of compound **3b** (Note 19). The major diastereomer is found in fractions 18-30, which are similarly concentrated to afford 2.56–2.57 g (51%) of compound **3a** (Notes 19 and 20).

Set-up for Step A   Set-up for Step B   Set-up for Step C

1. Benzyl bromide (99%) was purchased from Alfa-Aesar, allylamine (98%) and potassium carbonate (99%, anhydrous, Redi-Dry™) from Sigma-Aldrich, and used as received. EtOAc and hexanes (both ACS grade) were purchased from Fisher and used as received.

2. A large excess of allylamine was used to ensure a complete conversion of benzyl bromide.

3. TLC was performed on Silica gel 60 F_{254} glass plates purchased from EMD Millipore and visualized with a permanganate stain (prepared from 2 g of $KMnO_4$, 13 g of K_2CO_3 and 200 mL of H_2O).

4. The progress of the reaction was followed by TLC analysis on silica gel with 15% EtOAc-hexanes as eluent and visualization with the $KMnO_4$ stain. Benzyl bromide, $R_f = 0.81$; allylamine $R_f = 0.00$; product **1** $R_f = 0.10$.

5. The submitters purified the product by flash chromatography on a column (8 x 40 cm) of 250 g of silica gel conditioned with EtOAc:cyclohexane (10:90) (Note 1) and eluted with 1.5 L of EtOAc:cyclohexane (10:90) followed by 2 L of EtOAc:cyclohexane (50:50) in 500 mL fractions. The desired product is obtained in fractions 3-6, which are concentrated by rotary evaporation (40 °C, 20 mmHg) and dried under vacuum (1.3 mmHg) for 2 h to give 17.34 g (70%) of **1** as yellow oil. Silica gel: Gerudan Si60 (40-63 µm) was purchased from Merck.

6. The submitters checked purity using GC analysis. GC conditions: Varian GC430 apparatus, column VF1-MS (15 m x 0.25 mm x 0.25 µm); vector gas: He; flow: 2 mL/min; injection temperature: 250 °C; temperature profile: initial temperature = 90 °C for 1 min, temperature gradient = 10 °C/min, final temperature = 250 °C for 5 min; detection: FID (250 °C).

7. N-Allyl-N-benzylamine (**1**) is bench-stable. Physical properties: GC Retention time: 2.96 min (Note 6); FT-IR (film): 3315, 3064, 3027, 2978, 2811, 1643, 1495, 1454, 1106, 918, 736, 698 cm^{-1}; HRMS (ESI-TOF): m/z calcd. for $C_{10}H_{14}N$ (M + H)$^+$ 148.1121, found 148.1116; ^1H NMR (250 MHz, CDCl$_3$) δ: 1.35 (br s, 1 H), 3.28 (dd, J=6.0, 1.5 Hz, 2 H), 3.79 (s, 2 H), 5.10 (dt, J=10.5, 1.5 Hz, 1 H), 5.19 (dt, J=17.5, 1.5 Hz, 1 H), 5.93 (ddt, J =17.5, 6.0, 1.5 Hz, 1 H), 7.23-7.27 (m, 1 H), 7.30-7.34 (m, 4H); ^{13}C NMR (63 MHz, CDCl$_3$): δ 51.7, 53.2, 115.9, 126.9, 128.1, 128.3, 136.8, 140.3.

Anal. calcd for $C_{10}H_{13}N$: C, 81.59; H, 8.90; N, 9.51. Found: C, 81.66; H, 8.92; N, 9.57.

8. *N*-Allyl-*N*-benzylamine (**1**) is also commercially available. The submitters and checkers only used the product prepared by the method described here.

9. 4-Dimethylaminopyridine (99%) was purchased from Alfa-Aesar, toluene (HPLC grade) from Fisher, and ethyl 2-oxocyclohexanecarboxylate (95%) from Sigma-Aldrich. All were used as received.

10. The progress of the reaction was followed by TLC analysis on silica gel with 15% EtOAc-hexanes as eluent. Ethyl 2-oxocyclohexanecarboxylate, R_f = 0.75 and 0.18 (the latter spot probably corresponds to the enol form); amine **1**, R_f = 0.19; product **2**, R_f = 0.34. The reaction does not reach full conversion after 2 days (~95% complete by NMR), but a longer reaction time (4-5 days) provides only slight increase in conversion (~ 2%).

11. Silica gel: SiliaFlash® P60 40-63μm (230-400 mesh) 60Å Irregular Silica Gel was purchased from Siliycle.

12. Unreacted starting material is found in fractions 2-3, which after concentration and drying under vacuum provides 0.66 g (7%) of starting ethyl 2-oxocyclohexanecarboxylate.

13. *N*-Allyl-*N*-benzyl-2-oxocyclohexanecarboxamide **2** is bench-stable. Physical properties: GC retention time: 12.21 min (see Note 6); FT-IR (film): 3063, 3029, 2940, 2865, 1709, 1648, 1448, 1180, 1128, 737, 700 cm^{-1}; HRMS (ESI-TOF): *m/z* calcd. for $C_{17}H_{22}NO_2$ (M + H)$^+$ 272.1645. Found: 272.1657; ^{1}H NMR (500 MHz, CDCl$_3$, mixture of rotomeric forms) δ: 1.54–1.86 (m, 4 H), 1.95–2.12 (m, 6 H), 2.22–2.37 (m, 4 H), 2.51–2.59 (m, 2 H), 3.52–3.82 (m, 6 H), 4.19–4.50 (m, 4 H), 5.08–5.28 (m, 4 H), 5.70–5.83 (m, 2 H), 7.15–7.38 (m, 10 H); ^{13}C NMR (125 MHz, CDCl$_3$, mixture of rotomeric forms) δ: 21.1, 23.4, 23.6, 26.7, 26.9, 50.1, 30.3, 30.4, 34.5, 41.8, 41.9, 48.0, 48.4, 49.0, 54.3, 54.4, 116.6, 117.1, 126.2, 127.1, 127.5, 127.8, 127.9, 128.4, 128.5, 128.8, 132.5, 133.1, 136.7, 137.1, 169.9, 170.1, 207.2, 207.4. Calcd for $C_{17}H_{21}NO_2$: C, 75.25; H, 7.80; N, 5.16. Found: C, 75.11; H, 7.86; N, 5.17.

14. Copper (II) triflate (99%) was purchased from Alfa-Aesar, JohnPhosAuCl ((2-biphenyl)-di-*tert*-butylphosphine gold chloride, 98%) from Strem Chemicals, and toluene (HPLC grade) from Fisher. All were used as received.

15. The solution of gold complex was obtained by dissolving 20.0 mg JohnPhosAuCl in 10.0 ml of toluene with stirring for 15 min.

16. The progress of the reaction was followed by TLC analysis on silica gel with 15% EtOAc-hexanes as eluent and visualization with $KMnO_4$. Amide **2**, R_f = 0.19; minor diastereomer of **3**, R_f = 0.34; major diastereomer of **3**, R_f = 0.16.

17. The initial clear green solution turned brown during the course of the reaction.

18. NMR spectrum of the crude product shows the presence of two diastereomers, formed in ca. 2:1 ratio.

19. Both diastereomers of *N*-benzyl-4-methyl-2-azaspiro[4.5]decane-1,6-dione **3** are bench-stable. Physical properties of minor diastereomer: GC Retention time = 12.52 min (see Note 6); FT-IR (film): 2938, 2865, 1705, 1682, 1494, 1428, 1262, 1230, 701 cm⁻¹; HRMS (ESI-TOF): [M + H]⁺ calcd for $C_{17}H_{22}NO_2$ (minor): 272.1645. Found: 272.1657. ¹H NMR (500 MHz, CDCl₃) (minor) δ: 1.10 (d, *J*=7.0 Hz, 3 H), 1.70–1.97 (m, 4 H), 2.15–2.34 (m, 4 H), 2.60 (ddd, *J*=16.0, 7.0, 6.5 Hz, 1 H), 3.09 (t, *J*=9.0 Hz, 1 H), 3.14 (dd, *J*=9.0, 8.0 Hz, 1 H), 4.45 (d, *J*=15.0 Hz, 1 H), 4.50 (d, *J*=15.0 Hz, 1 H), 7.23–7.35 (m, 5 H); ¹³C NMR (125 MHz, CDCl₃) (minor) δ: 13.4, 20.7, 26.9, 30.2, 31.6, 40.4, 46.8, 50.4, 61.5, 127.6, 128.0, 128.7, 136.3, 172.8, 208.4; Major diastereomer: GC Retention time = 12.39 min (see Note 6); FT-IR (film): 2937, 2867, 1705, 1682, 1440, 1264, 701 cm⁻¹; HRMS (ESI-TOF): [M + H]⁺ calcd for $C_{17}H_{22}NO_2$ (major): 272.1645. Found: 272.1657. ¹H NMR (500 MHz, CDCl₃) (major) δ: 0.92 (d, *J*=7.0 Hz, 3H), 1.66–1.76 (m, 3 H), 1.98–2.03 (m, 1 H), 2.06–2.12 (m, 1 H), 2.23–2.32 (m, 1 H), 2.48 (dt, *J*=14.0, 4.5 Hz, 1 H), 2.71 (dd, *J*=9.5, 7.0 Hz, 1 H), 2.98 (app. sex, *J*=7.5 Hz, 1 H), 3.09 (ddd, *J*=12, 11.5, 6.0 Hz, 1 H), 3.25 (dd, *J*=9.5, 7.5 Hz, 1 H), 4.38 (d, *J*=14.5 Hz, 1 H), 4.48 (d, *J*=14.5 Hz, 1 H), 7.20–7.21 (m, 2 H), 7.26–7.34 (m, 3 H); ¹³C NMR (125 MHz, CDCl₃) (major) δ: 14.0, 21.2, 24.5, 34.6, 39.6, 41.7, 46.8, 51.3, 61.8, 127.5, 127.8, 127.9, 128.6, 136.2, 174.1, 208.9.

20. In the submitter's original procedure, the reaction was carried out in air. Compounds **3a** and **3b** were obtained in 69% overall yield. The isolated yields reported in Table 1 are those obtained when the reaction is carried in air.

Working with Hazardous Chemicals

The procedures in *Organic Syntheses* are intended for use only by persons with proper training in experimental organic chemistry. All hazardous materials should be handled using the standard procedures for work with chemicals described in references such as "Prudent Practices in the Laboratory" (The National Academies Press, Washington, D.C., 2011; the full text can be accessed free of charge at http://www.nap.edu/catalog.php?record_id=12654). All chemical waste should be disposed of in accordance with local regulations. For general guidelines for the management of chemical waste, see Chapter 8 of Prudent Practices.

In some articles in *Organic Syntheses*, chemical-specific hazards are highlighted in red "Caution Notes" within a procedure. It is important to recognize that the absence of a caution note does not imply that no significant hazards are associated with the chemicals involved in that procedure. Prior to performing a reaction, a thorough risk assessment should be carried out that includes a review of the potential hazards associated with each chemical and experimental operation on the scale that is planned for the procedure. Guidelines for carrying out a risk assessment and for analyzing the hazards associated with chemicals can be found in Chapter 4 of Prudent Practices.

The procedures described in *Organic Syntheses* are provided as published and are conducted at one's own risk. *Organic Syntheses, Inc.*, its Editors, and its Board of Directors do not warrant or guarantee the safety of individuals using these procedures and hereby disclaim any liability for any injuries or damages claimed to have resulted from or related in any way to the procedures herein.

Discussion

Homogeneous Au(I)-catalysis has become an essential tool in organic chemistry. The majority of reactions involve $[LAu]^+Y^-$ as the active species where Y^- is a weakly coordinating anion ($Y^- = TfO^-$, Tf_2N^-, BF_4^-, PF_6^-, SbF_6^-, ...; $L = PR_3$, NHC, ...).[2] These electrophilic compounds act as soft Lewis acids, which can coordinate and activate unsaturated C–C bonds towards nucleophilic attack. They are usually generated by anion metathesis between LAuX (X= Cl, Br) and a silver salt (AgY), which ensures a fast and irreversible generation of $[LAu]^+Y^-$. However, these species may rapidly decay to give Au(0) (mirror, precipitate, or nanoparticles) and inactive $[L_2Au]^+Y^-$ under the reaction conditions.[3] Thus, some Au(I)-catalyzed reactions suffer from a limited scale (usually milligram scale) and restriction on temperature range (usually below 80 °C). A catalyst loading superior to 1 mol% is also often required.

Several groups have focused their efforts on the development of bulky ligands to circumvent these problems.[4] On our side, we decided to play on the anion metathesis itself. We have shown very recently that the use of copper salts as additives in Au(I)-catalyzed reactions allows the gradual delivery of $[LAu]^+$ from a reservoir of stable LAuX.[5] Thus, readily available, non-light-sensitive, and cheap Cu(II) salts can advantageously replace silver additives in Au(I)-catalyzed reactions. With the Au/Cu catalytic system, it becomes possible to carry out gram-scale reactions in a small amount of solvent, even at elevated temperature, without observing the formation of Au(0). Since then, we have worked on the scope of this Au/Cu catalytic system. The practical features of this method, including its operational simplicity, make it an expedient alternative to traditional methods.

In particular, it allows one to synthesize (spiro) lactams with low loadings of gold complex via hydroalkylation of unactivated alkenes. Lactams are valuable building blocks and are ubiquitous frameworks of compounds of biological interest. Che has reported similar transformations using 1 to 20 mol% of JohnPhosAuCl/AgOTf.[6] With some substrates, as shown in our preliminary communication, Che's procedure is not as efficient as the one we propose. With the Au/Cu catalytic system, the intramolecular hydroalkylation of readily accessible β-ketoamides can be performed on a 1 gram-scale using 0.1 mol% of JohnPhosAuCl and 10 mol% of copper(II) triflate at 110 °C with good yield (60-78%) (Table 1). In addition to substrate scope, this new Au/Cu system has many practical advantages. All the reactions can be carried out in standard glassware without precautions towards air and moisture and employs commercially available reagents and catalysts. This new catalytic system is also suitable to achieve highly selective transformations in a large scale.

Table 1. Gram-scale Hydroalkylation of Unactivated Alkenes

JohnPhosAuCl (0.1 mol%)
Cu(OTf)$_2$ (10 mol%)

PhMe, 110 °C

Entry	Substrate	Time (h)	Product	Conv (Yield) (%)	d.r.
1		1.5		100 (73)	66/34
2		1.5		100 (69)	67/33
3		1.0		100 (78)	95/5
4		24		100 (60)	60/40
5		24		100 (70)	n/a
6		3		80 (60)	90/10

rganic
Syntheses

References

1. ICMMO (UMR CNRS 8182), LabEx CHARMMMAT Université Paris-sud 91405 Orsay, France. Fax: (+)33 169 154 747. sophie.bezzenine@u-psud.fr; vincent.gandon@u-psud.fr. We thank ANR JCJC HAONA, UPS, CNRS, CSC, and IUF for financial support.
2. Modern Gold-Catalyzed Synthesis, Hashmi, A. S. K.; Toste, F. D. Eds; Wiley-VCH: Weinheim, 2012.
3. See Chapter 8 in ref 2 and: Wang, W.; Hammond, G. B.; Xu, B. *J. Am. Chem. Soc.* **2012**, *134*, 5697.
4. a) Hashmi, A. S. K.; Weyrauch, J. P.; Rudolph, M.; Kurpejovic, E. *Angew. Chem. Int. Ed.* **2004**, *43*, 6545; b) Marion, N.; Ramon, R. S.; Nolan, S. P. *J. Am. Chem. Soc.* **2009**, *131*, 448; c) Teller, H.; Corbet, M.; Mantilli, L.; Gopakumar, G.; Goddard, R.; Thiel, W.; Fürstner, A. *J. Am. Chem. Soc.* **2012**, *134*, 15331; d) Lavallo, V.; Wright II, J. H.; Tham, F. S.; Quinlivan, S. *Angew. Chem. Int. Ed.* **2013**, *52*, 3172.
5. a) Guérinot, A.; Fang, W.; Sircoglou, M.; Bour, C.; Bezzenine-Lafollée, S.; Gandon, V. *Angew. Chem. Int. Ed.* **2013**, *52*, 5848; b) Fang, W.; Presset, M.; Guérinot, A.; Bour, C.; Bezzenine-Lafollée, S.; Gandon, V. *Chem. Eur. J.* **2014**, *20*, 5439.
6. Zhou, C.-Y.; Che, C.-M. *J. Am. Chem. Soc.* **2007**, *129*, 5828.

Appendix
Chemical Abstracts Nomenclature (Registry Number)

Potassium carbonate: carbonic acid, potassium salt (1:2); (584-08-7)

Allylamine: 2-propen-1-amine; (107-11-9)

Benzyl bromide: benzene, (bromomethyl)-; (100-39-0)

4-Dimethylaminopyridine: 4-pyridinamine,N,N-dimethyl-; (1122-58-3)

Ethyl 2-oxocyclohexanecarboxylate: cyclohexanecarboxylic acid, 2-oxo-, ethyl ester; (1655-07-8)

Copper (II) triflate: methanesulfonic acid, 1,1,1-trifluoro-, copper(2+) salt (2:1); (34946-82-2)

JohnPhosAuCl: gold, [[1,1'-biphenyl]-2-ylbis(1,1-dimethylethyl)phosphine]chloro-; (854045-93-5)

Organic
Syntheses

Weizhen Fang was born in 1985 in Anhui, China. He received his B.Eng. and M.S. from Southwest Jiaotong University in 2008 and 2011 respectively. He is now pursuing his Ph.D. in the group of Prof. Vincent Gandon at Université Paris-Sud (2011-2014). His current research interests include two-component catalyst system (Gold(I) complexes and Lewis acids) for asymmetric hydroalkylation and the synthesis of natural products.

Marc Presset was born in Thonon-les-bains (France) in 1981. After joining the ENS Cachan, he passed the Agrégation de Sciences Physiques in 2007. He then moved to Aix- Marseille Université, where he obtained his Ph.D. under the supervision of Prof. Rodriguez and Dr. Coquerel in 2010. After previous postdoctoral experiences with Prof. Molander at the University of Pennsylvania (USA) and in Janssen (Belgium), he is currently working with Prof. Gandon at the Université Paris-Sud.

Amandine Guérinot was born in 1983 in Troyes (France). She received her engineer's diploma from ESPCI ParisTech in 2007 and then joined the organic chemistry laboratory at ESPCI where she prepared her Ph.D. under the supervision of Prof. Cossy and Dr. Reymond. After a one-year postdoctoral stay with Pr Canesi at UQAM (Montreal), she moved to Université Paris-Sud to work with Prof. Gandon. After an additional postodoctoral fellowship in Université Paris V in the group of Dr. Micouin, she was appointed associate professor in 2013 at ESPCI in the group of Pr Cossy.

Christophe Bour was born in Haguenau (France) in 1980. He studied chemistry at the University of Strasbourg and completed his Ph.D. with Dr. J. Suffert in 2006. He then joined the research group of Prof. Antonio Echavarren in Tarragona (Spain) as a post-doctoral fellow. In early 2009, he moved to the ECPM at the University of Strasbourg to work with Dr. G. Hanquet as a post-doctoral researcher. In September 2010, he was appointed associate professor at the Université Paris-Sud. His scientific interests include catalysis, coordination chemistry, and new synthetic methodologies.

Sophie Bezzenine-Lafollée was born in Paris, France in 1972. She received her Ph.D. prepared under the supervision of Dr. H. Rudler in 1998 in the University of Paris VI. She spent one post-doctoral year in Prof. Müller's group at the University of Geneva, Switzerland and then two years in the laboratory of Prof. J. Ardisson and Dr. A. Pancrazi at the University of Cergy-Pontoise. She became associate professor in 2001 in Orsay University. She worked with Dr. F. Guibé and then with Dr. J. Collin. Her main current research interests are enantioselective catalysis and applications in synthesis.

Vincent Gandon was born in Soissons, France, in 1973. He received his Ph.D. in 2002 from the University of Reims Champagne Ardenne (group of Prof. Jan Szymoniak). After a postdoctoral stay at the University of California, Riverside in the group of Prof. Guy Bertrand, he joined the faculty of the University Pierre et Marie Curie in Paris in 2003, as associate professor in the laboratory of Prof. Max Malacria. In 2009, he was appointed full Professor at the University of Paris-Sud. His research interests are focused on homogeneous catalysis using cobalt, gold, platinum, and gallium complexes.

Pavel K. Elkin was born in Kaliningrad, Russia, in 1991. He received his B.S. from D. Mendeleyev University of Chemical Technology of Russia in 2012 and M.S. from the University of Chicago in 2013. He is now pursuing his Ph.D. in the group of Prof. Viresh H. Rawal at the University of Chicago. His current research interests include development of novel dienes for asymmetric Diels-Alder reactions, new types of single-point hydrogen-bonding catalysts and the synthesis of natural products.

Ruthenium-Catalyzed Direct Oxidative Alkenylation of Arenes through Twofold C–H Bond Functionalization in Water: Synthesis of Ethyl (E)-3-(2-Acetamido-4-methylphenyl)acrylate

Lianhui Wang, Karsten Rauch, Alexander V. Lygin, Sergei I. Kozhushkov and Lutz Ackermann[*1]

Institut für Organische und Biomolekulare Chemie der Georg-August-Universität Göttingen, Tammannstrasse 2, 37077 Göttingen (Germany)

Checked by William Trieu and Margaret Faul

Ac~NH, Me, H, H~CO₂Et, [RuCl₂(p-cymene)]₂ (2.5 mol%), KPF₆ (20 mol%), Cu(OAc)₂·H₂O (1 equiv), H₂O, 105 °C, 20–26 h → Ac~NH, Me, CO₂Et

Procedure

A. *Ethyl (E)-3-(2-acetamido-4-methylphenyl)acrylate.* A 500-mL, two-necked, round-bottomed flask is equipped with a 2.5 cm rod-shaped, Teflon-coated magnetic stirring bar, rubber septum, reflux condenser and nitrogen inlet and outlet at the top of the reflux condenser. The flask is flushed with nitrogen (Note 1) and charged with N-*m*-tolylacetamide (5.00 g, 33.5 mmol), [RuCl₂(p-cymene)]₂ (513 mg, 0.84 mmol, 2.5 mol %), KPF₆ (1.233 g, 6.70 mmol, 20.0 mol %), Cu(OAc)₂·H₂O (6.690 g, 33.5 mmol, 1.0 equiv), (Note 2) and H₂O (100 mL) (Note 3). The reaction mixture is stirred for 10 min at ambient temperature, then ethyl acrylate (3.39 g, 3.6 mL, 33.86 mmol, 1.0 equiv) is added via syringe in one portion, and the rubber septum is changed to a glass stopper. After stirring for additional 15 min at ambient

Org. Synth. **2015**, *92*, 131-147
DOI: 10.15227/orgsyn.092.0131

Published on the Web 5/5/2015
© 2015 Organic Syntheses, Inc.

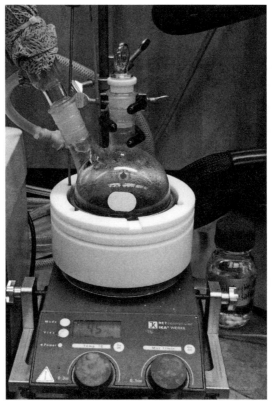

Experimental Set-up

temperature, the reaction mixture is vigorously stirred and heated in an oil bath (105 °C) (Notes 4 and 5). After 16 h of stirring at 105 °C, a second portion of ethyl acrylate (1.70 g, 1.8 mL, 16.95 mmol, 0.5 equiv) is added, and the mixture is stirred at 105 °C for an additional 10 h. After cooling to ambient temperature, the reaction mixture is dissolved in EtOAc (250 mL) and transferred to a 1-L Erlenmeyer flask. Saturated aqueous NH$_4$Cl and 25% aqueous NH$_3$ solutions (120 mL each) (Note 6) are added, the resulting mixture is vigorously stirred for 10 min with a 6 cm egg-shaped, Teflon-coated magnetic stirring bar and then transferred to a 1-L separatory funnel. The aqueous phase is extracted with EtOAc (2 × 150 mL). The combined organic phases are washed with brine (100 mL), dried over anhydrous MgSO$_4$ (80 g) for 30 min under stirring, and filtered. The solvent is removed from the filtrate under water-aspirator vacuum at 40 °C (20 mmHg). The residue (8.57 g) is purified by column chromatography on silica gel (Note 7) to furnish 6.68–6.73 g (81%, potency corrected) of ethyl (E)-3-(2-acetamido-4-methylphenyl)acrylate as a colorless solid (Note 8) with a purity ≥98%, as determined by quantitative ^{1}H NMR spectroscopy and GC analysis (Note 9 and 10).

Notes

1. This operation is performed by opening the nitrogen inlet from the condenser and flushing the vessel for 10 minutes.
2. N-m-Tolylacetamide, [RuCl$_2$(p-cymene)]$_2$, Cu(OAc)$_2$·H$_2$O, and ethyl acrylate are obtained from Sigma-Aldrich, (97–98% purity). KPF$_6$ is also obtained from Sigma-Aldrich and used as received. The N-m-tolyacetamide should be ground with a mortar and pestle if it is supplied as a hardened solid.
3. Water is distilled in a stream of nitrogen prior to its use.
4. The initially deep-green color of the reaction mixture is changed to red-brown after 30 min heating, and subsequently becomes dark after additional 2 h of heating. Heating was performed using an Al block heater in place of an oil bath.
5. The consumption of N-m-tolylacetamide is monitored by GC-MS. Small aliquots were withdrawn, worked-up by following the extraction procedures to the final EtOAc extraction, and then studied by a coupled gas chromatography/mass spectrometry instrument 7890A GC-System with mass detector 5975C (Triplex-Axis-Detector) from Agilent Technologies equipped with HP-5MS column (30 m × 0.25 mm, film 0.25 μm). Retention time t_{ret} = 4.65 min with helium flowrate of 3 mL/min, temperature profile T = 70 °C/1 min, then 70–150 °C/2.7 min, then 150 °C/1 min, then 150–250 °C/2.9 min, then 250 °C/3 min. The final conversion of the limiting substrate is 95–100%.
6. Technical grade NH$_4$Cl and aqueous NH$_3$ solutions are obtained from Teknova and SAFC respectively.
7. The residue is dissolved in CH$_2$Cl$_2$ (60 mL) and then charged onto a column (7 × 20 cm, 300 g of silica gel). The column is eluted with n-hexane/EtOAc/CH$_2$Cl$_2$ 2:2:1 (5 L) collecting the 200-mL fractions that contain material with R_f = 0.22. A total of eighteen 200-mL portions were collected, and the product was identified to be present in fractions 8 through 16.

8. Ethyl (E)-3-(2-acetamido-4-methylphenyl)acrylate has the following physicochemical and spectroscopic properties: mp = 155–156 °C; R_f = 0.06 (n-hexane/EtOAc 3:1); ¹H NMR (400 MHz, CDCl₃) δ: 1.32 (t, J=7.1 Hz, 3 H), 2.22 (s, 3 H), 2.34 (s, 3 H), 4.24 (q, J=7.0, 2 H), 6.34 (d, J=15.9 Hz, 1 H), 7.00 (d, J=7.63 Hz, 1 H), 7.44 (d, J=7.8 Hz, 2 H), 7.54 (br s, 1 H), 7.77 (d, J=15.9 Hz, 1 H); ¹³C NMR (100 MHz, CDCl₃) δ: 14.3, 21.3, 21.4, 23.7, 60.5, 118.6, 125.6, 126.5, 126.6, 127.0, 136.1, 139.1, 141.2, 167.1, 169.7; IR (neat): 3225, 1711, 1661, 1635, 1610, 1537, 1492, 1164, 983, 813 cm⁻¹; MS m/z [M]⁺: 247, 204, 160, 132, 117. HRMS [M + H⁺] calcd for $C_{14}H_{18}NO_3$: 248.12812, found: 248.12787; [M + Na⁺] calcd for $C_{14}H_{17}NNaO_3$: 270.11006, found: 270.10978

9. The analysis is performed applying coupled gas chromatography/mass spectrometry instrument *7890A GC-System* with mass detector *5975C (Triplex-Axis-Detector)* from Agilent Technologies equipped with *HP-5MS* column (30 m × 0.25 mm, film 0.25 μm). Retention time t_{ret} = 7.41 min with 3 mL·min⁻¹ of He, temperature profile T = 70 °C/1 min, then 70–150 °C/2.7 min, then 150 °C/1 min, then 150–250 °C/2.9 min, then 250 °C/3 min, then 250–290 °C/1 min.

10. Quantitative ¹H NMR was performed on each sample using benzyl benzoate (Sigma-Aldrich, part number 55177) as an internal standard. Analysis of each sample was performed in duplicate. Isolated products from the two replicates analyzed as 99.7 wt% and 98.8 wt%, respectively.

Working with Hazardous Chemicals

The procedures in *Organic Syntheses* are intended for use only by persons with proper training in experimental organic chemistry. All hazardous materials should be handled using the standard procedures for work with chemicals described in references such as "Prudent Practices in the Laboratory" (The National Academies Press, Washington, D.C., 2011; the full text can be accessed free of charge at http://www.nap.edu/catalog.php?record_id=12654). All chemical waste should be disposed of in accordance with local regulations. For general guidelines for the management of chemical waste, see Chapter 8 of Prudent Practices.

Organic
Syntheses

In some articles in *Organic Syntheses*, chemical-specific hazards are highlighted in red "Caution Notes" within a procedure. It is important to recognize that the absence of a caution note does not imply that no significant hazards are associated with the chemicals involved in that procedure. Prior to performing a reaction, a thorough risk assessment should be carried out that includes a review of the potential hazards associated with each chemical and experimental operation on the scale that is planned for the procedure. Guidelines for carrying out a risk assessment and for analyzing the hazards associated with chemicals can be found in Chapter 4 of Prudent Practices.

The procedures described in *Organic Syntheses* are provided as published and are conducted at one's own risk. *Organic Syntheses, Inc.*, its Editors, and its Board of Directors do not warrant or guarantee the safety of individuals using these procedures and hereby disclaim any liability for any injuries or damages claimed to have resulted from or related in any way to the procedures herein.

Discussion

Conventional palladium-catalyzed[2] cross-coupling reactions have matured to being among the most reliable tools for the formation of $C_{sp}2$–$C_{sp}2$ bonds for the preparation of styrene derivatives, which present useful intermediates in synthetic organic chemistry. Based on the pioneering studies by Mizoroki[3a] and by Heck,[3b] regioselective syntheses of styrenes[4] – including naturally occurring products[5] – have predominantly exploited palladium catalysts for reactions between prefunctionalized aryl (pseudo)halides and alkenes (Figure 1a).[6]

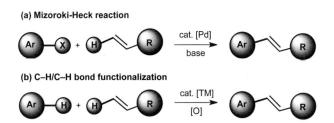

(a) Mizoroki-Heck reaction

(b) C–H/C–H bond functionalization

Figure 1. Strategies for transition metal-catalyzed styrene syntheses

Despite its remarkable importance and the thus achieved considerable advances in organic synthesis, the Mizoroki-Heck reaction is unfortunately accompanied by the formation of stoichiometric amounts of potentially hazardous halide salts which can cause a significant environmental pollution. For this reason, recent research interest has shifted towards the development of more environmentally-friendly halide-free alkenylations. In 1967, Fujiwara and Moritani thus reported the first example of catalyzed direct oxidative coupling of arenes with styrene through a twofold C–H bond activation approach, wherein the C–H bond of the alkene was replaced with the aromatic moiety in the presence of a palladium catalyst.[7] This approach is not only advantageous with respect to the overall minimization of by-product formation (atom-economy),[8a,b] but also enables a streamlining of organic syntheses by significantly reducing the overall number of required reaction steps (step-economy).[8c] As a consequence, a plethora of synthetically useful protocols for palladium-catalyzed direct oxidative couplings between arenes and alkenes (Figure 1b) was elaborated during the last decade.[9] Furthermore, relatively expensive rhodium catalysts were also developed for oxidative alkenylations in recent years.[10,11] Conversely, fourteen times less expensive[12] ruthenium[13] complexes have only recently been exploited as catalysts for direct C–H bond alkenylations on arenes.[14]

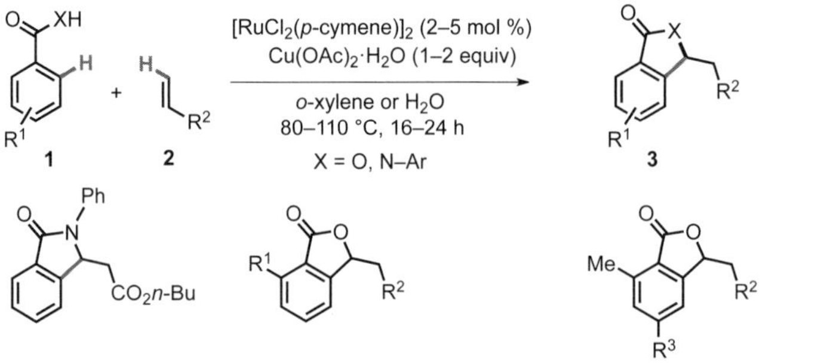

[RuCl$_2$(p-cymene)]$_2$ (2–5 mol %)
Cu(OAc)$_2$·H$_2$O (1–2 equiv)

o-xylene or H$_2$O
80–110 °C, 16–24 h

X = O, N–Ar

3a, 80%

3b, R^1 = OMe, R^2 = CO$_2$Et: 89%
3c, R^1 = OMe, R^2 = CN: 76%
3d, R^1 = F, R^2 = CO$_2$Et: 75%
3e, R^1 = F, R^2 = CN: 76%
3f, R^1 = Me, R^2 = CO$_2$n-Bu: 90%

3g, R^2 = CO$_2$Et, R^3 = Me: 89%
3h, R^2 = CN, R^3 = Me: 97%
3i, R^2 = CO$_2$Et, R^3 = Br: 67%

Scheme 1. Ruthenium(II)-catalyzed oxidative alkenylation/aza(oxa)-Michael reaction sequence with benzanilides and benzoic acids 1[16]

Recent years have witnessed a tremendous development in direct olefinations through twofold C–H bond functionalization of arenes and heteroarenes with readily accessible, selective, and rather inexpensive ruthenium catalysts.[15] Thus, oxidative alkenylations of arenes with electron-withdrawing coordinating substituents proved viable with ruthenium(II) complexes. For instance, ruthenium(II)-catalyzed oxidative alkenylations of benzanilides and benzoic acids **1** with acrylates or acrylonitriles **2** delivered bicycles **3** through intermolecular oxidative alkenylation and subsequent intramolecular aza- or oxa-Michael addition (Scheme 1).[16] Importantly, benzoic acids underwent this transformation with water as an environmentally benign, nontoxic reaction medium.[17]

Two-fold C–H bond functionalizations with *N,N*-dialkylbenzamides and their derivatives were, on the contrary, accomplished by Satoh, Miura and co-workers utilizing [Ru(*p*-cymene)Cl₂]₂ as the catalyst, along with AgSbF₆ as the additive (Scheme 2a).[18] Notably, the reaction did not proceed in the absence of the silver salt. In independent studies, our research group simultaneously found that the use of less expensive KPF₆ instead of AgSbF₆ as the co-catalytic additive enabled the twofold C–H bond functionalizations of *N*-monoalkylated aromatic amides with ample scope and comparable efficacy in water as the reaction media (Scheme 2b).[19] Alternatively, ruthenium-catalyzed C–H bond olefination of benzamides can be realized applying pre-functionalized starting materials bearing an internal oxidizing directing group, such as found in *N*-methoxybenzamides. Their ruthenium-catalyzed C–H bond alkenylations afforded olefinated *N*–H-free benzamides **6** in MeOH as the solvent (Scheme 2c).[20]

[RuCl₂(p-cymene)]₂ (2–5 mol %)
AgSbF₆ or KPF₆ (10–20 mol %)
Cu(OAc)₂·H₂O
solvent, T

(a) N,N-Dialkylbenzamides (FG = CONR₂)[a]

4a, R² = CO₂n-Bu: 98%
4b, R² = CO₂Cy: 82%
4c, R² = CO₂Et: 92%
4d R² = Ph: 62%

4e, n = 1: 75%
4f, n = 2: 82%

(b) N-Alkylbenzamides (FG = CONHR)[b]

5a, R¹ = OMe: 65%
5b, R¹ = H: 71%
5c, R¹ = F: 72%

5d, R¹ = Me: 69%
5e, R¹ = CF₃: 51%

(c) N-Methoxybenzamides (FG = CONH–OMe)[c]

6a, R¹ = H: 87%; 6b, R¹ = Me: 95%; 6c, R¹ = OMe: 70%;
6d, R¹ = CO₂Me: 45%; 6e, R¹ = NO₂: 65%

(d) Esters (FG = CO₂R)[d]

7a, R¹ = OMe: 62%
7b, R¹ = OH: 45%
7c, R¹ = I: 59%

7a, R¹ = H: 48%
7b, R¹ = Me: 57%

(e) Phenones (FG = COR)[e]

8a, R¹ = H: 86%
8b, R¹ = F: 89%
8c, R¹ = CO₂Me: 85%

8d: 75%

8e: 83%*

(f) Benzaldehydes (FG = CHO)[e]

9a, R¹ = OMe: 60%
9b, R¹ = NMe₂: 39%

9c, R³ = Me: 84%
9d, R³ = t-Bu: 68%*
9e, R³ = H: 62%

Reaction conditions: [a]Alkene (2 equiv), [RuCl₂(p-cymene)]₂ (5.2 mol%), AgSbF₆ (20 mol%)
Cu(OAc)₂·H₂O (2 equiv), t-AmOH, 100 °C, 4 h.[18a] [b]Alkene (1.5 equiv), [RuCl₂(p-cymene)]₂ (2.5–
5.0 mol%), KPF₆ (10–20 mol%), Cu(OAc)₂·H₂O (1 equiv), H₂O, 100 °C, 20 h.[19] [c]Alk
(1.8 equiv), [RuCl₂(p-cymene)]₂ (5.0 mol%), NaOAc (30 mol%), MeOH, 60 °C, 4–24 h.
[d][RuCl₂(p-cymene)]₂ (3–5 mol%), AgSbF₆ (20–40 mol%), Cu(OAc)₂·H₂O (30 mol%), DCE, air
100 °C, 12–16 h.[21] [e][RuCl₂(p-cymene)]₂ (2–3 mol %), AgSbF₆ (10–20 mol%), Cu(OAc)₂·H₂O (25–
50 mol%), DCE, air, 100–110 °C, 12–16 h.[22] *Reaction in *tert*-butanol.

**Scheme 2. Ruthenium(II)-catalyzed oxidative alkenylations with
electron-withdrawing coordinating substituents**

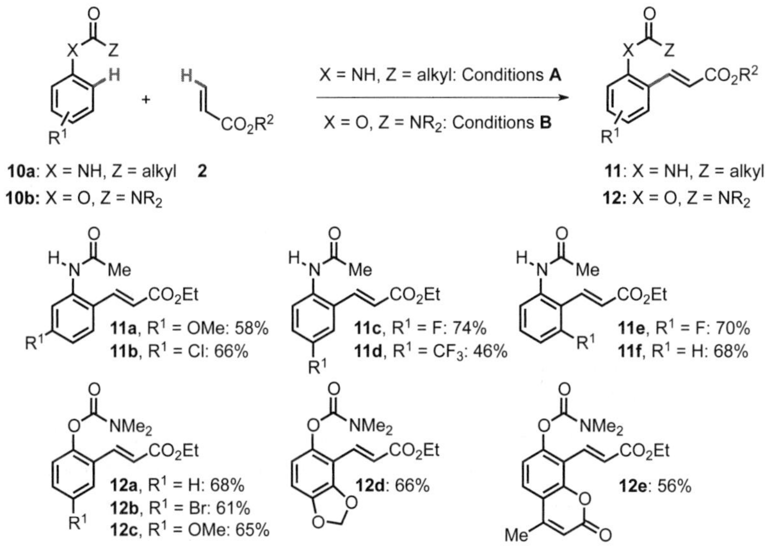

Scheme showing reaction:

10a: X = NH, Z = alkyl 2
10b: X = O, Z = NR₂

X = NH, Z = alkyl: Conditions **A**
X = O, Z = NR₂: Conditions **B**

11: X = NH, Z = alkyl
12: X = O, Z = NR₂

11a, R¹ = OMe: 58%
11b, R¹ = Cl: 66%

11c, R¹ = F: 74%
11d, R¹ = CF₃: 46%

11e, R¹ = F: 70%
11f, R¹ = H: 68%

12a, R¹ = H: 68%
12b, R¹ = Br: 61%
12c, R¹ = OMe: 65%

12d: 66%

12e: 56%

Conditions **A**: [RuCl$_2$(p-cymene)]$_2$ (5.0 mol%), KPF$_6$ (20 mol%), Cu(OAc)$_2$·H$_2$O (1 equiv), H$_2$O, 120 °C, 20 h.[19] Conditions **B**: [RuCl$_2$(p-cymene)]$_2$ (2.5 mol%), AgSbF$_6$ (10 mol%), Cu(OAc)$_2$·H$_2$O (30 mol%), DME, 110 °C, 24 h.[23]

Scheme 3. Ruthenium(II)-catalyzed oxidative alkenylations with electron-donating coordinating substituents

In contrast to these chelation-assisted alkenylations of benzamides, analogous ruthenium-catalyzed oxidative functionalizations of readily available, yet only weakly coordinating esters have until recent proven elusive. Yet, the research groups of Ackermann[21a] and Jeganmohan[21b] independently disclosed reaction conditions for versatile oxidative direct functionalization of diversely decorated esters **7** (Scheme 2d). Notably, the catalytic system consisting of [RuCl$_2$(p-cymene)]$_2$, AgSbF$_6$ and Cu(OAc)$_2$·H$_2$O was also found to be effective for alkenylations of phenones (Scheme 2e)[22a] and substituted benzaldehydes (Scheme 2f).[22b] Indeed, a ruthenium-catalyzed C–H bond functionalization of aromatic ketones provided alkenylated products **8** in 75–89% and 55–62% yield, when using substituted acrylates and styrenes, respectively, while alkenylated benzaldehydes **9** were obtained in slightly lower isolated yields.

Until recently, ruthenium-catalyzed oxidative alkenylations through twofold C–H bond functionalizations have proven to be limited to (hetero)arenes bearing electron-withdrawing groups (*vide supra*). Challenging oxidative olefinations with electron-rich arenes **10a**, on the contrary, were elaborated very recently with [RuCl$_2$(*p*-cymene)]$_2$, KPF$_6$ and Cu(OAc)$_2$·H$_2$O as the catalytic system in water as the reaction medium (Scheme 3).[19] Moreover, a cationic ruthenium (II) catalyst derived from [RuCl$_2$(*p*-cymene)]$_2$ and AgSbF$_6$ enabled highly efficient oxidative alkenylations of electron-rich aryl carbamates with weakly coordinating and removable directing groups.[23] These catalytic conditions allowed for highly productive cross-dehydrogenative C–H bond functionalizations of **10b** in a highly chemo-, diastereo- and site-selective fashion, affording diversely decorated phenol derivatives **12** (Scheme 3).[23]

Moreover, ruthenium-catalyzed oxidative alkenylations of arenes *N*-arylpyrazoles **13**, 2-aryl-1*H*-imidazoles, 2-aryl-1*H*-benzo[*d*]imidazoles, 2-arylbenzo[*d*]thiazoles **14** and 2-aryl-4,5-dihydrooxazoles **15** (Figure 2*a*) with heterocyclic directing groups have recently been reported by Dixneuf and coworkers[24] as well as Satoh, Miura and coworkers.[16a,18] Hence, substrates **13** were directly alkenylated with acrylates and acrylamides **2** employing [RuCl$_2$(*p*-cymene)]$_2$/Cu(OAc)$_2$·H$_2$O or [Ru(OAc)$_2$(*p*-cymene)]/Cu(OAc)$_2$·H$_2$O as the catalysts.[15]

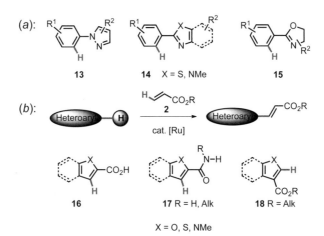

Figure 2. Ruthenium(II)-catalyzed oxidative alkenylations of (*a*) substrates bearing heterocyclic directing groups and (*b*) heteroarenes.

Ruthenium catalysts for oxidative C–H/C–H alkenylation reactions of heteroarenes have hitherto been less explored as compared to palladium- or rhodium-catalyzed analogous transformations. Yet, ruthenium-catalyzed alkenylations of furan-, thiophene-, 1-methyl-1*H*-pyrrole-, benzo[*b*]-thiophene-, benzofuran- or 1-methyl-1*H*-indole-2-carboxylic acids **16**, 2-carboxamides **17** and 3-alkoxycarbonyls **18** (Figure 2*b*) were achieved recently.[19,20,21,22b,25]

In summary, ruthenium(II) complexes allowed for challenging direct double C–H/C–H bond alkenylations of arenes with ample scope. Considering the practical importance of atom- and step-economical C–H bond alkenylations for natural product synthesis, drug discovery and crop protection, along with the unique features of the robust and selective ruthenium catalysts, significant further progress is expected in this rapidly evolving research area.

References

1. Institut für Organische und Biomolekulare Chemie der Georg-August-Universität Göttingen, Tammannstrasse 2, 37077 Göttingen (Germany). E-mail: lutz.ackermann@chemie.uni-goettingen.de. This study was partly funded by the DFG, the Fonds der Chemischen Industrie, AstraZeneca, and the Ministry for Science and Culture of Lower Saxony. Support by the Chinese Scholarship Council (fellowship to L.W.) is furthermore gratefully acknowledged.

2. (a) Handbook of Organopalladium Chemistry for Organic Synthesis (Ed.: Negishi, E.), Wiley-Interscience, New York, **2002**; (b) Metal-Catalyzed Cross-Coupling Reactions (Eds.: de Meijere, A.; Diederich, F.), Wiley-VCH, Weinheim, 2nd edn, **2004**; (c) Transition Metals for Organic Synthesis (Eds.: Beller, M.; Bolm, C.), Wiley-VCH, Weinheim, 2nd edn, **2004**.

3. (a) Mizoroki, T.; Mori, K.; A. Ozaki A. *Bull. Chem. Soc. Jpn.* **1971**, *44*, 581; (b) Heck, R. F. Nolley, J. P. *J. Org. Chem.* **1972**, *37*, 2320–2322.

4. Selected reviews: (a). Ruan, J.; Xiao, J. *Acc. Chem. Res.* **2011**, *44*, 614–626; (b) Mc Cartney, D.; Guiry, P. J. *Chem. Soc. Rev.* **2011**, *40*, 5122–5150; (c) Bellina, F.; Chiappe, C. *Molecules* **2010**, *15*, 2211–2245; (d) Heravi, M. M.; Fazeli, A. *Heterocycles* **2010**, *81*, 1079–2026; (e) The Mizoroki–Heck Reaction (Ed.: Oestreich, M.), Wiley, Chichester, **2009**; (f) Polshettiwar,

V.; Molnar, A. *Tetrahedron* **2007**, *63*, 6949–6976; (g) Corbet, J.-P.; Mignani, G. *Chem. Rev.* **2006**, *106*, 2651–2710; (h) Amatore, C.; Jutand, A. *Acc. Chem. Res.* **2000**, *33*, 314–321; (i) Beletskaya, I. P.; Cheprakov, A. V. *Chem. Rev.* **2000**, *100*, 3009–3066.

5. An illustrative review: Dounay, A. B.; Overman, L. E. *Chem. Rev.* **2003**, *103*, 2945–2964.

6. The use of copper, nickel, platinum, cobalt, rhodium or iridium complexes in the Mizoroki-Heck reactions is significantly less developed; see: Ackermann, L.; Born, R. Mizoroki-Heck Reactions with Metals Other than Palladium. In: The Mizoroki–Heck Reaction (Ed.: Oestreich, M.), Wiley, Chichester, **2009**, pp. 383–403.

7. (a) Moritani, I.; Fujiwara, Y. *Tetrahedron Lett.* **1967**, *8*, 1119–1122; (b) Fujiwara, Y.; Moritani, I.; Matsuda, M. *Tetrahedron* **1968**, *24*, 4819–4824.

8. (a) Trost, B. M. *Science* **1991**, *254*, 1471–1477; (b) Trost, B. M. *Acc. Chem. Res.* **2002**, *35*, 695–705.

9. Reviews: (a) Le Bras, J.; Muzart, J. *Chem. Rev.* **2011**, *111*, 1170–1214; (b) Yeung, C. S. Dong, V. M. *Chem. Rev.* **2011**, *111*, 1215–1292; (c) Karimi, B.; Behzadnia, H.; Elhamifar, D.; Akhavan, P. F.; Esfahani, F. K. *Synthesis* **2010**, 1399–1427; (d) Ferreira, E. M.; Zhang, H.; Stolz, B. M. Oxidative Heck-Type Reactions (Fujiwara-Moritani Reactions). In: The Mizoroki–Heck Reaction (Ed.: Oestreich, M.), Wiley, Chichester, **2009**, pp. 345–382. For selected recent communications, see: (e) Min, M.; Kim, Y.; Hong, S. *Chem. Commun.* **2013**, *49*, 196–198; (f) Cong, X.; You, J.; Gao, G.; Lan, J. *Chem. Commun.* **2013**, *49*, 662–664; (g) Harada, S.; Yano, H.; Obora, Y. *ChemCatChem* **2013**, *5*, 121–125; (h) Vasseur, A.; Harakat, D.; Muzart, J.; Le Bras, J. *Adv. Synth. Catal.* **2013**, *355*, 59–67; (k) Li, G.; Leow, D.; Wan, L.; Yu, J.-Q. *Angew. Chem. Int. Ed.* **2013**, *52*, 1245–1247.

10. Pioneering reports: (a) Hong P.; Yamazaki, H. *Chem. Lett.* **1979**, *8*, 1335–1336; (b) Matsumoto, T.; Yoshida, H. *Chem. Lett.* **2000**, *29*, 1064–1065; (c) Matsumoto, T.; Periana, R. A.; Taube, D. J.; Yoshida, H. *J. Catal.* **2002**, *206*, 272–280, and references cited therein.

11. For selected recent examples, see: (a) Yang, X.-F.; Hu, X.-H.; Feng, C.; Loh, T.-P. *Chem. Commun.* **2015**, *51*, 2532–2535; (b) Algarra, A. G.; Davies, D. L.; Khamker, Q.; Macgregor, S. A.; McMullin, C. L.; Singh, K.; Villa-Marcos, B. *Chem. Eur. J.* **2015**, *21*, 3087–3096; (c) Unoh, Y.; Hirano, K.; Satoh, T.; Miura, M. *Org. Lett.* **2014**, *17*, 704–707; (d) Becker, P.; Priebbenow, D. L.; Pirwerdjan, R.; Bolm, C. *Angew. Chem. Int. Ed.* **2014**, *53*, 269–271; (e) Parthasarathy, K.; Bolm, C. *Chem. Eur. J.* **2014**, *20*, 4896–4900; (f) Vora, H. U.; Silvestri, A. P.; Engelin, C. J.; Yu, J.-Q. *Angew.*

Chem. Int. Ed. **2014**, *53*, 2683–2686; (g) Nobushige, K.; Hirano, K.; Satoh, T.; Miura, M. *Org. Lett.* **2014**, *16*, 1188–1191; (h) Suzuki, C.; Morimoto, K.; Hirano, K.; Satoh, T.; Miura, M. *Adv. Synth. Cat.* **2014**, *356*, 1521–1526; (i) Ye, B.; Cramer, N. *J. Am. Chem. Soc.* **2013**, *135*, 636–639; (j) Patureau, F. W.; Besset, T.; Glorius, F. *Org. Synth.* **2013**, *90*, 41–51, and references cited therein; (k) Zheng, L.; Wang, J. *Chem. Eur. J.* **2012**, *18*, 9699–9704; (l) Wencel-Delord, J.; Nimphius, C.; Patureau, F. W.; Glorius, F. *Angew. Chem. Int Ed.* **2012**, *51*, 2247–2251; (m) Tsai, A. S.; Brasse, M.; Bergman, R. G.; Ellman, J. A. *Org. Lett.* **2011**, *13*, 540–542; (n) Park, S. H.; Kim, J. Y.; Chang, S. *Org. Lett.* **2011**, *13*, 2372–2375; (o) Li, X.; Gong, X.; Zhao, M.; Song, G.; Deng, J.; X. Li, X. *Org. Lett.* **2011**, *13*, 5808–5811; (p) Wang, F.; Song, G.; Du, Z.; Li, X. *J. Org. Chem.* **2011**, *76*, 2926 –2932; (q) Mochida, S.; Hirano, K.; Satoh, T.; Miura, M. *J. Org. Chem.* **2011**, *76*, 3024–3033; (r) Ueura, K.; Satoh, T.; Miura, M. *J. Org. Chem.* **2007**, *72*, 5362–5367; (s) Ueura, K.; Satoh, T.; Miura, M. *Org. Lett.* **2007**, *9*, 1407–1409.

12. Thus, in April 2015, the prices of gold, platinum, rhodium, iridium, palladium and ruthenium were 1210, 1161, 1165, 580, 779 and 50 US$ per troy oz, respectively. See: http://www.platinumgroupmetals.org/

13. For recent reviews on ruthenium-catalyzed C–H bond functionalization, see: (a) Ackermann, L. *Org. Process Res. Dev.* **2015**, *18*, 260–269; (b) Bruneau, C. *Top. Organomet. Chem.* **2014**, *48*, 195–236; (c) De Sarkar, S.; Liu, W.; Kozhushkov, S. I.; Ackermann, L. *Adv. Synth. Cat.* **2014**, *356*, 1461–1479; (d) Thirunavukkarasu, V. S.; Kozhushkov, S. I.; Ackermann, L. *Chem. Commun.* **2014**, *50*, 29–39; (e) Li, B.; Dixneuf, P. H. *Chem. Soc. Rev.* **2013**, *42*, 5744–5767; (f) Arockiam, P. B.; Bruneau, C.; Dixneuf, P. H. *Chem. Rev.* **2012**, *112*, 5879–5918; (g) Ackermann, L. *Pure Appl. Chem.* **2010**, *82*, 1403–1413; (h) Ackermann, L.; Vicente, R. *Top. Curr. Chem.* **2010**, *292*, 211–229.

14. For a review on related ruthenium-catalyzed oxidative annulations of alkynes through initial C–H bond functionalizations, see: L. Ackermann, L. *Acc. Chem. Res.* **2014**, *47*, 281–295;

15. A review: Kozhushkov, S. I.; Ackermann, L. *Chem. Sci.* **2013**, *4*, 886–896.

16. (a) Hashimoto, Y.; Ueyama, T.; Fukutani, T.; Hirano, K.; Satoh, T.; Miura, M. *Chem. Lett.* **2011**, *40*, 1165–1166; (b) Ackermann, L.; Pospech, J. *Org. Lett.* **2011**, *13*, 4153–4155. Recently reported results: (c) Manoharan, R.; M. Jeganmohan, M. *Chem. Commun.* **2015**, *51*, 2929–2932; (d) Li, J.; John, M.; Ackermann, L. *Chem. Eur. J.* **2014**, *20*, 5403–5408. For the related ruthenium(II)-catalyzed oxidative C–H akenylations of sulfonic

acids, sulfonamides and phenols, see: (e) Fabry, D. C.; Ronge, M. A.; Zoller, J.; Rueping, M. *Angew. Chem. Int. Ed.* **2015**, *54*, 2801–2805. (f) Ma, W.; Mei, R.; Tenti, G.; Ackermann, L. *Chem. Eur. J.* **2014**, *20*, 15248–15251; (g) Ma, W.; Ackermann, L. *Chem. Eur. J.* **2013**, *19*, 13925–13928.

17. Reviews: (a) Simon, M.-O.; Li, C.-J. *Chem. Soc. Rev.* **2012**, *41*, 1415–1427; (b) Organic Reactions in Water (Ed.: Lindstorm, U. M.) Wiley-Blackwell, New York, **2007**.

18. (a) Hashimoto, Y.; Ortloff, T.; Hirano, K.; Satoh, T.; Bolm, C.; Miura, M. *Chem. Lett.* **2012**, *41*, 151–153; (b) Zhao, P.; Niu, R.; Wang, F.; Han, K.; Li, X. *Org. Lett.* **2012**, *14*, 4166–4169.

19. Ackermann, L.; Wang, L.; Wolfram, R.; Lygin, A. V. *Org. Lett.* **2012**, *14*, 728–731.

20. (a) Li, B.; Ma, J.; Wang, N.; Feng, H.; Xu, S.; Wang, B. *Org. Lett.* **2012**, *14*, 736–739. See also: (b) Ackermann, L.; Fenner, S. *Org. Lett.* **2011**, *13*, 6548–6551.

21. (a) Graczyk, K.; Ma, W.; Ackermann, L. *Org. Lett.* **2012**, *14*, 4110–4113. See also: (b) Padala, K.; Pimparkar, S.; Madasamy, P.; Jeganmohan, M. *Chem. Commun.* **2012**, *48*, 7140–7142.

22. (a) Padala, K.; Jeganmohan, M. *Org. Lett.* **2011**, *13*, 6144–6147; (b) Padala, K.; Jeganmohan, M. *Org. Lett.* **2012**, *14*, 1134–1137.

23. Li, J.; Kornhaaß, C.; Ackermann, L. *Chem. Commun.* **2012**, *48*, 11343–11345.

24. (a) Arockiam, P. B.; Fischmeister, C.; Bruneau, C.; Dixneuf, P. H. *Green Chem.* **2011**, *13*, 3075–3078; (b) Li, B.; Devaraj, K.; Darcel, C.; Dixneuf, P. H. *Green Chem.* **2012**, *14*, 2706–2709.

25. Ueyama, T.; Mochida, S.; Fukutani, T.; Hirano, K.; Satoh, T.; Miura, M. *Org. Lett.* **2011**, *13*, 706–708.

Appendix
Chemical Abstracts Nomenclature (Registry Number)

Ethyl (*E*)-3-(2-Acetamido-4-methylphenyl)acrylate: 2-Propenoic acid, 3-[2-(acetylamino)-4-methylphenyl]-, ethyl ester, (2*E*); (1355020-48-2)
N-*m*-Tolylacetamide: m-Methyl-acetanilide (537-92-8)
[RuCl$_2$(*p*-cymene)]$_2$ (52462-29-0)
KPF$_6$: Potassium hexafluorophosphate; (17084-13-8)
Cu(OAc)$_2$·H$_2$O: Copper(II) acetate monohydrate; (6046-93-1)
Ethyl acrylate: Ethyl propenoate; (140-88-5)

Lianhui Wang was born in Henan, People's Republic of China, in 1983. He received his B. S. in chemistry in 2007 and completed his M. Sc. in chemistry on palladium-catalyzed cross-coupling reactions in 2010 under the supervision of Prof. Dr. Xiuling Cui and Prof. Dr. Zhiwu Zhu from Zhengzhou University (P. R. China). In 2010 he was awarded with the China Scholarship Council doctoral fellowship and started his PhD studies on ruthenium-catalyzed oxidative C–H activation reactions in the group of Prof. Dr. Lutz Ackermann (University of Göttingen, Germany). Upon completion of his Ph.D. in 2014, he initiated his academic career at the Institute of Molecular Medicine and School of Biomedical Sciences at Huaqiao University, Xiamen, P. R. of China.

Karsten Rauch was born in 1963 in Göttingen, Germany. In 1982, he completed his training as a technician in chemistry (University of Göttingen). From 1983 to 1984, he worked in the group of Prof. Dr. Hans Kuhn (MPI for Biophysical Chemistry, Göttingen) on monolayers and amino acid chemistry. Since 1985, he worked in the Department for Organic Chemistry (University of Göttingen) in the group of Prof. Dr. Lüttke on synthetic chemistry, since 1989, in the group of Prof. Dr. Armin de Meijere on synthesis and catalysis. Since 2007, he is working in the group of Prof. Dr. Lutz Ackermann on catalysis.

Alexander V. Lygin was born in 1984 in Krasnokamensk, Russia. He studied chemistry at the M.V. Lomonosov Moscow State University (Moscow, Russia) in 2001–2006 and completed his diploma on metallocene chemistry in 2006 under the supervision of Dr. Alexander Z. Voskoboynikov. He completed his PhD studies on the synthesis of heterocycles from isocyanides in the group of Prof. Dr. Armin de Meijere (University of Göttingen, Germany) and obtained his doctoral degree in 2009. He was a postdoctoral coworker in the laboratory of Prof. Dr. Lutz Ackermann at the University of Göttingen in 2011–2012. Since 2012 he is employed at EVONIC Industries, Darmstadt, Germany. His research interests include organometallic chemistry, chemistry of heterocycles and catalysis.

Sergei I. Kozhushkov was born in 1956 in Kharkov, USSR. He studied chemistry at Lomonosov Moscow State University, where he obtained his doctoral degree in 1983 under the supervision of Professor N. S. Zefirov and performed his "Habilitation" in 1998. From 1983 to 1991, he worked at Moscow State University and then at Zelinsky Institute of Organic Chemistry. In 1991, he joined the research group of Professor A. de Meijere (Georg-August-Universität Göttingen, Germany) as an Alexander von Humboldt Research Fellow. Since 2001 he has a permanent position as a Senior Scientist at the Georg-August-University of Göttingen. Since 2007, he is working in the research group of Professor L. Ackermann (Georg-August-Universität Göttingen, Germany). His current research interests focus on the chemistry of highly strained small ring compounds under transition metals catalysis.

Organic
Syntheses

Lutz Ackermann (1972) studied Chemistry at the Christian-Albrechts-University Kiel, Germany, and received his Ph.D. from the University of Dortmund in 2001 for research under the supervision of Alois Fürstner at the Max-Plank-Institut für Kohlenforschung in Mülheim / Ruhr. He was a postdoctoral coworker in the laboratory of Robert G. Bergman at the UC Berkeley before initiating his independent career in 2003 at the Ludwig-Maximilians-University München. In 2007, he became Full Professor at Georg-August-University Göttingen, and serves as the Dean at the Georg-August-University Göttingen since 04.2011. His recent awards and distinctions include a JSPS visiting professor fellowship (2009), an AstraZeneca Excellence in Chemistry Award (2011) and an ERC Consolidates Grant (2012). The development of novel concepts for sustainable catalysis and their application to organic synthesis constitute his major current research interests.

William Trieu received his B.S in chemical engineering in 2007 at the University of California, Berkeley where he worked for Prof. Jay Keasling on the production of an antimalarial drug precursor. He worked at Merck supporting Gardasil fermentation operations prior to joining Amgen in 2008. At Amgen he has worked on a number of projects in which he has optimized chemical processes by utilizing chemical engineering principles. He received his M.S in chemical engineering in 2012 at the University of Southern California.

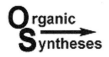

Preparation of 3,5-Dibromo-2-pyrone from Coumalic Acid

Hyun-Kyu Cho and Cheon-Gyu Cho[1]*

Department of Chemistry, Hanyang University, 222 Wangsimni-ro, Seongdong-gu, Seoul 133-791, Korea

Checked by Arata Nishii and Keisuke Suzuki

[Reaction scheme: Coumalic acid **1** (2-pyrone with CO_2H) treated with NBS (2.5 equiv), Bu_4NBr, $CHCl_3$, 50 °C, 12 h, gives 3,5-dibromo-2-pyrone **2** (51–53%) and byproduct **3**]

1 → **2 (51–53%)** | **3 (byproduct)**

Procedure

3,5-Dibromo-2-pyrone (2). A 200-mL two-necked round-bottomed flask, equipped with a 4.5-cm Teflon-coated magnetic stir bar and a reflux condenser, is sequentially charged with *N*-bromosuccinimide (15.8 g, 88.9 mmol, 2.5 equiv) (Note 1), tetrabutylammonium bromide (571 mg, 1.77 mmol, 0.05 equiv) (Note 2), and chloroform (50 mL) (Note 3). Coumalic acid (**1**) (5.00 g, 35.7 mmol) (Notes 4 and 5) is added to the stirring solution, which is heated by using an oil bath at 50 °C for 12 h (Note 6). After cooling to rt, hexane (100 mL) is added. The resultant two-phase mixture becomes one-phase after vigorous stirring. To remove the succinimide byproduct, the resulting mixture is filtered through a short plug of silica gel (50 g) eluting with 1400 mL of 1:1 dichloromethane–hexane, until TLC analysis shows that the product **2** is no longer detected in the eluent (Note 7). The filtrate is concentrated with a rotary evaporator (30 °C, 100 mmHg) (Note 8), and the resulting crude oil is purified by flash silica-gel column chromatography, eluting with hexane–dichloromethane (3:2) (Note 9). The combined eluents are concentrated with a rotary evaporator (30 °C, 100 mmHg) and further

Org. Synth. **2015**, *92*, 148-155
DOI: 10.15227/orgsyn.092.0148

Published on the Web 5/29/2015
© 2015 Organic Syntheses, Inc.

evaporated with an oil vacuum pump at room temperature to afford 3,5-dibromo-2-pyrone (**2**, 4.66–4.73 g, 51–53%) as a pale yellow solid (Note 10).

Notes

1. *N*-Bromosuccinimide (99%) was purchased from Sigma-Aldrich Co. (checkers), Alfa Aesar (submitters) and used as received.
2. Tetrabutylammonium bromide ($\geq$98%) was purchased from Kanto Chemical Co., Inc. (checkers), Daejung Chemicals (submitters) and used as received.
3. The checkers: Chloroform (99%) was purchased from Nacalai Tesque, Inc., hexane ($\geq$95%) from Kanto Chemical Co., Inc., dichloromethane ($\geq$ 99%) from ASAHI GLASS Co., Ltd. The submitters: Chloroform (99.8%) and hexane (CP) were purchased from Samchun Chemical, dichloromethane (EP) from Duksan Company.
4. The checkers and the submitters: Coumalic acid (>97.0%) was purchased from Tokyo Chemical Industry Co., Ltd. The checkers used coumalic acid as received. The submitters purified coumalic acid by the method of Wiley and Smith.[2] Coumalic acid **1** (25 g) was dissolved in methanol (200 mL) with heating. After dissolution, the solution was cooled in an ice bath. The precipitated solids were collected on a Büchner funnel with filter paper (No 20. 5 µm) and washed with 25 mL of cold methanol, which afforded pure coumalic acid (18 g).
5. After adding coumalic acid, the color of the resulting suspension is yellow (Fig. 1).

Figure 1. After 15 minutes   **Figure 2.** After 12 hours

6. After the reaction temperature increases to 50 °C, the color of the suspension turns to a reddish-orange (Fig. 2). The submitters report the observation of carbon dioxide evolution.

7. TLC analysis was performed with silica-gel plates (1.5 cm x 4 cm, glass-backed, Merck in Darmstadt, Germany), using hexane:ethyl acetate (5:1) as the eluent; $R_f = 0.40$. Plates were visualized by UV and a potassium permanganate stain solution.

8. Bromine is generated and evaporated during the concentration.

9. For silica-gel column chromatography, the checkers employed 0.063–0.210 mm particle size silica gel (Kanto Chemical Co., Inc., Japan), while the submitters employed 230–400 mesh, 0.040–0.063 mm particle size silica gel (Merck in Darmstadt, Germany). The crude residue was dissolved in 2:3 dichloromethane–hexane (10 mL), and the solution was charged onto a column (diameter = 6 cm) of silica gel (101 g). The column was eluted with ca. 300 mL of dichloromethane–hexane (2:3). At this point, fraction collection (100 mL fractions) was begun, and elution was continued with 2.3 L of dichloromethane–hexane (2:3). 3,5-Dibromo-2-pyrone (**2**) was obtained in fractions 4–21 (compound **2**: $R_f = 0.40$, hexane:ethyl acetate = 5:1). Fractions were combined and evaporated (30 °C, 100 mmHg). The fraction collection needed to be carefully performed, particularly with respect to the later fractions; a byproduct, 5-bromo-2-pyrone (**3**), runs slower on silica-gel chromatography ($R_f = 0.30$, hexane:ethyl acetate = 5:1), and it may contaminate the later fractions.

10. 3,5-Dibromo-2-pyrone has the following physical and spectroscopic properties: mp 63.0–64.7 °C; ^1H NMR (600 MHz, CDCl$_3$) δ: 7.59 (d, J = 2.4 Hz, 1 H), 7.74 (d, J = 2.4 Hz, 1 H); ^{13}C NMR (150 MHz, CDCl$_3$) δ: 99.9, 113.5, 146.6, 149.5, 156.4; IR (ATR): 3120, 3079, 1718, 1602, 1516, 1362, 1311, 1204, 1064, 973, 852 cm^{-1}; $R_f = 0.40$ (hexane:ethyl acetate = 5:1); Anal. Calcd. for C$_5$H$_2$Br$_2$O$_2$: C, 23.66; H, 0.79. Found C, 23.88; H, 0.73. The product (**2**) gradually turns to yellow when stored at rt, although the ^1H NMR spectrum shows no decomposition peaks. No color change is observed when the product (**2**) is stored in a refrigerator for weeks.

Working with Hazardous Chemicals

The procedures in *Organic Syntheses* are intended for use only by persons with proper training in experimental organic chemistry. All hazardous materials should be handled using the standard procedures for work with chemicals described in references such as "Prudent Practices in the Laboratory" (The National Academies Press, Washington, D.C., 2011; the full text can be accessed free of charge at http://www.nap.edu/catalog.php?record_id=12654). All chemical waste should be disposed of in accordance with local regulations. For general guidelines for the management of chemical waste, see Chapter 8 of Prudent Practices.

In some articles in *Organic Syntheses*, chemical-specific hazards are highlighted in red "Caution Notes" within a procedure. It is important to recognize that the absence of a caution note does not imply that no significant hazards are associated with the chemicals involved in that procedure. Prior to performing a reaction, a thorough risk assessment should be carried out that includes a review of the potential hazards associated with each chemical and experimental operation on the scale that is planned for the procedure. Guidelines for carrying out a risk assessment and for analyzing the hazards associated with chemicals can be found in Chapter 4 of Prudent Practices.

The procedures described in *Organic Syntheses* are provided as published and are conducted at one's own risk. *Organic Syntheses, Inc.*, its Editors, and its Board of Directors do not warrant or guarantee the safety of individuals using these procedures and hereby disclaim any liability for any injuries or damages claimed to have resulted from or related in any way to the procedures herein.

Discussion

The synthesis of 3,5-dibromo-2-pyrone was first reported by Pirkle and coworkers in 1969.[3] It was prepared from 2-pyrone (prepared from coumalic acid via thermal decarboxylation) by following either a three-step sequence that involved two successive brominations and HBr elimination or a four-step sequence that involved a bromination, HBr elimination, and photochemical bromination, followed by another HBr elimination. This complex and even cumbersome process would the best alternative yet as there is no other method available in the literature. In this reaction, coumalic acid underwent an electrophilic aromatic bromination at C3 and bromo-decarboxylation at C5. 3,5-Dibromo-2-pyrone is a potent neutral diene that can react with both electron-poor and –rich dienophiles via either normal- or inverse-demand Diels-Alder cycloaddition reactions in good to excellent chemical yield and diastereoselectivity.[4]

Table 1. Diels-Alder Reactions

entry	dienophile	conditions	*endo*-adduct	yield (*endo:exo*)
1	$\overset{O}{\underset{}{\|}}$ OCH$_3$	toluene 100 °C, 5 h	Br, Br, O, OCH$_3$	84% (94:6)
2	$\overset{O}{\underset{}{\|}}$ Me	toluene 100 °C, 5 h	Br, Br, Me	84% (94:6)
3	CN	toluene 100 °C, 12 h	Br, Br, CN	90% (76:24)
4	$\overset{O}{\underset{Me}{\|}}$ OCH$_3$	CH$_2$Cl$_2$, 100 °C, 24 h	Br, Br, OCH$_3$, CH$_3$	84% (86:14)
5	OBn	toluene 100 °C, 3 d	Br, Br, OBn	69% (100:0)
6	OTMS	toluene 100 °C, 2 d	Br, Br, OTMS	73% (99:1)

Also disclosed was that either of the two C-Br groups of 3,5-dibromo-2-pyrone could be selectively mono-functionalized into the corresponding 3-substituted-5-bromo-2-pyrones **6** or 5-substituted-3-bromo-2-pyrones **7**.[5]

Table 2. Regioselective Coupling Reactions

entry	nucleophile	conditions	6 (yield)	7 (yield)
1	Bu₃Sn (phenyl)	condition **A**	94%	trace
		condition **B**	trace	75%
2	Bu₃Sn (aryl-NC)	condition **A**	80%	trace
		condition **B**	trace	79%
3	Bu₃Sn (aryl-CO₂Me)	condition **A**	79%	trace
		condition **B**	trace	75%
4	Bu₃Sn (aryl-OMe)	condition **A**	61%	trace
		condition **B**	trace	55%
5	Bu₃Sn (furyl)	condition **A**	61%	trace
		condition **B**	trace	60%
6	Bu₃Sn (thienyl)	condition **A**	72%	trace
		condition **B**	trace	68%
7	TMS-alkyne	condition **C**	83%	trace
8	H₂N (aniline)	condition **D**	61%	trace

A: Pd(PPh₃)₄, CuI (0.1 eq), PhMe, 100 °C. **B**: Pd(PPh₃)₄, CuI (1.0 eq), DMF, 50 °C.
C: Pd(PPh₃)₂Cl₂, CuI (0.05 eq), dioxane, rt, **D**: Pd(OAc)₂, xantphos, Cs₂CO₃, PhMe, 110 °C

Our group has been utilizing the potent diene reactivity of both parent 3,5-dibromo-2-pyrone and the aforementioned 2-pyrone derivatives toward the synthesis of various alkaloid natural products.[6]

References

1. Department of Chemistry, Hanyang University, 222 Wangsimni-ro, Seongdong-gu, Seoul 133-791, Korea. E-mail: ccho@hanyang.ac.kr
2. Wiley, R.; Smith, N. *Organic Syntheses*; Wiley & Sons: New York, 1963; Collect. Vol. No. IV, pp 201
3. Pirkle, W. H.; Dines, M. J. *Org. Chem.* **1969**, *34*, 2239–2244.
4. (a) Cho, C.-G.; Kim, Y.-W.; Kim, Y.-K. *Tetrahedron Lett.* **2001**, *42*, 8193–8195. (b) Cho, C.-G.; Park, J.-S.; Jung, I.-H.; Lee, H.-W. *Tetrahedron Lett.* **2001**, *42*, 1065–1067. (c) Cho, C.-G.; Kim, Y.-W.; Lim, Y.-K.; Park, J.-S.; Lee, H.-W.; Koo, S.-M. J. *Org. Chem.* **2002**, *67*, 290–293.
5. (a) Lee, J.-H.; Park, J.-S.; Cho, C.-G. *Org Lett.* **2002**, *4*, 1171–1173. (b) Lee, J.-H.; Cho, C.-G. *Tetrahedron Lett.* **2003**, *44*, 65–67. (c) Kim, W.-S.; Kim, H.-J.; Cho, C.-G. J. *Am. Chem. Soc.* **2003**, *125*, 14288–14289. (d) Ryu, K.-M.; Arun, K. G.; Han, J.-W.; Oh, C.-H.; Cho, C.-G. *Synlett* **2004**, *12*, 2197–2199.
6. (a) Tam. N. T.; Cho, C.-G. *Org. Lett.* **2007**, *9*, 3391–3392, (b) Shin, I.-J.; Choi, E.-S.; Cho, C.-G. *Angew. Chem. Int. Ed.* **2007**, *46*, 2303–2305. (c) Chang, J. H.; Kang, H.-U.; Jung, I.-H.; Cho, C.-G. *Org. Lett.* **2010**, *12*, 2016–2018. (d) Jung, Y.-G.; Kang, H.-U.; Cho, H.-K.; Cho, C.-G. *Org. Lett.* **2011**, *13*, 5890–5892. (e) Jung, Y.-G.; Lee, S.-C.; Cho, H.-K.; Nitin B. D.; Song, J.-Y.; Cho, C.-G. *Org. Lett.* **2013**, *15*, 132–135. (f) Cho, H.-K.; Lim, H.-Y.; Cho, C.-G. *Org. Lett.* **2013**, *15*, 5806–5809. (g) Shin, H.-S.; Jung, Y.-G.; Cho, H.-K.; Park, Y.-G.; Cho, C.G. *Org. Lett.* **2014**, *16*, 5718-5720.

Appendix
Chemical Abstracts Nomenclature (Registry Number)

Coumalic acid: 2*H*-Pyran-5-carboxylic acid, 2-oxo-; (**1**) (500-05-0)
3,5-Dibromo-2-pyrone: 2*H*-Pyran-2-one, 3,5-dibromo-; (**2**) (19978-41-7)
5-Bromo-2-pyrone: 2*H*-Pyran-2-one, 5-bromo-; (**3**) (19978-33-7)
N-Bromosuccinimide (128-08-5)
Tetrabutylammonium bromide (1643-19-2)

Org. Synth. **2015**, *92*, 148-155   **154**   DOI10.15227/orgsyn.092.0148

Organic
Syntheses

Hyun-Kyu Cho was born in Seoul, Korea in 1987. He graduated in 2006 with a B.S. degree in chemistry from Hanyang University. He is currently pursuing his doctoral studies under the supervision of Prof. Cheon-Gyu Cho at the same university. His research is focused on the target-oriented total syntheses of bioactive natural products using 3,5-dibromo-2-pyrone.

Cheon-Gyu Cho was born in Seoul, Korea in 1962. He graduated in 1984 with a B.S. degree in industrial chemistry from Hanyang University. He obtained his Ph.D. from the Johns Hopkins University in 1993 (Advisor: Prof. G. H. Posner). From 1993-1996 he was a postdoctoral fellow with Prof. P. T. Lansbury at MIT. From 1996-1997 he was an instructor at Harvard Medical School. He began his independent academic career at Hanyang University in 1997. From 2004-2005 he worked with Prof. A. B. Smith at UPenn as a visiting professor. His research interests include the total synthesis of bioactive natural products and the development of new synthetic methods.

Arata Nishii was born in 1992 in Mie, Japan. He received his B.S. degree from Tokyo Institute of Technology in 2014, and is continuing his graduate studies with Professor Keisuke Suzuki. His study focuses on the synthetic method of natural products.

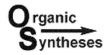

Synthesis of Phosphoryl Ynamides by Copper-Catalyzed Alkynylation of Phosphoramidates. Preparation of Diethyl Benzyl(oct-1-yn-1-yl)phosphoramidate

John M. Read, Yu-Pu Wang, and Rick L. Danheiser[*1]

Department of Chemistry, Massachusetts Institute of Technology, Cambridge, MA 02139

Checked by Erik Daa Funder and Erick M. Carreira

Procedure

A. Diethyl benzylphosphoramidate (1). A 250-mL, two-necked, heart-shaped flask (Note 1) equipped with a 32 x 16 mm, Teflon-coated, oval magnetic stir bar, a Liebig condenser fitted with a nitrogen inlet adapter, and a glass stopper is charged sequentially via syringe with benzylamine (3.19 mL, 3.13 g, 29.3 mmol, 2.0 equiv) (Note 2), 50 mL of diethyl ether

(Note 3), and diethyl phosphite (1.91 mL, 2.06 g, 14.6 mmol, 1.0 equiv). Iodoform (5.76 g, 14.6 mmol, 1.0 equiv) (Note 4) is added to the reaction mixture in three equal portions by temporarily removing the glass stopper. A vigorous bubbling starts after the first addition. Before the bubbling completely settles, the next portion of iodoform is added in order to maintain the momentum of the reaction. The portions are added approximately within 1 minute. The initiation of the reaction is indicated by an increase in temperature and spontaneous reflux of the solvent that begins a few seconds after the addition of iodoform (Note 5). During the course of the reaction the color of the heterogeneous reaction mixture changes from light yellow to white (see photos below).

After 2.5 h, TLC analysis indicates the complete disappearance of diethyl phosphite (Note 6). The reaction mixture is then poured into a 250-mL, round-bottomed flask. The original two-necked flask is rinsed with 50 mL of CH$_2$Cl$_2$, which is added to the round-bottmed flask, and the solution is concentrated by rotary evaporation (20 °C, 20 mmHg) to afford 10.7 g of a thick light yellow suspension. This material is diluted with 40 mL of chloroform (Note 7), and the homogeneous solution is transferred to a 125-mL separatory funnel and washed with water (2 x 30 mL), 0.5% aqueous acetic acid (2 x 30 mL), water (2 x 50 mL), and saturated NaCl solution (1 x 50 mL). The light peach-colored organic layer is dried over 4 g of MgSO$_4$ and filtered through a 30-mL sintered glass Büchner funnel

(medium porosity, 30 mm diameter). The MgSO$_4$ is washed with chloroform (3 x 10 mL) and the combined filtrate is concentrated by rotary evaporation (20 °C, 20 mmHg) to afford 6.3 g of a light yellow oil.

This material is dissolved in a minimum amount of 5:1 EtOAc-CH$_2$Cl$_2$ (ca. 8 mL) and loaded onto a column (75 mm diameter) of 160 g of silica gel (Note 8) prepared as a slurry in 5:1 EtOAc-CH$_2$Cl$_2$. Elution with 5:1 EtOAc-CH$_2$Cl$_2$ (30 mL fractions collected in test tubes) affords the product in fractions 18-130. These fractions are combined and the solvent is removed by rotary evaporation (20 °C, 20 mmHg). Further concentration at 20 °C, 0.5 mmHg over 16 h provides 3.18 g (89%) of phosphoramidate **1** as a viscous light yellow oil (Notes 9 and 10).

B. *1-Bromooct-1-yne (2)*. A 250-mL, one-necked, round-bottomed flask (Note 1) equipped with a 32 x 16 mm, Teflon-coated, oval magnetic stir bar, rubber septum, and nitrogen inlet needle is charged with 1-octyne (2.90 mL, 2.17 g, 19.7 mmol, 1.0 equiv) (Note 11) and 65 mL of acetone (Note 12). *N*-Bromosuccinimide (NBS) (3.86 g, 21.7 mmol, 1.1 equiv) (Note 13) is then added in one portion by temporarily removing the septum. The reaction mixture is stirred for 2 min to allow all of the NBS to dissolve. Silver(I) nitrate (0.335 g, 1.97 mmol, 0.1 equiv) (Note 14) is added in one portion by temporarily removing the septum, the flask is wrapped with aluminum foil, and the colorless suspension is stirred at room temperature for 3 h at which point TLC analysis shows complete consumption of octyne (Note 15).

The resulting white suspension is transferred to a 250-mL separatory funnel and diluted with 65 mL of cooled deionized water (4 °C) and 70 mL of pentane. The organic phase is separated and washed with a saturated aqueous Na$_2$S$_2$O$_3$ solution (3 x 25 mL) and brine (50 mL), dried over 5 g of MgSO$_4$, and filtered through a 150-mL sintered glass Büchner funnel (fine porosity, 60 mm diameter). The MgSO$_4$ is washed with pentane (3 x 30 mL) and the filtrate is concentrated by rotary evaporation (20 °C, 20 mmHg) to yield 3.41–3.58 g (92–96%) of **2** as a light yellow oil, which is used in the next step without further purification (Notes 16 and 17).

C. *Diethyl Benzyl(oct-1-yn-1-yl)phosphoramidate (3)*. A 100-mL, one-necked, round-bottomed flask (Note 1) equipped with a 20 x 9 mm, Teflon-coated, oval magnetic stir bar, rubber septum, and nitrogen inlet needle is charged with diethyl benzylphosphoramidate (**1**) (2.70 g, 11.1 mmol, 1.0 equiv), copper(II) sulfate pentahydrate (0.416 g, 1.67 mmol, 0.15 equiv) (Note 18), 1,10-phenanthroline (0.600 g, 3.33 mmol, 0.30 equiv), and potassium phosphate (4.71 g, 22.2 mmol, 2.0 equiv) (Note 19). The flask is evacuated to 0.5 mmHg and then backfilled with nitrogen (repeated two more times), and then 5 mL of toluene (Note 20) is added. A separate 50-mL, one-necked, round-bottomed flask is charged with 1-bromooct-1-yne (**2**) (2.73 g, 14.4 mmol, 1.3 equiv) and fitted with a rubber septum and nitrogen inlet needle. Toluene (20 mL) is added and the solution of bromoalkyne is rapidly transferred into the 100-mL flask via a metal cannula and nitrogen pressure. The 50-mL flask is rinsed with two 1.5-mL portions of toluene. The reaction mixture is placed under reduced pressure via the vacuum manifold until bubbling ensues and then backfilled with argon; this is performed four times. The rubber septum is replaced with an 11-cm Liebig condenser fitted with a rubber septum and nitrogen inlet needle. The system is evacuated and backfilled with nitrogen twice after which the heterogeneous brown reaction mixture (see photo) is heated in a 95 °C oil bath for 24 h (750 rpm stirring) (Note 21) at which point TLC analysis indicates complete consumption of phosphoramidate **1** (Note 22).

The reaction mixture is cooled to room temperature and filtered through 5 g of Celite in a 30-mL sintered glass Büchner funnel (medium porosity, 30 mm diameter). The Celite is washed with 150 mL of EtOAc and the filtrate is concentrated to afford ca. 9 to 10 g of an orange oil. This material is dissolved in a minimum amount of CH_2Cl_2 (ca. 15 mL) and loaded onto a column (75 mm diameter) of 200 g of silica gel prepared as a slurry in 10:5:85 EtOAc-Et$_3$N-hexanes. Elution with 10:5:85 EtOAc-Et$_3$N-hexanes (30 mL fractions) affords the product in fractions 26-80. These fractions are combined, and the solvent is removed by rotary evaporation (20 °C, 20 mmHg). Further concentration at 20 °C, 0.5 mmHg over 16 h provides 3.16–3.27 g (81–84%) of ynamide **3** as a viscous yellow oil (Note 22).

Notes

1. All reaction glassware was flame-dried under vacuum (0.5 mmHg), back-filled with argon while hot, and then maintained under the inert atmosphere during the course of the reaction. The checkers used an atmosphere of nitrogen in all reactions described. The submitters used a 250-mL round-bottomed flask for Step A with a rubber septum equipped with a thermocouple probe in place of the glass stopper employed by the checkers.

2. The submitters used benzylamine (99%), which was purchased from Acros Organics, and diethyl phosphite (technical grade, 94%) purchased from Aldrich Chemical Company. Both were used as received. If the benzylamine appeared yellow or showed any impurities by ^{1}H NMR analysis, distillation from CaH$_2$ was necessary to obtain optimal results. The checkers used benzylamine (>99%) purchased from TCI and distilled it from CaH$_2$ under an atmosphere of nitrogen. Diethyl phosphite (98%) was purchased from Acros Organics and used as received.

3. The submitters used Et$_2$O (ultra low water) that was purchased from J.T. Baker and purified by pressure filtration through activated alumina prior to use. The checkers used Et$_2$O purchased from Sigma Aldrich containing BHT as a stabilizer. Before use the solvent was first distilled

and then passed through an activated alumina column embedded in a solvent purification system provided by LC Technology Solutions. The solvent was finally further dried overnight under argon using 4 Å microwave activated molecular sieves purchased from Sigma Aldrich.

4. The submitters used iodoform (99%) that was purchased from Aldrich Chemical Company and used as received. The checkers purchased iodoform (99%) from Fluka, and it was used as received.

5. *CAUTION: The rapid addition of iodoform results in a vigorous exothermic reaction. Care needs to be taken in this step. The checkers chose to add the iodoform in portions, while ensuring the mixture starts to reflux in order to initiate the reaction.*

6. TLC analysis was performed with silica gel plates (2 cm x 5 cm, glass backed, purchased from EMD Chemicals) with ethyl acetate as eluent and visualization with Seebach's stain (2.5 g of phosphomolybdic acid, 1 g of Ce(SO$_4$)$_2$, and 6 mL conc H$_2$SO$_4$ dissolved in 94 mL of H$_2$O). Benzylamine: R$_f$ = 0.08, diethyl benzylphosphoramidate (1): R$_f$ = 0.20, diethyl phosphite: R$_f$ = 0.33. Iodoform and diiodomethane do not stain. The checkers found that it was helpful to elute the TLC plate several times to obtain clear resolution of diethyl phosphite and diethyl phosphoramidate **1**.

7. Chloroform (ACS grade) was purchased from Mallinckrodt Chemicals and used as received, but the solvent should be checked for acidity. A sample of chloroform (ca. 5 mL) is shaken with an equal volume of deionized water and the aqueous layer is tested with pH paper. If a pH of less than 5 is obtained, then the chloroform is washed with saturated aqueous NaHCO$_3$, dried over MgSO$_4$, and filtered before use.

8. The submitters used silica gel (40-63 µm) that was purchased from Sorbent Technologies and used as received. The checkers used high purity grade silica gel with a pore size of 60 Å and a 230-400 mesh particle size purchased from Fluka. The submitters collected 30-mL fractions and obtained the product in fractions 13-74.

9. A second run on similar scale provided the product in 88% yield.

10. Diethyl benzylphosphoramidate (**1**) has the following spectroscopic properties: ^{1}H NMR (400 MHz, CDCl$_3$) δ: 1.31 (td, *J* = 7.1, 0.9 Hz, 6 H), 2.87 (br s, 1 H), 3.96–4.17 (m, 6 H), 7.27 – 7.37 (m, 5 H); ^{13}C NMR (100 MHz, CDCl$_3$) δ: 16.4 (d, *J* = 7.1 Hz), 45.6, 62.6 (d, *J* = 5.3 Hz), 127.45, 127.53, 128.7, 139.8 (d, *J* = 6.5 Hz); ^{31}P NMR (162 MHz, CDCl$_3$) δ: 8.39;

HRMS (ESI) [M + H]⁺ calcd for $C_{11}H_{19}NO_3P$: 244.1097. Found: 244.1097; IR (neat): 3218, 2981, 2931, 2905, 1454, 1226, 1024, 958, 697, 494 cm⁻¹; Anal. Calcd for $C_{11}H_{18}NO_3P$: C, 54.32; H, 7.46; N, 5.76. Found: C, 54.13; H, 7.65; N, 5.63.

11. The submitters used 1-octyne (98%) that was purchased from Alfa Aesar and used as received. The checkers used 1-octyne (97%) purchased from Aldrich and used as received.

12. The submitters used acetone (histological grade) that was purchased from Mallinckrodt Chemicals and used as received. The checkers used a new bottle of acetone purchased from Sigma Aldrich Chromasolv (>99.9%).

13. The submitters used N-bromosuccinimide (NBS) (99%) that was purchased from Alfa Aesar and used as received. If the material was dark yellow in color, NBS was recrystallized from water as recommended using the procedure described previously.[2] The checkers used N-bromosuccinimide (99%) purchased from Acros Organics and used it as received.

14. The submitters used silver(I) nitrate (99.9+% metal basis) which was purchased from Alfa Aesar and used as received. The checkers used silver(I) nitrate (99%) purchased from Acros Organics and used as received.

15. TLC analysis was performed with silica gel plates (2 cm x 5 cm, glass backed, purchased from EMD Chemicals) with hexanes as the eluent and visualization with $KMnO_4$. 1-Octyne: $R_f = 0.47$, 1-bromooct-1-yne: $R_f = 0.58$. The checkers observed: 1-bromooct-1-yne: $R_f = 0.64$

16. The bromoalkyne product has a very distinct smell and the checkers recommend that all manipulations (especially evaporations) of the material be performed in a well ventilated hood.

17. This material was found to be of high purity provided that pure 1-octyne and NBS were used as starting materials. 1-Bromooct-1-yne (**2**) has the following spectroscopic properties: ¹H NMR (400 MHz, CDCl₃) δ: 0.89 (t, $J = 6.9$ Hz, 3 H), 1.21 – 1.43 (m, 6 H), 1.51 (app quint, $J = 7.1$ Hz, 2 H), 2.20 (t, $J = 7.1$ Hz, 2 H); ¹³C NMR (100 MHz, CDCl₃) δ: 14.2, 19.8, 22.7, 28.4, 28.6, 31.4, 37.6, 80.6; IR (neat): 2960, 2934, 2860, 1468, 1459, 1379, 1327, 725 cm⁻¹; Anal. Calcd for $C_8H_{13}Br$: C, 50.81; H, 6.93; Br, 42.26. Found: C, 50.62; H, 7.08; Br, 42.09.

18. The submitters used copper(II) sulfate pentahydrate which was purchased from Mallinckrodt Chemicals and ground to a fine powder with a mortar and pestle before use. The checkers used copper(II) sulfate pentahydrate purchased from Merck and ground to a fine light blue powder before use (see picture).

19. The submitters used 1,10-phenanthroline (≥99%) purchased from Aldrich Chemical Company and potassium phosphate (97%, anhydrous) purchased from Acros Organics. Both were used as received. The checkers used 1,10-phenanthroline (99%) purchased from Lancaster and newly purchased potassium phosphate (97%, anhydrous) from Acros Organics.

20. The checkers used toluene (ACS grade) purchased from J.T. Baker and purified by pressure filtration through activated alumina prior to use. The submitters used toluene purchased from Fisher. Before use the solvent was passed through an activated alumina column embedded in a solvent purification system provided by LC Technology Solutions.

21. Very rapid stirring is crucial to achieve complete reaction due to the heterogeneity of the reaction mixture. The checkers used 750 rpm.

22. TLC analysis was performed with silica gel plates (2 cm x 5 cm, glass backed, purchased from EMD Chemicals) with 4:1:1 EtOAc-CH$_2$Cl$_2$-hexanes as eluent and visualization with KMnO$_4$. Diethyl benzylphosphoramidate (1): R$_f$ = 0.29, 1-bromooct-1-yne (2): R$_f$ = 0.79, ynamide 3: R$_f$ = 0.64. The checkers observed the following values: Diethyl benzylphosphoramidate (1): R$_f$ = 0.14, 1-bromooct-1-yne (2): R$_f$ = 0.83, ynamide 3: R$_f$ = 0.57

23. Diethyl benzyl(oct-1-yn-1-yl)phosphoramidate (3) has the following spectroscopic properties: ^{1}H NMR (400 MHz, CDCl$_3$) δ: 0.87 (t, J = 6.9 Hz, 3 H), 1.13 – 1.33 (m, 12 H), 1.40 (quint, J = 6.9 Hz, 2 H), 2.16 (td,

J = 6.8, 2.9 Hz, 2 H), 3.95 – 4.17 (m, 4 H), 4.39 (d, J = 8.8 Hz, 2 H), 7.27 –
7.38 (m, 3 H), 7.39 – 7.46 (m, 2 H); ^{13}C NMR (100 MHz, CDCl$_3$) δ:14.2,
16.2 (d, J = 7.2 Hz), 18.5 (d, J = 1.4 Hz), 22.7, 28.5, 29.3 (d, J = 1.2 Hz), 31.5,
55.1 (d, J = 5.8 Hz), 63.6 (d, J = 5.6 Hz), 65.0 (d, J = 5.0 Hz), 76.0 (d, J =
5.0 Hz), 128.0, 128.4, 128.9, 137.4 (d, J = 1.8 Hz); ^{31}P NMR (162 MHz,
CDCl$_3$) δ: 4.64; IR (neat): 2986, 2964, 2929, 2858, 2253, 1497, 1455, 1265,
1024, 974, 700, 596, 545 cm^{-1}; HRMS (ESI) [M + H]$^+$ calcd for C$_{19}$H$_{31}$NO$_3$P:
352.2036. Found: 352.2037; Anal. Calcd for C$_{19}$H$_{30}$NO$_3$P: C, 64.94; H,
8.60; N, 3.99. Found: C, 64.85; H, 8.53; N, 3.97.

Working with Hazardous Chemicals

The procedures in *Organic Syntheses* are intended for use only by
persons with proper training in experimental organic chemistry. All
hazardous materials should be handled using the standard procedures for
work with chemicals described in references such as "Prudent Practices in
the Laboratory" (The National Academies Press, Washington, D.C., 2011;
the full text can be accessed free of charge at
http://www.nap.edu/catalog.php?record_id=12654). All chemical waste
should be disposed of in accordance with local regulations. For general
guidelines for the management of chemical waste, see Chapter 8 of Prudent
Practices.

In some articles in *Organic Syntheses*, chemical-specific hazards are
highlighted in red "Caution Notes" within a procedure. It is important to
recognize that the absence of a caution note does not imply that no
significant hazards are associated with the chemicals involved in that
procedure. Prior to performing a reaction, a thorough risk assessment
should be carried out that includes a review of the potential hazards
associated with each chemical and experimental operation on the scale that
is planned for the procedure. Guidelines for carrying out a risk assessment
and for analyzing the hazards associated with chemicals can be found in
Chapter 4 of Prudent Practices.

The procedures described in *Organic Syntheses* are provided as
published and are conducted at one's own risk. *Organic Syntheses, Inc.*, its
Editors, and its Board of Directors do not warrant or guarantee the safety of
individuals using these procedures and hereby disclaim any liability for any

injuries or damages claimed to have resulted from or related in any way to the procedures herein.

Discussion

Alkynes substituted with electron-donating nitrogen functional groups exhibit unique reactivity in a variety of polar and pericyclic reactions. In recent years, ynamides have emerged as an exceptionally valuable class of building blocks for organic synthesis.[3] Ynamides are significantly more stable than simple ynamines, are more easily stored and handled, and are more resistant to hydrolysis and polymerization. While most ynamides studied to date are carbonyl and sulfonyl derivatives of ynamines, very recently *N*-phosphoryl ynamides have attracted attention as especially useful substrates for a variety of synthetic transformations.[4,5] Research in our laboratory has focused on the application of *N*-phosphoryl ynamides in [2 + 2] cycloadditions with ketenes where we have observed their reaction to occur at a rate as much as an order of magnitude faster than that of the corresponding *N*-carbomethoxy ynamides.[5] As a result of this exceptional reactivity, *N*-phosphoryl ynamides function as particularly useful alkyne partners in our vinylketene-based benzannulation strategy for the synthesis of highly substituted carboaromatic and heteroaromatic compounds.[5]

Scheme 1 outlines the general method for the synthesis of *N*-phosphoryl ynamides employed in this article. In 2003, concurrent research in our laboratory[6] and that of Hsung[7] led to the development of complementary methods for the synthesis of ynamides via copper-promoted coupling of haloalkynes with several classes of amide derivatives. The extension of this chemistry to the *N*-alkynylation of phosphoramidates (Scheme 1) was introduced by Hsung and coworkers in 2011[4a] and provides a convenient and efficient method for the preparation of this class of ynamides.

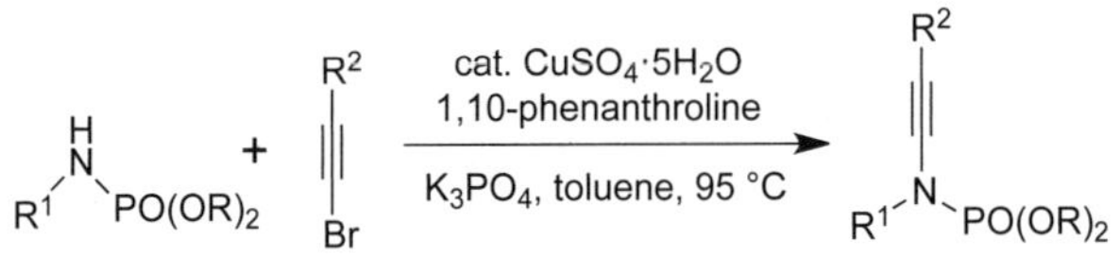

$$\underset{\substack{R^1}}{\overset{\substack{H\\N}}{}}PO(OR)_2 \quad + \quad \underset{Br}{\overset{R^2}{\vert\vert\vert}} \quad \xrightarrow[\substack{K_3PO_4,\ toluene,\ 95\ ^\circ C}]{\substack{cat.\ CuSO_4\cdot5H_2O\\1,10\text{-phenanthroline}}} \quad \underset{\substack{R^1}}{\overset{\substack{R^2\\\vert\vert\vert\\N}}{}}PO(OR)_2$$

Scheme 1. Synthesis of *N*-Phosphoryl Ynamides

Org. Synth. **2015**, *92*, 156-170   **165**   DOI: 10.15227/orgsyn.092.0156

A variety of methods are available for the preparation of the phosphoramidate reaction partners required in this strategy. In addition to the direct reaction of the corresponding amines with phosphoryl halides, phosphoramidates can be synthesized via methods involving electrosynthesis,[8] copper-catalyzed oxidative coupling,[9] the Todd-Atherton reaction,[10] the Staudinger phosphite reaction,[11] and a recently reported method involving the reaction of aldehydes with phosphoryl nitrenoids.[12]

Herein we describe the synthesis of phosphoramidate **1** by reaction of benzylamine with an iodophosphate generated in situ via the modified Todd-Atherton procedure introduced by Mielniczak and Łopusinski.[13] This approach avoids the use of toxic diethyl chlorophosphate which is employed in the more popular direct phosphorylation route,[14] and is competitive in terms of cost since diethyl phosphite is very inexpensive.

The mechanism of the Todd-Atherton and related reactions has been the subject of several mechanistic studies.[15,16] A general mechanism for the transformation described in this article is outlined in the following scheme. Overall, two equivalents of the amine are required for reaction, though in principle the amine can be recovered in cases where it is valuable.

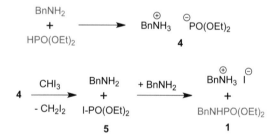

Scheme 2. Mechanism of the Formation of Phosphoramidate 1

The alkynyl bromide **2** employed for the *N*-alkynylation of phosphoramidate **1** is conveniently prepared via silver-catalyzed bromination of octyne with NBS following the method originally reported by Hofmeister and coworkers.[17] This convenient method typically provides bromoalkynes in high yield without the need to deprotonate the terminal alkyne with strong base. Several examples of the application of this method have previously been described in *Organic Syntheses*.[6b,18] In the case of bromooctyne **2**, the crude product of the bromination reaction is obtained in high purity and can be used in the subsequent alkynylation step without a need for further purification.

The synthesis of ynamide **3** described here follows the general procedure described by Hsung and coworkers.[4a] Despite the high reaction temperature and relatively high catalyst loading, this method provides a convenient and reliable procedure for the synthesis of N-phosphoryl ynamides provided that they do not incorporate heat-sensitive functional groups in their structure. As in the case of other metal-catalyzed coupling reactions, it is important to note that the use of anhydrous K_3PO_4 is crucial to obtain optimal yields.[19]

In summary, the reactions described in this article provide an efficient and economical synthetic route to **3**, and also serve to illustrate an excellent general method for the preparation of N-phosphoryl ynamides, an important emerging class of synthetic building blocks.

References

1. Department of Chemistry, Massachusetts Institute of Technology, Cambridge, MA 02139. Email: danheisr@mit.edu. We thank the National Science Foundation (CHE-1111567) for generous financial support.
2. Kohnen, A. L.; Dunetz, J. R.; Danheiser, R. L. *Org. Synth.* **2007**, *84*, 88.
3. For recent reviews on the synthesis and transformations of ynamides, see: (a) Evano, G.; Jouvin, K.; Coste, A. *Synthesis* **2013**, *45*, 17–26. (b) DeKorver, K. A.; Li, H.; Lohse, A. G.; Hayashi, R.; Lu, Z.; Zhang, Y.; Hsung, R. P. *Chem. Rev.* **2010**, *110*, 5064–5106. (c) Evano, G.; Coste, A.; Jouvin, K. *Angew. Chem., Int. Ed.* **2010**, *49*, 2840–2859. (d) Wang, X.-N.; Yeom, H.-S.; Fang, L.-C.; He, S.; Ma, Z.-X.; Kedrowski, B. L.; Hsung, R. P. *Acc. Chem. Res.* **2014**, *47*, 560–578.
4. (a) DeKorver, K. A.; Walton, M. C.; North, T. D.; Hsung, R. P. *Org. Lett.* **2011**, *13*, 4862–4865. (b) Wang, X.-N.; Winston-McPherson, G. N.; Walton, M. C.; Zhang, Y.; Hsung, R. P.; DeKorver, K. A. *J. Org. Chem.* **2013**, *78*, 6233–6244. (c) DeKorver, K. A.; Hsung, R. P.; Song, W.-Z.; Wang, X.-N.; Walton, M. C. *Org. Lett.* **2012**, *14*, 3214–3217. (d) DeKorver, K. A.; Wang, X.-N.; Walton, M. C.; Hsung, R. P. *Org. Lett.* **2012**, *14*, 1768–1771.

5. Willumstad, T. P.; Haze, O.; Mak, X. Y.; Lam, T. Y.; Wang, Y.-P.; Danheiser, R. L. *J. Org. Chem.* **2013**, *78*, 11450–11469.

6. (a) Dunetz, J. R.; Danheiser, R. L. *Org. Lett.* **2003**, *5*, 4011–4014. (b) Kohnen, A. L.; Dunetz, J. R.; Danheiser, R. L. *Org. Synth.* **2007**, *84*, 88–101.

7. (a) Frederick, M. O.; Mulder, J. A.; Tracey, M. R.; Hsung, R. P.; Huang, J.; Kurtz, K. C. M.; Shen, L.; Douglas, C. J. *J. Am. Chem. Soc.* **2003**, *125*, 2368–2369. (b) Zhang, Y.; Hsung, R. P.; Tracey, M. R.; Kurtz, K. C. M.; Vera, E. L. *Org. Lett.* **2004**, *6*, 1151–1154.

8. Torii, S.; Sayo, N.; Tanaka, H. *Tetrahedron Lett.* **1979**, *46*, 4471–4474.

9. Fraser, J.; Wilson, L. J.; Blundell, R. K.; Hayes, C. J. *Chem. Commun.* **2013**, *49*, 8919–8921.

10. (a) Atherton, F. R.; Openshaw, H. T.; Todd, A. R. *J. Chem. Soc.* **1945**, 660–663. (b) Atherton, F.R.; Todd, A.R. *J. Chem. Soc.* **1947**, 674–678.

11. (a) Letsinger, R. L.; Schott, M. E. *J. Am. Chem. Soc.* **1981**, *103*, 7394–7396. (b) Nielsen, J.; Caruthers, M. H. *J. Am. Chem. Soc.* **1988**, *110*, 6275–6276. (c) Wilkening, I.; del Signore, G.; Hackenberger, C. P. R. *Chem. Commun.* **2008**, 2932–2934. (d) Serwa, R.; Majkut, P.; Horstmann, B.; Swiecicki, J.-M.; Gerrits, M.; Krause, E.; Hackenberger, C. P. R. *Chem. Sci.* **2010**, *1*, 596–602.

12. Xiao, W.; Zhou, C.-Y.; Che, C.-M. *Chem. Commun.* **2012**, *48*, 5871–5873.

13. Mielniczak, G.; Łopusinski, A. *Synth. Commun.* **2003**, *33*, 3851–3859.

14. For the synthesis of **1** via reaction of benzylamine with ClPO(OEt)$_2$, see (a) Hammerschmidt, F.; Hanbauer, M. *J. Org. Chem.* **2000**, *65*, 6121–6131. (b) Kumar, G. D. K.; Saenz, D.; Lokesh, G. L.; Natarajan, A. *Tetrahedron Lett.* **2006**, *47*, 6281–6284.

15. Kong, A.; Engel, R. *Bull. Chem. Soc. Jpn.* **1985**, *58*, 3671–3672.

16. Troev, K.; Kirilov, E. M. G.; Roundhill, D. M. *Bull. Chem. Soc. Jpn.* **1990**, *63*, 1284–1285.

17. Hofmeister, H.; Annen, K.; Laurent, H.; Wiechert, R. *Angew. Chem. Int. Ed.* **1984**, *23*, 727–729

18. Leroy, J. *Org. Synth.* **1997**, *74*, 212–215.

19. Dooleweerdt, K.; Birkedal, H.; Ruhland, T.; Skrydstrup, T. *J. Org. Chem.* **2008**, *73*, 9447–9450.

rganic
yntheses

Appendix
Chemical Abstracts Nomenclature (Registry Number)

Diethyl benzylphosphoramidate: Phosphoramidic acid, *N*-(phenylmethyl)-, diethyl ester; (53640-96-3)
Benzylamine: Benzenemethanamine; (100-46-9)
Diethyl phosphite: Phosphonic acid, diethyl ester; (762-04-9)
Iodoform: Methane, triiodo-; (75-47-8)
1-Bromooct-1-yne: 1-Octyne, 1-bromo-; (38761-0)
N-Bromosuccinimide: 2,5-Pyrrolidinedione, 1-bromo-; (128-08-5)
1-Octyne; (629-05-0)
Silver(I) nitrate: Nitric acid silver(1+) salt (1:1); (7761-88-8)
Diethyl Benzyl(oct-1-yn-1-yl)phosphoramidate: Phosphoramidic acid, *N*-1-octyn-1-yl-*N*-(phenylmethyl)-, diethyl ester; (1332480-36-0)
Copper(II) sulfate pentahydrate: Sulfuric acid copper(2+) salt (1:1), hydrate (1:5); (7758-99-8)
Potassium phosphate: Phosphoric acid, potassium salt (1:3); (7778-53-2)
1,10-Phenanthroline (66-71-7)

John M. Read was born in Laredo, Texas, in 1994. He is currently an undergraduate at the Massachusetts Institute of Technology and expects to receive his B.S. degree in chemistry in 2016. John joined the laboratory of Professor Rick Danheiser in 2013, and his research has focused on the synthesis of highly substituted polycyclic compounds and requisite precursors.

Organic
Syntheses

Yu-Pu Wang was born in Taipei, Taiwan. He received a B.S. degree in Chemistry in 2009 from Rice University working in the laboratory of Professor James M. Tour. He is currently pursuing a Ph.D. degree at the Massachusetts Institute of Technology in the research group of Professor Rick L. Danheiser and his work involves the development of new methods for the synthesis of highly substituted indoles and their application to the synthesis of natural products and polycyclic systems with interesting electronic properties.

Rick L. Danheiser received his undergraduate education at Columbia where he carried out research in the laboratory of Professor Gilbert Stork. He received his Ph.D. at Harvard in 1978 working under the direction of E. J. Corey on the total synthesis of gibberellic acid. Dr. Danheiser is the A. C. Cope Professor of Chemistry at MIT where his research focuses on the design and invention of new annulation and cycloaddition reactions, and their application in the total synthesis of biologically active compounds.

Erik Daa Funder obtained his Ph.D. from Aarhus University, Denmark in 2013. During his Ph.D. studies he worked under the supervision of Prof. Kurt V. Gothelf dealing with the target synthesis of small molecules and the development of new reactions. The Ph.D. studies included a six-month stay in the group of Prof. Phil S. Baran at the Scripps Research Institute, La Jolla, CA, USA working on C-H activation. Currently, as a postdoctroral associate in the group of Prof. Erick M. Carreira, he is pursuing the synthesis of hydroxylated steroids, as well as the development and optimization of new reactions.

Synthesis of 2-Azido-1,3-dimethylimidazolinium Hexafluorophosphate (ADMP).

Mitsuru Kitamura[*1] and Kento Murakami

Department of Applied Chemistry, Kyushu Institute of Technology, 1-1 Sensuicho, Tobata, Kitakyushu, 804-8550, Japan

Checked by Raul Leal and Richmond Sarpong

Procedure

CAUTION: *Although 2-azido-1,3-dimethylimidazolinium hexafluorophosphate (ADMP) was found to be safe and stable in the safety tests (impact sensitivity test, friction sensitivity test, DSC), it is potentially explosive. It must be handled with care. This preparation should be carried out in a well-ventilated hood and should be conducted behind a safety shield.*

A. *2-Chloro-1,3-dimethylimidazolinium Hexafluorophosphate (2).* An oven-dried 250-mL round-bottomed flask equipped with a 3-cm Teflon-coated magnetic stirbar and a rubber septum (containing nitrogen inlet and outlet needles) is charged with 2-chloro-1,3-dimethylimidazolinium chloride (1) (Note 1) (21.2 g, 125 mmol, 1.0 equiv) and potassium hexafluorophosphate (Notes 2 and 3) (23.1 g, 126 mmol, 1.0 equiv) by temporary removal of the septum, and 60 mL of acetonitrile (Note 4) is added by syringe through the septum. The mixture is stirred for 10 min at room temperature then vacuum filtered through a pad of dry Celite (2 g) packed in a 60 mL sintered glass filter funnel (Note 5). The filter cake is washed with acetonitrile (3 x 20 mL), then the filtrate is concentrated on a rotary evaporator (heating bath temp < 50 °C, 15 mmHg). The residue is dissolved in a small amount of acetonitrile (15 mL) and the solution is poured into a 300 mL Erlenmeyer flask with diethyl ether (100 mL), under stirring, over the course of 2–3 min, to form an off-white precipitate. The suspension is vacuum-filtered through a 150 mL Büchner funnel with medium porosity filter paper. The solids are washed with diethyl ether (3 x 10 mL), then collected and dried under vacuum (22 °C, 0.2 mmHg) for 12 h to afford 2-chloro-1,3-dimethylimidazolinium hexafluorophosphate (**2**) (Note 6) (31.9–32.2 g, 91%).

Reaction Setup	First Filtration	Second Filtration

B. *2-Azido-1,3-dimethylimidazolinium Hexafluorophosphate (ADMP, 3).* An oven-dried 250-mL round-bottomed flask equipped with a 3-cm Teflon-coated magnetic stirbar and a rubber septum (containing nitrogen inlet and outlet needles) is charged with 2-chloro-1,3-dimethylimidazolinium hexafluorophosphate (**2**) (31.4 g, 113 mmol, 1.0 equiv) and sodium azide (Note 7) (11.0 g, 169 mmol, 1.5 equiv) by temporary removal of the septum. The flask is placed in a room temperature water bath and 115 mL of acetonitrile is added by syringe through the septum (Note 8). The mixture is stirred for 1 h at room temperature, then filtered through a pad of dry Celite (2 g) packed in a 60 mL sintered glass filter funnel (Note 5). The filter cake (Note 9) is washed with acetonitrile (3 x 30 mL), and the filtrate is concentrated on a rotary evaporator (heating bath temp < 50 °C). The residue is dissolved in a small amount of acetonitrile (13 mL) and the solution is poured into a 300 mL Erlenmeyer flask with diethyl ether (100 mL), under stirring, over the course of 2–3 min, to form a white precipitate. The suspension is vacuum-filtered through a 150 mL Büchner funnel with medium porosity filter paper. The solids are washed with diethyl ether (3 x 10 mL) then collected and dried under vacuum (22 °C, 0.2 mmHg) for 12 h to afford 2-azide-1,3-dimethylimidazolinium hexafluorophosphate (**3**) (30.1–30.5 g, 94–95%) Further purification of the title compound **3** is achieved by recrystallization (Note 10) from toluene and acetone to afford pure 2-azido-1,3-dimethylimidazolinium hexafluorophosphate (**3**) as white crystals (26.8–27.5 g, 83–85%) (Note 11).

Reaction Setup   First Filtration   Second Filtration

ADMP (**3**) Crystals

Notes

1. Both submitters and checkers purchased 2-chloro-1,3-dimethylimidazolinium chloride (> 80.0 %) from Tokyo Chemical Industry (TCI) and used it as received. This material is very hygroscopic. Material was used each time from a new bottle, and weighed quickly under air.

2. The submitters and the checkers purchased potassium hexafluorophosphate (>97%) from Wako Pure Chemical Industries, and used it as received.

3. Kiso *et al.* reported the synthesis of 2-chloro-1,3-dimethylimidazolinium hexafluorophosphate by the reaction of 2-chloro-1,3-dimethylimidazolinium chloride and ammonium hexafluorophosphate, see: Akaji, K.; Kuriyama, N.; Kimura, T.; Fujiwara, Y.; Kiso, Y. *Tetrahedron Lett.* **1992**, *33*, 3177; Kiso, Y.; Fujiwara, Y.; Kimura, T.; Nishitani A.; Akaji, K. *Int. J. Peptide Protein Res.* **1992**, *40*, 308. The submitters chose to use potassium hexafluorophosphate for the anion exchange, because it is a cheaper alternative compared to ammonium hexafluorophosphate.

4. Acetonitrile (99.8%) was purchased from Sigma-Aldrich in a Sure/Seal bottle and was used as received.

5. Celite 545 was purchased from Wako Pure Chemical Industries. The Celite was packed tightly into the funnel using the bottom of a beaker.

6. The isolated 2-chloro-1,3-dimethylimidazolinium hexafluorophosphate (**2**) has the following physicochemical properties: mp 231–232 °C (decomp.); IR (ATR, uncorrected) cm^{-1}: 1637, 1548, 1299, 819, 620, 554; ^{1}H NMR (600 MHz, CD$_3$CN) δ: 3.11 (s, 6 H), 3.93 (s, 4 H); ^{13}C NMR (151 MHz, CD$_3$CN) δ: 35.2, 50.8, 157.1; ^{19}F NMR (376 MHz, CD$_3$CN) δ: –71.71 (d, *J* = 706.6 Hz); Anal. calcd for C$_5$H$_{10}$ClF$_6$N$_2$P: C, 21.56; H, 3.62; N, 10.06. Found: C, 21.24; H, 3.49; N, 9.90; HRMS (*m/z*) calcd for [C$_5$H$_{10}$N$_2$Cl]$^+$: 133.0527. Found: 133.0526.

7. Sodium azide (>98%) was purchased from Wako Pure Chemical Industries and was used as received for both submitters and checkers.

8. The submitters reported running the reaction at 0 °C for 30 min. The reaction was found to be sluggish at 0 °C, but went to completion at 22 °C within 1 h. The internal temperature of the reaction was monitored with a thermometer, and no change (>2 °C) in internal

temperature was noted during addition of acetonitrile and over the course of the reaction.

9. The filter cake contains excess sodium azide used in the reaction. The cake was diluted with water and disposed as waste liquid, keeping the waste separate from other waste streams that contain metals, oxidizing reagents, and chlorinated solvents.

10. While the obtained material is of sufficient purity for efficient diazo-transfer reactions, the 2-azido-1,3-dimethylimidazolinium hexafluoro-phosphate (ADMP, **3**) was recrystallized by dissolving the crude product in the minimum amount of 1:1 toluene:acetone necessary (160 mL) at 50 °C (heated using a water bath). The material was allowed to cool to room temperature, and then placed in a –20 °C freezer overnight. The crystals were collected by vacuum filtration through a Büchner funnel with medium porosity filter paper, and were washed with cold (–20 °C) 1:1 toluene:acetone (2 x 10 mL). The solids were collected and dried under vacuum (22 °C, 0.2 mmHg) for 12 h. The filtrate was concentrated on a rotary evaporator. This concentrated residue was recrystallized twice more using the same method using 70 mL and 15 mL of 1:1 toluene:acetone, respectively. The white crystals were combined and analyzed for the properties listed in Note 11.

11. The 2-azido-1,3-dimethylimidazolinium hexafluorophosphate (ADMP, **3**) has the following physicochemical properties: mp 202–204 °C (decomp); IR (ATR, uncorrected) cm^{-1}: 2172, 1638, 1579, 1307, 1289, 819, 668, 554; ^{1}H NMR (600 MHz, CD$_3$CN) δ: 3.05 (s, 6 H), 3.79 (s, 4 H); ^{13}C NMR (151 MHz, CD$_3$CN) δ: 33.8, 50.0, 156.5; ^{19}F NMR (376 MHz, CD$_3$CN) δ: –71.81 (d, J = 706.6 Hz); Anal. calcd for C$_5$H$_{10}$F$_6$N$_5$P: C, 21.06; H, 3.54; N, 24.56. Found: C, 21.08; H, 3.63; N, 24.39; HRMS (m/z) calcd for [C$_5$H$_{10}$N$_5$]$^{+}$: 140.0931. Found: 140.0929.

Working with Hazardous Chemicals

The procedures in *Organic Syntheses* are intended for use only by persons with proper training in experimental organic chemistry. All hazardous materials should be handled using the standard procedures for work with chemicals described in references such as "Prudent Practices in the Laboratory" (The National Academies Press, Washington, D.C., 2011; the full text can be accessed free of charge at

). All chemical waste should be disposed of in accordance with local regulations. For general guidelines for the management of chemical waste, see Chapter 8 of Prudent Practices.

In some articles in *Organic Syntheses*, chemical-specific hazards are highlighted in red "Caution Notes" within a procedure. It is important to recognize that the absence of a caution note does not imply that no significant hazards are associated with the chemicals involved in that procedure. Prior to performing a reaction, a thorough risk assessment should be carried out that includes a review of the potential hazards associated with each chemical and experimental operation on the scale that is planned for the procedure. Guidelines for carrying out a risk assessment and for analyzing the hazards associated with chemicals can be found in Chapter 4 of Prudent Practices.

The procedures described in *Organic Syntheses* are provided as published and are conducted at one's own risk. *Organic Syntheses, Inc.*, its Editors, and its Board of Directors do not warrant or guarantee the safety of individuals using these procedures and hereby disclaim any liability for any injuries or damages claimed to have resulted from or related in any way to the procedures herein.

Discussion

Organic azides are generally synthesized by the nucleophilic substitution of a leaving group on an organic compound with an azide ion.[2] The diazo-transfer reaction of primary amines is an alternative method for synthesizing organic azides. Trifluoromethanesulfonyl azide (TfN_3) has been used as a diazo-transfer reagent for primary amines,[3] and the diazo-transfer reaction proceeds efficiently with the help of a metal catalyst such as Cu^{II} salt.[3c-8] However, the reaction has several drawbacks. The safety and stability of the diazo-transfer reagent is a concern; TfN_3 has an explosive nature and requires very careful treatment.[3b,4] Furthermore, it has poor reactivity toward poorly nucleophilic primary amines such as anilines with electron-withdrawing groups.[3e] Several low-explosive reagents have recently been reported for diazo-transfer to primary amines, such as nonafluorobutanesulfonyl azide,[5] imidazole-1-sulfonyl azide,[6] benzotriazol-1-yl-sulfonyl azide.[7]

2-Azido-1,3-dimethylimidazolinium hexafluorophosphate (ADMP) is safe and stable crystalline solid, and show efficient diazo-transfer ability to primary amines.[8] ADMP is not very hygroscopic and can be stored in a freezer (at −10 °C) for at least two months. Impact sensitivity tests[9] and friction sensitivity tests[10] showed that ADMP is not explosive in the test ranges. The exothermic decomposition of ADMP was observed through differential scanning calorimetry experiments from approximately 200 °C.[8a] These results suggest that ADMP can be safely used below its decomposition temperature, preferably below 100 °C in order to allow an appropriate safety margin.

The diazo-transfer reaction of ADMP to primary amines proceeded smoothly in the presence of 4-(N,N-dimethyl)aminopyridine (DMAP) as a base and the reaction did not require the addition of metal catalyst (Scheme 1).[11] Even though low nucleophilic anilines substituted with strong electron-withdrawing groups such as acetyl, cyano and nitro groups were employed, the diazo-transfer reaction proceeded at 50 °C.

Scheme 1. Synthesis of organic azides by the diazo-transfer reaction of ADMP to primary amines

$$R-NH_2 \ + \ \underset{\text{ADMP}}{\underset{\text{MeN} \overset{+}{\frown} \text{NMe}}{PF_6^- \ N_3}} \ \xrightarrow[\substack{CH_3CN \\ 30\ °C}]{DMAP} \ R-N_3$$

R = OMe	87%
R = Me	86%
R = n-Bu	84%
R = Cl	87%[a,b]
R = Ac	87%[a,b]
R = CN	70%[a,c]
R = NO₂	63%[a,c]

$PhCH_2CH_2-N_3$ 94%[d]

cyclo-Hex-N₃   85%[d]

$Ph \overset{}{\diagup} N_3$ 90%

Ph_3C-N_3 96%

R = Me	97%[a,c]
R = Cl	15%[c]

   90%

[a] Reaction was carried out at 50 °C. [b] THF was used as solvent.
[c] C_6H_5Cl was used as solvent. [d] Et₂NH was used instead of DMAP.

Compared to low nucleophilic amines, the diazo-transfer reaction to nucleophilic amines proceeded easier at low temperature (30 °C). However, for the diazo-transfer to less-hindered primary alkyl amines, DMAP was not suitable as a base, but a stronger base such as alkylamine or DBU was appropriate.

In addition to diazo-transfer to primary amines, ADMP and its corresponding chloride (ADMC), which was prepared by the reaction of 2-chloro-1,3-dimethylimidazolinium chloride (DMC) and sodium azide, could be used for various nitrogen functionalization (Scheme 2).[12]

ADMP and ADMC reacted with 1,3-dicarbonyl compounds under mild basic conditions to give 2-diazo-1,3-dicarbonyl compounds in high yields, which are easily isolated because the by-products are highly soluble in water.[13] Naphthols also reacted with ADMC to give corresponding diazonaphthoquinones in good to high yields.[14] In addition, 2-azido-1,3-dimethylimidazolinium salts were found to be employed as azide-transfer[15] and migratory amination reagents.[16]

Scheme 2. Reactions of ADMP and ADMC

Organic Syntheses

References

1. Department of Applied Chemistry, Kyushu Institute of Technology, 1-1 Sensuicho, Tobata, Kitakyushu, 804-8550, Japan. kita@che.kyutech.ac.jp
2. (a) Scriven, E. F. V.; Turnbull, K. *Chem. Rev.* **1988**, *88*, 297; (b) Bräse, S.; Gil, C.; Knepper, K.; Zimmermann, V. *Angew. Chem., Int. Ed.* **2005**, *44*, 5188. (c) Bräse, S.; Banert, K. *Organic Azides*; Wiley, Wiltshire, **2010**.
3. (a) Cavender, C. J.; Shiner, V. J.; Jr, *J. Org. Chem.* **1972**, *37*, 3567. (b) Zaloom, J.; Roberts, D. C. *J. Org. Chem.* **1981**, *46*, 5173. (c) Alper, P. B.; Hung, S.-C.; Wong, C.-H. *Tetrahedron Lett.* **1996**, *37*, 6029. (d) Nyffeler, P. T.; Liang, C.-H.; Koeller, K. M.; Wong, C.-H. *J. Am. Chem. Soc.* **2002**, *124*, 10773. (e) Liu, Q.; Tor, Y. *Org. Lett.* **2003**, *5*, 2571. (f) Yan, R.-B.; Yang, F.; Wu, Y.; Zhang, L.-H.; Ye, X.-S. *Tetrahedron Lett.* **2005**, *46*, 8993. (g) Titz, A.; Radic, Z.; Schwardt, O.; Ernst, B. *Tetrahedron Lett.* **2006**, *47*, 2383.
4. For the stablity of sulfonyl azides, see: (a) Hazen, G. G.; Weinstock, L. M.; Connell, R.; Bollinger, F. W. *Synth. Commun.* **1981**, *11*, 947. (b) F. W. Bollinger, F. W.; Tuma, L. D. *Synlett* 1996, 407.
5. (a) Yekta, S.; Prisyazhnyuk, V.; Reissig, H.-U. *Synlett*, **2007**, 2069. (b) Al-Harrasi, A.; Pfrengle, F. ; Prisyazhnyuk, V.; Yekta,S.; Kóos, P.; Reissig, H.-U. *Chem. Eur. J.* **2009**, *15*, 11632. (c) Suárez, J. R.; Trastoy, B.; Pérez-Ojeda, M. E.; Marín-Barrios, R.; Chiara, J. L. *Adv. Synth. Catal.* **2010**, *352*, 2515.
6. Goddard-Borger, E. D.; Stick, R. V. *Org. Lett.* **2007**, *9*, 3797. Fischer, N.; Goddard-Borger, E. D.; Greiner, R.; Klapötke, T. M.; Skelton, B. W.; Stierstorfer, J. *J. Org. Chem.* **2012**, *77*, 1760.
7. Katritzky, A. R.; Khatib, M. EI.; Bol'shakov, O.; Khelashvili, L.; P. J. Steel, P. J. *Org. Chem.* **2010**, *75*, 6532. Although benzotriazol-1-yl-sulfonyl azide is not explosive in nature, an accident caused by a mistake made during its preparation has been reported, see: Katritzky, A. R.; Khatib, M. EI. *Chem. Eng. News* **2012**, *90*, 4.
8. (a) Kitamura, M.; Yano, M.; Tashiro, N.; Miyagawa, S.; Sando, M.; Okauchi, T. *Eur. J. Org. Chem.* **2011**, 458. (b) Kitamura, M.; Kato, S.; Yano, M.; Tashiro, N.; Shiratake, Y.; Sando, M.; Okauchi, T. *Org. Biomol. Chem.* **2014**, *12*, 4397.
9. Impact sensitivity was >25 [Nm] by German Federal Institute for Materials Research and Testing (BAM) procedure.
10. Friction sensitivity was >360 [N] by German Federal Institute for Materials Research and Testing (BAM) procedure.

11. Although the diazo-transfer from ADMP to primary amines proceeded smoothly in CH_2Cl_2 as shown in preliminary communication,[8a] CH_2Cl_2 should not be used in the reaction to avoid the risk of the formation of diazidomethane,[8b] which is very explosive and formed by the reaction of CH_2Cl_2 with N_3^-. For the formation of diazidomethane, see: R. E. Conrow, R. E.; Dean, W. D. *Org. Process Res. Dev.* **2008**, *12*, 1285 and references cited therein.

12. For a review, see: Kitamura, M. *J. Synth. Org. Chem., Jpn.* **2014**, *72*, 14.

13. (a) Kitamura, M.; Tashiro, N.; Okauchi, T.; *Synlett* **2009**, 2943. (b) Kitamura, M.; Tashiro, N.; Miyagawa, S.; Okauchi, T. *Synthesis* **2011**, 1037.

14. (a) Kitamura, M.; Tashiro, N.; Sakata, R.; Okauchi, T. *Synlett* **2010**, 2503. (b) Kitamura, M.; Sakata, R.; Tashiro, N.; Ikegami, A.; Okauchi, T. *Bull. Chem. Soc. Jpn.* **2015**, doi:10.1246/bcsj.20150021.

15. (a) Kitamura, M.; Tashiro, N.; Takamoto, Y.; Okauchi, T. *Chem. Lett.* **2010**, *39*, 732. (b) Kitamura, M.; Koga, T.; Yano M.; Okauchi, T. *Synlett*, **2012**, 1335.

16. (a) Kitamura, M.; Miyagawa, S.; Okauchi, T. *Tetrahedron Lett.* **2011**, *52*, 3158. (b) Kitamura, M.; Murakami, K.; Shiratake, Y.; Okauchi, T. *Chem. Lett.* **2013**, *42*, 691.

Appendix
Chemical Abstracts Nomenclature (Registry Number)

2-Chloro-1,3-dimethylimidazolinium chloride: 1*H*-Imidazolium, 2-chloro-4,5-dihydro-1,3-dimethyl-, chloride (1:1); (37091-73-9)
Potassium hexafluorophosphate: Phosphate(1-), hexafluoro-, potassium (1:1); (17084-13-8)
2-Chloro-1,3-dimethylimidazolinium hexafluorophosphate: 1*H*-Imidazolium, 2-chloro-4,5-dihydro-1,3-dimethyl-, hexafluorophosphate(1-) (1:1); (101385-69-7)
Sodium azide: Sodium azide (Na(N_3)); (26628-22-8)
2-Azido-1,3-dimethylimidazolinium hexafluorophosphate: 1*H*-Imidazolium, 2-azido-4,5-dihydro-1,3-dimethyl-, hexafluorophosphate(1-) (1:1); (1266134-54-6)

Mitsuru Kitamura was born in Takamatsu, Japan, in 1971 and received his B.Sc. in 1994 and M. Sc. in 1996 from Keio University. He received his Ph.D. in 1999 from Tokyo Institute of Technology under the direction of Professor K. Suzuki. Then, he was appointed as research associate in Professor K. Narasaka's group at the University of Tokyo. In 2005, he moved to Kyushu Institute of Technology as associate professor, and promoted to full professor in 2015. He was a recipient of the Chemical Society of Japan Award for Young Chemists (2005) and Award for a Creative Young Chemist, Kyushu-Yamaguchi Branch, the Society of Synthetic Organic Chemistry, Japan (2011).

Kento Murakami was born in Matsuyama, Japan, in 1990. He received his B. Eng. in 2013 from Kyushu Institute of Technology. In the same year, he began his graduate studies at the same institute under the guidance of Professor Mitsuru Kitamura.

Raul Leal was born in Chesapeake, Virginia in 1988. He received his Bachelor of Science in 2010 from the State University of New York at Binghamton. He is currently pursuing graduate studies in the area of natural product synthesis under the mentorship of Professor Richmond Sarpong at the University of California, Berkeley.

Asymmetric Synthesis of All-Carbon Benzylic Quaternary Stereocenters via Conjugate Addition to Alkylidene Meldrum's Acids

Eric Beaton and Eric Fillion[*1]

Department of Chemistry, University of Waterloo, Waterloo, Ontario, Canada, N2L 3G1

Checked by Li Zhang and Chris Senanayake

Published on the Web 6/18/2015
© 2015 Organic Syntheses, Inc.

rganic
yntheses

Procedure

Caution! Neat diethylzinc is extremely pyrophoric and moisture sensitive. Handle under argon or nitrogen at all times. Fires should be extinguished using a dry powder extinguisher.

A. *5-(1-(4-Chlorophenyl)ethylidene)-2,2-dimethyl-1,3-dioxane-4,6-dione* (**1**).[2,3] An oven-dried 1000 mL, single-necked, round-bottomed flask, equipped with a 3 cm oval Teflon®-coated magnetic stirring bar and a rubber septum, is purged with nitrogen. The flask is charged with THF (300 mL) and cooled to 0 °C (bath temp) in an ice bath (Note 1). An oven-dried 100 mL, single-necked, round-bottomed flask equipped with a rubber septum is charged with dichloromethane (50 mL), and then titanium (IV) chloride (18.0 mL, 31.1 g, 164 mmol, 2.13 equiv) is added and the flask is swirled for 10 s (Notes 2 and 3). The TiCl$_4$ solution is added to the flask containing THF in four portions via a 20 mL Norm-ject plastic syringe at a rate of 4.3 mL/min with stirring (Note 4). During this addition smoke forms inside the reaction flask and the mixture becomes a cloudy yellow suspension. An oven-dried 250 mL, single-necked, round-bottomed flask is charged with 2,2-dimethyl-1,3-dioxane-4,6-dione (11.1 g, 77.0 mmol, 1.00 equiv) and THF (80 mL). The flask is sealed with a rubber septum and purged with nitrogen before 4'-chloroacetophenone (10.1 mL, 12.0 g, 77.8 mol, 1.01 equiv) is added via syringe. The flask is swirled until the solid has dissolved (<2 min) (Notes 5 and 6). The Meldrum's acid/4'-chloroacetophenone solution is added to the reaction flask containing TiCl$_4$ via cannula over a period of 10 min. During the addition, the mixture changes color from bright yellow to dark yellow/brown. The 250 mL flask is rinsed with 5 mL of THF, which is rapidly transferred to the reaction via cannula. Pyridine (31.5 mL, 30.7 g, 388 mmol, 5.04 equiv) is then added via a 20 mL syringe in 2 portions at a rate of 4.3 mL/min (Note 7). During this addition, a red color is observed initially and the mixture becomes a muddy brown suspension by the end of the addition. The vigorously stirred reaction is kept at 0 °C for 1 h and then allowed to stir at room temperature for 24 h. The mixture is cooled to 0 °C using an ice-water bath and 150 mL of deionized water is added. The stir bar is removed and the reaction is transferred to a 1000 mL separatory funnel with the assistance of 50 mL of deionized water and 50 mL of ethyl acetate. The layers are partitioned and the aqueous layer is extracted with ethyl acetate (3 x 150 mL). The combined organic layers are dried over MgSO$_4$ (14 g, <2 min), filtered through a 7 cm diameter filter funnel packed with cotton into a 2 L round-bottomed flask, and concentrated by rotary evaporation (30 °C, 20 mmHg). The resulting dark yellow oil is dissolved in ethyl acetate (200 mL), which is washed sequentially with saturated NaHCO$_3$, (2 x 100 mL) and brine (150 mL) and then dried over MgSO$_4$ (6 g, <2 min) (Note 8). The mixture is filtered through a 7 cm diameter filter funnel packed with cotton and concentrated by rotary evaporation (30 °C, 20 mmHg) to obtain a bright yellow solid (Note 9). The solid is dissolved in ethyl acetate (50 mL) and transferred, with rinsing, to a 250 mL, single-necked round-bottomed flask. The solvent is removed by rotary evaporation (30 °C, 20 mmHg) (Note 9). Methanol (60 mL) and a 1 cm rod-shaped magnetic stirring bar are added and the mixture is placed in an oil bath heated to 50 °C with stirring until the solid dissolves (20 min) (Note 10). The flask is removed from the oil bath and swirled gently to dissolve any solid on the walls of the flask, and then the solution is allowed to cool to room temperature (1 h) before being placed in an ice-water bath for 1 h. The solid is collected by suction filtration (room temperature, ~20 mmHg) through a Büchner funnel equipped with a filter paper and is washed with 30 mL of methanol that has been cooled to –25 °C. The nearly colorless needles are pulverized using a mortar and pestle before being dried under reduced pressure (1 mmHg, 1 h, room temperature) to afford 13.2–13.8 g (61–64%) of 5-(1-(4-chlorophenyl)ethylidene)-2,2-dimethyl-1,3-dioxane-4,6-dione [1] as an off-white powder (Notes 11, 12, 13, and 14).

B. *(R)-5-(2-(4-Chlorophenyl)butan-2-yl)-2,2-dimethyl-1,3-dioxane-4,6-dione* **(2)**.[3] An oven-dried 250 mL, two-necked round-bottomed flask equipped with a 2 cm rod-shaped Teflon®-coated magnetic stirring bar and a thermocouple is charged with copper (II) trifluoromethanesulfonate (322 mg, 0.890 mmol, 2.5 mol %) then connected to a vacuum line and placed under reduced pressure (0.5 mmHg) before being placed in an oil bath heated to 100 °C (Note 15). The flask is held at this temperature for 3 h before the flask is removed from the oil bath and allowed to cool to room temperature (30 min) while still under reduced pressure. The flask is opened to air and (*S*)-2,2′-binaphthoyl-(*R,R*)-di(1-phenylethyl)aminoyl-phosphine[4] (961 mg, 1.78 mmol, 5.0 mol%) is quickly added before the flask is sealed with a rubber septum and purged with argon (Note 16). 1,2-Dimethoxyethane (45 mL) is then added via syringe and the mixture is stirred at room temperature for 30 min before being placed in an isopropanol bath maintained at -40 °C using an immersion cooler

(Notes 17 and 18). Diethylzinc (7.3 mL, 8.8 g, 71 mmol, 2.0 equiv) is added via a 10 mL gas-tight syringe under argon over 10 min and the reaction is allowed to stir for 15 min (Note 19). An oven-dried 100 mL, single-necked

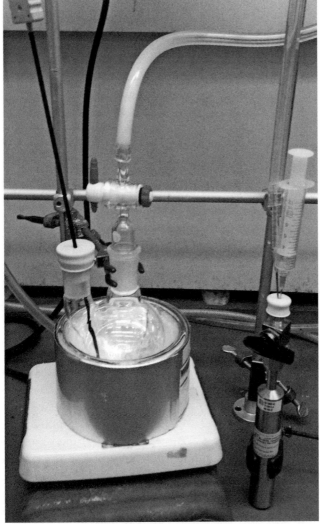

Set-up for Step B

round-bottomed flask equipped with a rubber septum is purged with argon and charged with 5-(1-(4-chlorophenyl)ethylidene)-2,2-dimethyl-1,3-

dioxane-4,6-dione (**1**) (10.0 g, 35.6 mmol, 1.00 equiv) and 1,2-dimethoxyethane (35 mL). The flask is swirled to dissolve the solid (1 min) before being transferred to the reaction by cannula over a period of 10 min. 1,2-Dimethoxyethane (5 mL) is used to rinse the 100 mL flask, which is transferred to the reaction by cannula in 30 s and the reaction is allowed to stir for 24 h. While still at –40 °C, hydrochloric acid (100 mL, 2 M) is added to the reaction over 2 min (Note 20). Then ethyl acetate (50 mL) is added. The cooling bath is removed and the reaction is allowed to stir for 10 min in a room temperature water bath. The mixture is transferred to a 500 mL separatory funnel with the assistance of deionized water (50 mL) and ethyl acetate (50 mL). The layers are partitioned and the aqueous layer is extracted with ethyl acetate (3 x 50 mL). The combined organic layers are washed once with 100 mL of brine and dried over $MgSO_4$ (6 g, <2 min), filtered through a 7 cm diameter filter funnel packed with cotton and concentrated by rotary evaporation (35 °C, 20 mmHg). The resulting white solid is mixed with dichloromethane (30 mL) to form a suspension, then subjected to flash chromatography (350 g silica, 60 cm x 5 cm inner diameter, 200 mL fractions) using 6 L of 9:1 hexanes/ethyl acetate as the eluent (Notes 21 and 22). The product is found in fractions 11-24, which are concentrated by rotary evaporation (35 °C, 20 mmHg) followed by removal of residual ethyl acetate under reduced pressure (23 °C, 5 mmHg, 30 min) to obtain 8.8–8.9 g (79–80% yield, 97:3–98:2 er) of (*R*)-5-(2-(4-chlorophenyl)butan-2-yl)-2,2-dimethyl-1,3-dioxane-4,6-dione as a white solid (Notes 23, 24, and 25).

C. (*R*)-3-(4-Chlorophenyl)-3-methylpentanoic acid (**3**).[3] A 100 mL, single-necked round-bottomed flask equipped with a 1 cm oval Teflon®-coated magnetic stirring bar and a reflux condenser is charged with (*R*)-5-(2-(4-chlorophenyl)butan-2-yl)-2,2-dimethyl-1,3-dioxane-4,6-dione (8.00 g, 25.7 mmol, 1.00 equiv), pyridine (47.0 mL, 45.8 g, 579 mmol, 22.5 equiv) and deionized water (4.7 mL, 4.7 g, 0.26 mol, 10 equiv) then the condenser is equipped with a rubber septum and the apparatus is purged with nitrogen (Note 7). The apparatus is placed in an oil bath preheated to 118 °C and the brownish yellow solution is allowed to stir for 4 h under nitrogen. The reaction is removed from the oil bath and allowed to cool for 10 min before being concentrated by rotary evaporation (50 °C, 20 mmHg) to obtain a brownish yellow oil. The residue is cooled to 0 °C in an ice-water bath before being treated with 3M hydrochloric acid (40 mL) and allowed to stir for 5 min (Note 20). The resulting biphasic solution is diluted with methyl *t*-butyl ether (40 mL) and transferred to a 125 mL separatory funnel. The layers are partitioned and the aqueous layer is extracted with methyl *t*-butyl ether (3 x 40 mL). The combined organic layers are dried over $MgSO_4$ (6 g, <2 min) and filtered through a 4 cm diameter filter funnel packed with cotton before being concentrated by rotary evaporation (25 °C, 20 mmHg). The resulting oil is subjected to flash chromatography (100 g silica, 60 cm x 5 cm inner diameter, 200 mL fractions) using 2 L of hexanes:ethyl acetate (7:1) as the eluent (Note 22). The product is collected from fractions 6-15, which are concentrated by rotary evaporation (35 °C, 20 mmHg) followed by removal of residual ethyl acetate under reduced pressure (23 °C, 5 mmHg, 30 min) to obtain 5.5–5.7 g (94–98%) of (*R*)-3-(4-chlorophenyl)-3-methylpentanoic acid [**3**] as a pale yellow oil (Notes 26, 27 and 28).

Notes

1. ACS Reagent tetrahydrofuran (99.0%, contains 250 ppm BHT) was obtained from Sigma-Aldrich Company and distilled from sodium-benzophenone ketyl immediately prior to use.
2. Titanium (IV) chloride, ReagentPlus®, 99.9%, was obtained from Sigma-Aldrich Company and used without further purification.
3. Dichloromethane (99.9%, HPLC grade, contains 15-200 ppm amylene) was dried by percolation through two columns packed with neutral alumina under a positive pressure of nitrogen.
4. In some cases the $TiCl_4$ solution causes corrosion of the plastic syringe during one of the additions. If this occurs, the current addition is allowed to complete and a new syringe is used for the remaining additions.
5. 4'-Chloroacetophenone, 98%, was obtained from Oakwood Products, Inc. and used without further purification.
6. 2,2-Dimethyl-1,3-dioxane-4,6-dione, 99%, was obtained from Oakwood Products, Inc. and used without further purification.
7. Pyridine, 99%, was obtained from Caledon Laboratories Ltd. and used without further purification. In step C, an excess of pyridine is used because it is the solvent for the reaction.
8. A large amount of CO_2 gas evolves when the $NaHCO_3$ is added.
9. If the resulting residue is an oil, methanol (10 mL) is added and the flask is swirled for <2 min to precipitate the solid.

10. Methanol (99.8%, HPLC grade) was obtained from Caledon Laboratories Ltd. and used without further purification.

11. Whatman® 70 mm filter papers are used.

12. 5-(1-(4-Chlorophenyl)ethylidene)-2,2-dimethyl-1,3-dioxane-4,6-dione is stable in excess of six months when stored in the freezer.

13. The product displayed the following physiochemical properties: mp 72–74 °C; ^{1}H NMR (500 MHz, CDCl$_3$) δ: 1.83 (s, 6 H), 2.70 (s, 3 H), 7.12 (d, J = 8.5 Hz, 2 H), 7.38 (d, J = 8.0 Hz, 2 H); ^{13}C NMR (125 MHz, CDCl$_3$) δ: 26.3, 27.4, 104.0, 117.2, 127.4, 128.8, 135.6, 140.0, 160.2, 161.0, 171.7; IR (thin film, CH$_2$Cl$_2$) ν_{max} (cm^{-1}) 1765, 1726; HRMS (DART) m/z calcd. for C$_{14}$H$_{13}$^{35}ClO$_4$•Na$^+$: 303.0400, found: 303.0386. The purity of product (97–99%) is determined by quantitative HPLC (Halo-C18 0.46 cm × 15 cm column) using acetonitrile (A) and water (B, contains 0.2% H$_3$PO$_4$ and 60 mM NH$_4$PF$_6$) as eluent (1.3 mL/min, percentage of A starts from 35%, changes to 45% in 4 min, increases to 98% in another 3 min and keeps at 98% for another 2 min). t_R = 7.00 min. Purity of the quantification standard is determined by NMR assay using dimethyl fumarate as internal standard.

14. Sometimes it is possible to obtain a second crop from the filtrate. The filtrate is transferred to a 100 mL single-necked round-bottomed flask and concentrated by rotary evaporation (30 °C, 20 mmHg). Methanol (5 mL) is added to the resulting dark orange oil and the flask is swirled for 1 min to form a uniform solution. The flask is placed in a –25 °C freezer for 18 h. The solid is collected by suction filtration (room temperature, ~20 mmHg) through a Büchner funnel equipped with a filter paper and are washed with 10 mL of methanol at –25 °C to obtain an additional 0.98 g (5%) of **1** as a pale yellow solid contaminated with traces of Meldrum's acid and 2,2-dimethyl-5-(propan-2-ylidene)-1,3-dioxane-4,6-dione.

15. Copper (II) trifluoromethanesulfonate, 98%, was purchased from Strem Chemicals, Inc. and used without further purification. It was stored in a desiccator containing Drierite® under nitrogen.

16. (S)-2,2'-Binaphthoyl-(R,R)-di(1-phenylethyl)aminoyl-phosphine was prepared following literature procedure.[4]

17. 1,2-Dimethoxyethane (99.0%) was obtained from TCI America, Inc., purified by distillation from sodium-benzophenone ketyl, and degassed using the freeze-pump-thaw method (3 cycles).

18. A Kinetics Flexi-Cool 100 immersion cooler was used.

19. Diethylzinc, 95%, was purchased from Strem Chemicals, Inc. and used without further purification.

20. Hydrochloric acid, 37%, was obtained from Sigma-Aldrich Co. and used as received. The 2 M solution was prepared by diluting 82 mL of concentrate in deionized water to a 500 mL solution. The 3 M solution was prepared by diluting 10 mL of concentrate in 30 mL of deionized water.

21. The crude white solid starts to turn yellow if allowed to stand overnight.

22. SiliaFlash® F60 (40-63 μm, 230-400 mesh) silica gel was used.

23. Occasionally, fractions 8-10 contained some product that was contaminated with an unknown compound that is UV active (R_f = 0.28 in 9:1 hexanes/ethyl acetate). These fractions can be purified via flash chromatography as follows. After concentration by rotary evaporation (35 °C, ~20 mmHg), the mixed fractions were dissolved in dichloromethane (3 mL) and loaded onto a 40 cm x 3 cm inner diameter column containing 60 g of silica gel. The product was eluted with 750 mL of hexanes/ethyl acetate (9:1) and collected in 30 mL fractions. Fractions 11-24 were concentrated using rotary evaporation (35 °C, ~20 mmHg) before drying under reduced pressure (23 °C, 0.5 mmHg, 0.5 h) to obtain 0.5–1.0 g of **2** (5–9% yield).

24. (R)-5-(2-(4-Chlorophenyl)butan-2-yl)-2,2-dimethyl-1,3-dioxane-4,6-dione is stable in excess of six months when stored in the freezer.

25. The product displayed the following physiochemical properties: mp 89–92 °C; ^{1}H NMR (500 MHz, CDCl$_3$) δ: 0.73 (t, J = 7.6 Hz, 3 H), 1.36 (s, 3 H), 1.60 (s, 3 H), 1.65 (s, 3 H), 2.11 (q, J = 7.5 Hz, 2 H), 3.62 (s, 1 H), 7.22–7.24 (m, 2 H), 7.29-7.32 (m, 2 H); ^{13}C NMR (125 MHz, CDCl$_3$) δ: 8.6, 21.4, 27.5, 29.0, 32.7, 45.7, 57.0, 105.1, 128.3, 128.4, 128.7, 132.8, 141.3, 163.8, 164.3; IR (thin film, CH$_2$Cl$_2$) v_{max} (cm^{-1}) 2982, 1742; HRMS (DART) m/z calcd. for C$_{16}$H$_{20}$^{35}ClO$_4$ $^+$: 311.1050, found: 311.1038. [α]$^{20}_D$ = +16.1 (c 0.57, MeOH) (er 97 (R) : 3 (S)). Enantiomeric ratio (er) (97:3–98:2 (R:S)) determined by HPLC (Chiralcel OD-H 0.46 cm × 25 cm column) using 1% 2-propanol and 0.1% trifluoroacetic acid in hexanes as eluent (1 mL/min, prepared by diluting 2-propanol (5 mL) and trifluoroacetic acid (0.5 mL) with hexanes to make a 500 mL solution). t_{R1} = 11.1 min, t_{R2} = 14.3 min. The purity of product (97–99%) is determined by quantitative HPLC (Halo-C18 0.46 cm × 15 cm column) using acetonitrile (A) and water (B, contains 0.2% H$_3$PO$_4$ and 60 mM NH$_4$PF$_6$) as eluent (1.3 mL/min, percentage of A starts from 35%, changes to 45%

in 4 min, increases to 98% in another 3 min and keeps at 98% for another 2 min). $t_R = 7.94$ min. Purity of the quantification standard is determined by NMR assay using dimethyl fumarate as internal standard.

26. The submitters report that under vacuum the pale yellow oil slowly (>24 h) crystallizes to a white solid (mp = 39-41 °C); however, the checkers did not observe crystallization.

27. (R)-3-(4-Chlorophenyl)-3-methylpentanoic acid is stable in excess of six months when stored in the freezer.

28. The product displayed the following physiochemical properties: ^{1}H NMR (400 MHz, CDCl$_3$) δ: 0.66 (t, $J = 7.6$ Hz, 3 H), 1.45 (s, 3 H), 1.64–1.73 (m, 1 H), 1.75–1.83 (m, 1 H), 2.55 (d, $J = 14.4$ Hz, 1 H), 2.69 (d, $J = 14.4$ Hz, 1 H), 7.21 (d, $J = 8.8$ Hz, 2 H), 7.26 (d, $J = 8.8$ Hz, 2 H); ^{13}C NMR (100 MHz, CDCl$_3$) δ: 8.5, 23.8, 35.4, 40.2, 46.3, 127.6, 128.2, 131.7, 144.5, 176.5; IR (thin film, neat) v_{max} (cm^{-1}) 2968 (br), 1701; HRMS (DART) m/z calcd. for C$_{12}$H$_{15}$^{35}ClO$_2$•Na$^+$: 249.0658, found: 249.0649. $[\alpha]^{20}_D = -11.8$ (c 0.67, MeOH) (er 97 (R):3 (S)). Enantiomeric ratio (er) (97:3–98:2 (R:S)) determined by HPLC (Chiralcel AD-H 0.46 cm × 25 cm column) using 1% 2-propanol and 0.1% trifluoroacetic acid in hexanes as eluent (1 mL/min, prepared by diluting 2-propanol (5 mL) and trifluoroacetic acid (0.5 mL) with hexanes to make a 500 mL solution). $t_{R1} = 28.8$ min, $t_{R2} = 30.2$ min. The purity of product (96–98%) is determined by quantitative HPLC (Halo-C18 0.46cm × 15 cm column) using acetonitrile (A) and water (B, contains 0.2% H$_3$PO$_4$ and 60 mM NH$_4$PF$_6$) as eluent (1.3 mL/min, percentage of A starts from 35%, changes to 45% in 4 min, increases to 98% in another 3 min and keeps at 98% for another 2 min). $t_R = 7.01$ min. Purity of the quantification standard is determined by NMR assay using dimethyl fumarate as internal standard.

Working with Hazardous Chemicals

The procedures in *Organic Syntheses* are intended for use only by persons with proper training in experimental organic chemistry. All hazardous materials should be handled using the standard procedures for work with chemicals described in references such as "Prudent Practices in the Laboratory" (The National Academies Press, Washington, D.C., 2011;

the full text can be accessed free of charge at
http://www.nap.edu/catalog.php?record_id=12654). All chemical waste
should be disposed of in accordance with local regulations. For general
guidelines for the management of chemical waste, see Chapter 8 of Prudent
Practices.

In some articles in *Organic Syntheses*, chemical-specific hazards are
highlighted in red "Caution Notes" within a procedure. It is important to
recognize that the absence of a caution note does not imply that no
significant hazards are associated with the chemicals involved in that
procedure. Prior to performing a reaction, a thorough risk assessment
should be carried out that includes a review of the potential hazards
associated with each chemical and experimental operation on the scale that
is planned for the procedure. Guidelines for carrying out a risk assessment
and for analyzing the hazards associated with chemicals can be found in
Chapter 4 of Prudent Practices.

The procedures described in *Organic Syntheses* are provided as
published and are conducted at one's own risk. *Organic Syntheses, Inc.*, its
Editors, and its Board of Directors do not warrant or guarantee the safety of
individuals using these procedures and hereby disclaim any liability for any
injuries or damages claimed to have resulted from or related in any way to
the procedures herein.

Discussion

The catalytic asymmetric formation of all carbon benzylic quaternary
centres is an important goal in organic chemistry as these moieties are
ubiquitous in natural products and pharmaceuticals. Conjugate addition of
organometallic reagents to tri- and tetrasubstituted alkenes activated by
carbonyl, nitro or sulfone groups is an efficient method that can be used to
synthesize these motifs. The conjugate addition of dialkylzinc reagents to 5-
ylidene Meldrum's acid derivatives presented herein is a flexible and
convenient method to perform this transformation because the substrates
are readily prepared by Knoevenagel condensation[2] of various
commercially available aryl ketones and the conjugate addition products
may undergo a wide variety of transformations. In addition, the substrate
scope is broad as the reaction tolerates numerous functional groups.[3,5-8]

Table 1. Asymmetric Conjugate Addition Reactions to Alkylidene Meldrum's Acids[3,7]

Cu(OTf)$_2$ (5 mol %)
R'$_2$Zn (2 equiv)
L (10 mol %)
DME
-40 °C to rt, 48 h

entry	Ar	R	R'	yield (%)	er (R:S)
1	C$_6$H$_5$	Me	Et	95	92:8
2	2-naphthyl	Me	Et	66	97.5:2.5
3	2-furyl	Me	Et	97	95.5:4.5
4	4-MeC$_6$H$_4$	Me	Et	82	94.5:5.5
5	4-PhC$_6$H$_4$	Me	Et	76	97.5:2.5
6	4-ClC$_6$H$_4$	Me	Et	88	97.5:2.5
7	4-BrC$_6$H$_4$	Me	Et	84	96:4
8	4-FC$_6$H$_4$	Me	Et	83	96:4
9	4-(F$_3$C)C$_6$H$_4$	Me	Et	87	96:4
10	4-(BnO)C$_6$H$_4$	Me	Et	75	96.5:3.5
11	3-MeC$_6$H$_4$	Me	Et	93	89:11
12	3-ClC$_6$H$_4$	Me	Et	96	87:13
13	3-(BnO)C$_6$H$_4$	Me	Et	97	89.5:10.5
14	3,4-Cl$_2$C$_6$H$_3$	Me	Et	98	92.5:7.5
15	2-MeC$_6$H$_4$	Me	Et	NR	N/A
16	2-ClC$_6$H$_4$	Me	Et	NR	N/A
17	2-(BnO)C$_6$H$_4$	Me	Et	NR	N/A
18	4-ClC$_6$H$_4$	n-Bu	Et	78	97:3
19	C$_6$H$_5$.	i-Pr	Et	NR	N/A
20	4-ClC$_6$H$_4$	Et	Me	N/A	N/A
21	4-ClC$_6$H$_4$	Me	n-Bu	87	93.5:6.5

As shown in Table 1, asymmetric conjugate addition reactions of 5-(1-arylalk-1-ylidene) Meldrum's acid derivatives occur in good to excellent yield and excellent enantiomeric excess for 3- and 4-substituted aryl groups, although no reaction is obtained for 2-substituted aryl groups due to steric reasons.[3,7] The reaction is successful when R is a primary alkyl group but no addition occurs for secondary alkyl groups; however, both primary and secondary alkyls are tolerated as the nucleophile, although poor results

rganic
yntheses

have been obtained for dimethylzinc under these conditions (it has been shown that dimethylzinc adds effectively under modified conditions).[6]

References

1. Department of Chemistry, University of Waterloo, Waterloo, Ontario, Canada, N2L 3G1, efillion@uwaterloo.ca. This work was supported by the Natural Sciences and Engineering Research Council of Canada (NSERC) and the University of Waterloo. E. B. thanks NSERC for USRA scholarship (2012) and CGS-M scholarship (2014) and the Government of Ontario for OGS scholarship (2013).
2. Baxter, G. J.; Brown, R. F. C. *Aust. J. Chem.* **1975**, *28*, 1551–1557.
3. Fillion, E.; Wilsily, A. *J. Am. Chem. Soc.* **2006**, *128*, 2774–2775.
4. Smith, C.; Mans, D.; RajanBabu, T. V. *Org. Synth.* **2008**, *85*, 238–245.
5. Wilsily, A.; Fillion, E. *Org. Lett.* **2008**, *10*, 2801–2804.
6. Wilsily, A.; Lou, T.; Fillion, E. *Synthesis* **2009**, 2066–2072.
7. Wilsily, A.; Fillion, E. *J. Org. Chem.* **2009**, 74, 8583–8594.
8. Dumas, A. M.; Fillion, E. *Acc. Chem. Res.* **2010**, *43*, 440–454.

Appendix
Chemical Abstracts Nomenclature (Registry Number)

1-(4-Chlorophenyl)ethanone: 4'-chloroacetophenone; (99-91-2)
2,2-Dimethyl-1,3-dioxane-4,6-dione: Meldrum's acid; (2033-24-1)
Titanium (IV) chloride; (7550-45-0)
Pyridine; (110-86-1)
5-(1-(4-Chlorophenyl)ethylidene)-2,2-dimethyl-1,3-dioxane-4,6-dione;
(882161-49-1)
Copper (II) trifluoromethanesulfonate; (34946-82-2)
Diethylzinc; (557-20-0)
(S)-2,2'-Binapthoyl-(R,R)-di(1-phenylethyl)aminoyl-phosphine; (712352-08-4)
1,2-dimethoxyethane; (110-71-4)
(R)-5-(2-(4-Chlorophenyl)butan-2-yl)-2,2-dimethyl-1,3-dioxane-4,6-dione;
(882161-62-8)
(R)-3-(4-Chlorophenyl)-3-methylpentanoic acid; (number not yet assigned)

Organic
Syntheses

Eric Beaton was born in 1988 in Kitchener, ON, Canada. He graduated with a B.Sc. in Chemical Physics in 2012 from the University of Waterloo where his honors thesis was done under the supervision of Professor Eric Fillion. In 2013, he started his M. Sc. studies under the direction of Eric Fillion where he is studying Lewis acid-promoted substitution reactions of benzyl Meldrum's acid derivatives.

Eric Fillion received his undergraduate degree in biochemistry at the Université de Sherbrooke. After completing his M. Sc. in medicinal chemistry at the Université de Montreal with Professor Denis Gravel, he began his doctoral studies at the University of Toronto under the direction of Professor Mark Lautens. From 1998-2000, he was an NSERC post-doctoral fellow in the laboratories of Professor Larry E. Overman at the University of California, Irvine. In August 2000, he joined the Department of Chemistry at the University of Waterloo, where he is currently a Professor of Chemistry.

Li Zhang received his undergraduate degree in chemistry at the Nanjing University in 1994. After completing his M. Sc. in organic chemistry at the Iowa State University with Professor Richard C. Larock in 1999, he joined Chemical Development at Boehringer Ingelheim Pharmaceuticals, Inc., where he is currently a senior scientist. He is the co-author of more than 30 papers and patents in synthetic organic chemistry and pharmaceutical industry.

Palladium-catalyzed Buchwald-Hartwig Amination and Suzuki-Miyaura Cross-coupling Reaction of Aryl Mesylates

Shun Man Wong, Pui Ying Choy, On Ying Yuen, Chau Ming So,* and Fuk Yee Kwong*[1]

State Key Laboratory of Chirosciences and Department of Applied Biology and Chemical Technology, The Hong Kong Polytechnic University, Hung Hom, Kowloon, Hong Kong

Checked by Kyohei Matsushita and Keisuke Suzuki

A. 4-(tert-Butyl)phenyl methanesulfonate + Et₃N, MsCl, DCM → ... [reaction scheme]

B. ... OMs + Me–N(H)–Ph → Pd(OAc)₂, CM-phos / K₂CO₃, t-BuOH, 120 °C → ... [reaction scheme]

C. ... OMs + PhB(OH)₂ → Pd(OAc)₂, CM-phos / K₃PO₄, t-BuOH, 120 °C → ... [reaction scheme]

Procedure

A. *4-(tert-Butyl)phenyl methanesulfonate.* An oven-dried 500-mL, single-necked, round-bottomed flask equipped with a Teflon-coated magnetic stirbar (oval, 25 mm × 7 mm) is charged with 4-*tert*-butylphenol (15.0 g, 100 mmol) (Note 1), dichloromethane (120 mL) (Note 2), and stirring is started. Triethylamine (42 mL, 30.5 g, 300 mmol, 3.0 equiv) (Note 3) is added slowly over 5 min, and the mixture is stirred at room temperature for 5 min. Methanesulfonyl chloride (15.5 mL, 22.94 g, 200 mmol, 2.0 equiv) (Note 4) is added dropwise through a dropping funnel (Note 5), and the mixture is stirred at room temperature under an atmosphere of air until the

4-*tert*-butylphenol has been completely consumed as judged by TLC analysis (Note 6). The reaction mixture is transferred into a 2-L, separatory funnel. The reaction flask is rinsed with ethyl acetate (2 × 30 mL), water (20 mL), and with ethyl acetate (30 mL). All rinses, additional ethyl acetate (500 mL) and water (200 mL) are added to the separatory funnel, the funnel is shaken, and the layers are separated. The organic layer is washed sequentially with water (200 mL), 3.5 M hydrochloric acid (2 × 200 mL) (Note 7), water (200 mL), saturated aqueous sodium hydrogen carbonate (2 × 200 mL), and brine (2 × 200 mL). The organic layer is dried over anhydrous sodium sulfate (5 g) and the mixture is filtered. The sodium sulfate is washed with ethyl acetate (50 mL) and filtered. The combined organic solution is concentrated by rotary evaporation (36 °C, 20 mmHg) to afford a yellow oil. The oil is charged on a column of 100 g of silica gel (5.5-cm diameter × 9-cm packed height) (Note 8) and eluted with 50 mL of ethyl acetate–hexane (1:15). At this point, fraction collection (100-mL fractions) is begun, and elution is continued with 250 mL of ethyl acetate–hexane (1:15) and then 1500 mL of ethyl acetate–hexane (1:9). The eluent containing the product, as identified by TLC analysis, is concentrated by rotary evaporation (33 °C, 36 mmHg) to afford a pale yellow oil. The oil is dissolved in 5 mL of dichloromethane, and 200 mL of hexane is added. The solution is cooled in a freezer (–14 °C) for 8 h, and the resulting crystals are collected by suction filtration on a Büchner funnel, washed with ice-cold hexane (4 × 25 mL), and then transferred to a 100-mL, round-bottomed flask and dried overnight at 0.01 mmHg to provide 4-(*tert*-butyl)phenyl methanesulfonate (18.0–18.7 g, 79–82%) (Note 9) as white shiny crystals.

B. *4-(tert-Butyl)-N-methyl-N-phenylaniline.* An oven-dried 250-mL, resealable Schlenk flask (Note 10) equipped with a Teflon-coated magnetic stirbar (cylindrical, 45 mm × 7 mm) is charged with palladium(II) acetate (0.0393 g, 0.175 mmol, 1.0 mol%) (Note 11) and CM-phos (0.282 g, 0.700 mmol, 4.0 mol%) (Note 12). The Schlenk flask is capped with a rubber septum and then evacuated and backfilled with nitrogen three times. Dichloromethane (18 mL) (Note 13) and triethylamine (1.8 mL) (Note 14) are added via syringe through the septum, and stirring is started. The septum is replaced with a Teflon screwcap, and the Schlenk flask is sealed. The resulting dark orange mixture is placed in a 50 °C pre-heated oil bath with stirring for 5 min to afford a yellow reaction mixture. The flask is removed from the oil bath, allowed to cool to room temperature, and then the volatiles are stripped off at 0.01 mmHg for 1 h to afford a yellow solid. The flask is charged with potassium carbonate (6.05 g, 43.8 mmol, 2.5 equiv)

(Note 15), phenylboronic acid (0.0427 g, 0.350 mmol, 0.2 equiv) (Note 16), and 4-(*tert*-butyl)phenyl methanesulfonate (3.99 g, 17.5 mmol, 1.0 equiv). The Schlenk flask is capped with a rubber septum and then evacuated and backfilled with nitrogen three times. *N*-Methylaniline (2.85 mL, 26.3 mmol) (Note 17) and *tert*-butyl alcohol (70 mL) (Note 18) are added via syringe through

the septum, and stirring is started. The septum is replaced with a Teflon screwcap, and the Schlenk flask is sealed. The reaction flask is stirred at room temperature for 10 min, then placed in a 120 °C pre-heated oil bath (Note 19) with stirring for 24 h (Note 20). The flask is removed from the oil bath, allowed to cool to room temperature, and the mixture is transferred into a 500-mL, separatory funnel. The reaction flask is rinsed with ether (2 × 50 mL), brine (2 × 50 mL), water (50 mL), and again with ether (50 mL). All rinses are added to the separatory funnel, the funnel is shaken, and the layers are separated. The aqueous layer is extracted with ether (2 × 50 mL), and the combined organic extracts are dried over anhydrous sodium sulfate (5 g). The organics are separated by filteration. The sodium sulfate is washed with ether (50 mL) and the ether separated by filtration. The combined organic phase is concentrated by rotary evaporation (40 °C,

12 mmHg) to afford brown oil. The oil is charged on a column of 130 g of silica gel (4.5-cm diameter × 16-cm packed height) (Note 7) and eluted with 130 mL hexane. At this point, fraction collection (100-mL fractions) is begun, and elution is continued with dichloromethane–hexane (1:49) until all the desired product is eluted. The eluent containing the product is concentrated by rotary evaporation (39 °C, 14 mmHg) to afford pale-yellow liquid. The liquid is dried for 1 h at 0.01 mmHg to provide 4-(*tert*-butyl)-*N*-methyl-*N*-phenylaniline (3.60–3.85 g, 86–92%, Note 21) as a pale-yellow liquid.

C. *4-(tert-Butyl)-1,1'-biphenyl*. An oven-dried 250-mL, resealable Schlenk flask (Note 10) equipped with a Teflon-coated magnetic stirbar (cylindrical, 45 mm × 7 mm) is charged with palladium(II) acetate (0.0393 g, 0.175 mmol, 1.0 mol%, Note 11) and CM-phos (0.282 g, 0.700 mmol, 4.0 mol%) (Note 12). The Schlenk flask is capped with a rubber septum and then evacuated and backfilled with nitrogen three times. Dichloromethane (18 mL) (Note 13) and triethylamine (1.8 mL, 1.31 g, 12.9 mmol, 0.74 equiv) (Note 14) are added *via* syringe through the septum, and stirring is started. The septum is replaced with a Teflon screwcap, and the Schlenk flask is sealed. The resulting dark orange mixture is placed in a 50 °C pre-heated oil bath with stirring for 5 min under reflux, affording a yellow reaction mixture. The flask is removed from the oil bath, allowed to cool to room temperature, then the volatiles are stripped off at 0.01 mmHg for 1 h to afford a yellow solid. The flask is charged with potassium phosphate (11.1 g, 52.5 mmol, 3.0 equiv) (Note 22), phenylboronic acid (4.27 g, 35.0 mmol, 2.0 equiv) (Note 16), and 4-(*tert*-butyl)phenyl methanesulfonate (3.99 g, 17.5 mmol, 1.0 equiv). The Schlenk flask is capped with a rubber septum and then evacuated and backfilled with nitrogen three times. *tert*-Butyl alcohol (88 mL) (Note 18) is added *via* syringe through the septum, and stirring is started. The septum is replaced with a Teflon screwcap and the Schlenk flask is sealed. The reaction flask is stirred at room temperature for 10 min, then placed in a 120 °C (Note 19) pre-heated oil bath with for 24 h (Note 20). The flask is removed from the oil bath, allowed to cool to room temperature, and the mixture is transferred into a 500-mL, separatory funnel. The reaction flask is rinsed with ether (2 × 50 mL), brine (2 × 50 mL), water (50 mL), and again with ether (50 mL). All rinses are added to the separatory funnel, the funnel is shaken, and the layers are separated. The aqueous layer is extracted with ether (2 × 50 mL), and the combined organic extracts are dried over anhydrous sodium sulfate (5 g). The mixture is filtered and the sodium sulfate is washed with ether (50 mL).

The organic solution is concentrated by rotary evaporation (40 °C, 15 mmHg) to afford a brown oil. The oil is charged on a column of 100 g of silica gel (4.5-cm diameter × 12-cm packed height) (Note 7) and eluted with 75 mL hexane. At this point, fraction collection (100-mL fractions) is begun, and elution is continued with hexane until all the desired product is eluted. The eluent containing the product is concentrated by rotary evaporation (40 °C, 15 mmHg) to afford a colorless oil. The oil is dried for 2 h at 0.01 mmHg to provide 4-(*tert*-butyl)-1,1'-biphenyl (2.20–2.90 g, 60–79%) (Note 23) as a white solid.

Notes

1. The checkers used 4-*tert*-butylphenol (>98%) as received from Tokyo Chemical Industry Co., Ltd. (TCI). The submitters used 4-*tert*-butylphenol (97%) as received from Acros Organics.
2. The checkers used dichloromethane (anhydrous) as received from KANTO CHEMICAL Co., Inc.. The submitters obtained dichloromethane (GR Grade) from DUKSAN and used it as received.
3. Triethylamine (≥99.5%) was obtained from Aldrich Co., Inc., and used as received.
4. Two equiv of methanesulfonyl chloride were used to ensure the reaction could be completed within 3 h. The checkers obtained methanesulfonyl chloride from Tokyo Chemical Industry Co., Ltd. (TCI) and used it as received. The submitters used methanesulfonyl chloride as obtained from Merck Millipore.
5. Upon addition of methanesulfonyl chloride, vapor and heat were released.
6. Ethyl acetate–dichloromethane–hexane (1:2:7) [SM (R_f = 0.43), product (R_f = 0.50)] Thin layer chromatography was performed on pre-coated TLC-plates (Merck Co., Inc. TLC silica gel 60 F_{254}, Art 5715, 0.25 mm). n-Hexane (≧95% grade) from Kanto Chemical Co. was used.
7. Hydrochloric acid (reagent grade, 35%–37%) was obtained from Koso Chemical Co., Inc. and diluted with water to the desired concentration.
8. The checkers used silica gel 60N (Spherical, neutral, 63–210 μm) obtained from KANTO CHEMICAL Co., Inc. The submitters used silica gel 60 (0.040–0.063 mm, 230–400 mesh ASTM) obtained from Merck Millipore.

9. The analytical data of 4-*(tert*-butyl)phenyl methanesulfonate are as follows mp = 53–54 °C; ^{1}H NMR (600 MHz, CDCl$_3$) δ: 1.32 (s, 9 H), 3.12 (s, 3 H), 7.20 (d, *J* = 8.1 Hz, 2 H), 7.42 (d, *J* = 8.1 Hz, 2 H); ^{13}C NMR (150 MHz, CDCl$_3$) δ: 31.5, 34.8, 37.3, 121.5, 127.1, 147.1, 150.7; MS (EI): *m/z* (relative intensity) 228 (M$^+$, 16), 213 (100), 135 (70), 91 (41), 79 (47); HRMS calcd. for C$_{11}$H$_{16}$O$_3$S$^+$: 228.0820, found 228.0811; Anal. calcd for C$_{11}$H$_{16}$O$_3$S: C, 57.87; H, 7.06; S, 14.04. Found: C, 57.86; H, 6.93; S, 13.88.

10. The resealable Schlenk flask used by the checkers was tailor-made with a Rotaflo stopcock and a 250–300-mL round-bottomed flask. The key bore of the Rotaflo stopcock was 10-mm. In order to have effective stirring during the reaction, Teflon-coated magnetic stirbar (cylindrical, 45 mm × 7 mm) was chosen. The resealable Schlenk flask used by the submitters was tailor-made with a Rotaflo stopcock and a 250-mL, round-bottomed flask. The key bore of the Rotaflo stopcock was 10-mm. In order to have effective stirring during reaction, Teflon-coated magnetic stirbar (cylindrical, 50 mm × 8 mm) was chosen. Photographs of the Schlenk flasks are shown below.

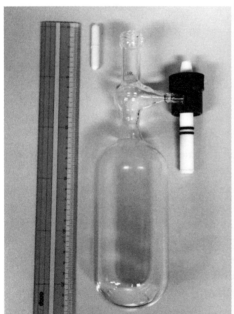

Checker's flask   Submitter's flask

11. Palladium(II) acetate (≥99.9% trace metals basis) was obtained from Aldrich Co., Inc., and used as received.
12. CM-Phos (98%) is available commercially from Strem Chemical, Co. and can be used for the reaction described above. The checkers used material produced in their laboratory based on a procedure that has been submitted for checking by *Organic Syntheses*.
13. The checkers used dichloromethane (anhydrous) as received from KANTO CHEMICAL Co., Inc. The submitters used dichloromethane (GR Grade) obtained from DUKSAN, and the solvent was distilled from calcium hydride under nitrogen prior to use.
14. Triethylamine (≥99.5%) was obtained from Aldrich Co., Inc., and distilled from potassium hydroxide under nitrogen prior to use.
15. Potassium carbonate (ACS reagent, ≥99.0%) was obtained from Aldrich Co., Inc., and used as received.
16. Two equiv of phenylboronic acid were used in order to obtain the desired yield of the coupling reaction. The checkers used phenylboronic acid from Aldrich Co., Inc., and the submitters used phenylboronic acid from Soochiral Chemical Science & Technology Co., Ltd. Phenylboronic acid was recrystallized from dichloromethane and hexane prior to use.
17. *N*-Methylaniline (98%) was obtained from Aldrich Co., Inc., and distilled under nitrogen.
18. *tert*-Butyl alcohol (ACS reagent, ≥99.0%) was obtained from Aldrich Co., Inc., and distilled from sodium under nitrogen.
19. The oil bath temperature was higher when run on a large scale than previously published examples because of the lower internal temperature of the large-scale reactions. A blast shield was used since the boiling point of *tert*-butyl alcohol is 82 °C and the reaction was run at 120 °C under a closed system.
20. Reaction times are longer when run on a large scale than the previously published reaction times for smaller scale reactions. Efficient stirring is very important for these large-scale reactions, otherwise, very sticky and glutinous reaction mixtures form, and lower product yields are obtained.
21. The analytical data of 4-(*tert*-butyl)-*N*-methyl-*N*-phenylaniline are as follows: ^{1}H NMR (600 MHz, CDCl$_3$) δ: 1.32 (s, 9 H), 3.30 (s, 3 H), 6.89 (t, *J* = 7.8 Hz, 1 H), 6.97 (d, *J* = 7.8 Hz, 2 H), 7.00 (brd, *J* = 8.7 Hz, 2 H), 7.24 (brt, 7.8 Hz, 2 H), 7.30 (brd, *J* = 8.7 Hz, 2 H); ^{13}C NMR (150 MHz, CDCl$_3$) δ: 31.6, 34.4, 40.4, 119.1, 120.4, 121.4, 126.2, 129.2, 145.0, 146.5, 149.4; MS (EI): *m/z* (relative intensity) 239 (M$^+$, 31), 224 (100); HRMS

calcd. for $C_{17}H_{21}NH^+$: 240.1752, found 240.1743; Anal. calcd for $C_{17}H_{21}N$: C, 85.30; H, 8.84; N, 5.85. Found: C, 85.21; H, 8.59; N, 5.92. R_f = 0.5, in dichloromethane–hexane (1:19) solvent system.

22. The checkers used potassium phosphate as received from Aldrich Co., Inc. The submitters use potassium phosphate (97%) as obtained from Strem Chemicals, Inc.

23. The analytical data of 4-(*tert*-butyl)-1,1'-biphenyl are as follows: mp = 47–49 °C. ^{1}H NMR (600 MHz, CDCl$_3$) δ: 1.36 (s, 9 H), 7.32 (t, J = 7.2 Hz, 1 H), 7.42 (t, J = 7.2 Hz, 2 H), 7.46 (d, J = 8.4 Hz, 2 H), 7.54 (d, J = 8.4 Hz, 2 H), 7.59 (d, J = 7.2 Hz, 2 H); ^{13}C NMR (150 MHz, CDCl$_3$) δ: 31.5, 34.7, 125.9, 126.9, 127.12, 127.17, 128.8, 138.5, 141.2, 150.4; MS (EI): m/z (relative intensity) 210 (M$^+$, 34), 195 (100), 167 (25); HRMS calcd. for $C_{16}H_{18}^+$: 210.1409, found 210.1402; Anal. calcd for $C_{16}H_{18}$: C, 91.37; H, 8.63. Found: C, 91.06; H, 8.47. R_f = 0.3, in hexane solvent system.

Working with Hazardous Chemicals

The procedures in *Organic Syntheses* are intended for use only by persons with proper training in experimental organic chemistry. All hazardous materials should be handled using the standard procedures for work with chemicals described in references such as "Prudent Practices in the Laboratory" (The National Academies Press, Washington, D.C., 2011; the full text can be accessed free of charge at http://www.nap.edu/catalog.php?record_id=12654). All chemical waste should be disposed of in accordance with local regulations. For general guidelines for the management of chemical waste, see Chapter 8 of Prudent Practices.

In some articles in *Organic Syntheses*, chemical-specific hazards are highlighted in red "Caution Notes" within a procedure. It is important to recognize that the absence of a caution note does not imply that no significant hazards are associated with the chemicals involved in that procedure. Prior to performing a reaction, a thorough risk assessment should be carried out that includes a review of the potential hazards associated with each chemical and experimental operation on the scale that is planned for the procedure. Guidelines for carrying out a risk assessment and for analyzing the hazards associated with chemicals can be found in Chapter 4 of Prudent Practices.

Discussion

Palladium-catalyzed cross-coupling reactions have become an extremely versatile tool in organic synthesis for the construction of carbon-carbon as well as carbon-heteroatom bonds.[2] Notably, it evolves into a synthetically attractive transformation in targeting pharmaceutically useful intermediates.[3] In particular, the Buchwald-Hartwig amination and Suzuki-Miyaura cross-coupling reaction represent an effective method for the construction of $C(sp^2)$-N and $C(sp^2)$-$C(sp^2)$ linkages, respectively.

In 2008, we reported the synthesis and applications of **CM-phos**,[4] which showed excellent catalytic activities towards the first palladium-catalyzed amination and Suzuki coupling reaction of aryl mesylates. Aryl mesylates can be easily accessed from phenols, owing to their lower molecular mass, and cross-coupling reactions utilizing these reagents have the advantage of higher atom economy than those employing the corresponding aryl tosylates.[5] However, their relatively inert leaving-group activity, with respect to tosylates, has limited their applications in coupling reactions. Thus, this area remains highly challenging as mesylates are regarded as the least active sulfonate leaving group.

The examples described here demonstrate that a variety of aryl mesylates with differing substitution patterns, electronic properties, and functional groups can be coupled with differing amines and arylboronic acids in high yield on large scale up to 40 mmol (Tables 1 and 2).

As shown in Tables 3 to 6, the Pd(OAc)$_2$/ **CM-phos** catalyst is highly effective for both amination and Suzuki coupling reactions of aryl mesylates. The examples showed that the catalytic system is applicable to couple a range of aryl/heteroaryl mesylate substrates with amine and arylboronic acid nucleophiles in high yield with low to moderate levels of catalyst loading (0.5-4 mol% Pd). Notably, the reactions require no special techniques and are amenable to large-scale synthesis.

Table 1. Palladium-catalyzed amination of aryl mesylates with amines or N-heterocycles on large scale[a]

Entry	ArOMs	Amine or N-heterocycle	Product	Pd (mol%)	Time (h)	Yield (%)[b]
1[c]	t-Bu—⬡—OMs	Me—N(H)—⬡	t-Bu—⬡—N(Me)—⬡	1	24	72
2	NC—⬡—OMs	Me—N(H)—⬡	NC—⬡—N(Me)—⬡	1	24	83
3	t-Bu—⬡—OMs	indole (H-N)	t-Bu—⬡—N-indole	1	24	74

[a] Reaction condtions: ArOMs (40 mmol), amine or N-heterocycle (60 mmol), K₂CO₃ (100 mmol), Pd(OAc)₂:**CM-phos** = 1:4 (mol% as indicated), PhB(OH)₂ (0.8 mmol), t-BuOH (160 mL), at 110 °C under N₂ for indicated period of time. [b] Yields of isolated product. [c] ArOMs (35 mmol), amine (52.5 mmol), K₂CO₃ (87.5 mmol), Pd(OAc)₂:**CM-phos** = 1:4 (mol% as indicated), PhB(OH)₂ (0.7 mmol), t-BuOH (140 mL), at 120 °C under N₂ for indicated period of time.

Table 2. Palladium-catalyzed Suzuki-Miyaura coupling of aryl/heteroaryl mesylates with arylboronic acids on large scale[a]

Entry	ArOMs or Het-OMs	Ar'B(OH)₂	Product	Pd (mol%)	Time (h)	Yield (%)[b]
1[c]	*t*-Bu—⬡—OMs	(HO)₂B—⬡	*t*-Bu—⬡—⬡	1	24	96
2	Ph—C(O)—⬡—OMs	(HO)₂B—⬡	Ph—C(O)—⬡—⬡	1	12	97
3	NC—⬡—OMs	(HO)₂B—⬡	NC—⬡—⬡	1	17	92
4	quinoline-OMs	(HO)₂B—⬡	quinoline-⬡	1	12	95

[a] Reaction conditions: ArOMs or Het-OMs (40 mmol), Ar'B(OH)₂ (80 mmol), K₃PO₄ (120 mmol), Pd(OAc)₂:**CM-phos** = 1:4 (mol% as indicated), *t*-BuOH (180 mL), at 110 °C under N₂ for indicated period of time. [b] Yields of isolated product. [c] Reaction conditions: ArOMs (35 mmol), Ar'B(OH)₂ (70 mmol), K₃PO₄ (105 mmol), Pd(OAc)₂:**CM-phos** = 1:4 (mol% as indicated), *t*-BuOH (175 mL), at 120 °C under N₂ for indicated period of time.

Table 3. Palladium-catalyzed amination of aryl mesylates[a]

Entry	ArOMs	Amine	Product	Pd (mol%)	Time (h)	Yield (%)[b]
1	t-Bu—⬡—OMs	Me-NH-Ph	t-Bu—⬡—N(Me)Ph	2	4	93
2				0.5	24	96
3	t-Bu—⬡—OMs	Me,Me-H₂N	t-Bu—⬡—NH	1	24	90
4	t-Bu—⬡—OMs	HN(Ph)Ph	t-Bu—⬡—N(Ph)Ph	4	24	80
5	t-Bu—⬡—OMs	HN-morpholine	t-Bu—⬡—N-morpholine	1	18	90
6[c]	t-Bu—⬡—OMs	HN-pyrrolidine	t-Bu—⬡—N-pyrrolidine	2	24	93
7[c]	t-Bu—⬡—OMs	HN(Bn)Bn	t-Bu—⬡—N(Bn)Bn	4	24	81
8	Me,Me—⬡—OMs	Me,Me-H₂N	Me,Me—⬡—NH	1	24	87
9	naphthyl-OMs	HN-pyrrolidine	naphthyl-N-pyrrolidine	2	24	85
10	MeO—⬡—OMs	Me-NH-Ph	MeO—⬡—N(Me)Ph	2	24	78

[a] Reaction condtions: ArOMs (1.0 mmol), amine (1.5 mmol), K_2CO_3 (2.5 mmol), Pd(OAc)$_2$:**CM-phos** = 1:4 (mol% as indicated), PhB(OH)$_2$ (0.04 mmol), t-BuOH (4.0 mL), at 110 °C under N_2 for indicated period of time. [b] Yields of isolated product. [c] K_3PO_4 was used instead of K_2CO_3.

Table 4. Palladium-catalyzed *N*-arylation of nitrogen heterocycles with aryl mesylates[a]

Entry	ArOMs	*N*-Heterocycle	Product	Pd (mol%)	Time (h)	Yield (%)[b]
1	*t*-Bu, OMs	indole (H)	*t*-Bu product	1	24	93
2	Me, Me, OMs	cyclopenta-fused indole (H)	Me, Me product	2	24	84
3	*t*-Bu, OMs	pyrrole (HN)	*t*-Bu product	1	24	80
4[c]	*t*-Bu, OMs	carbazole (H)	*t*-Bu product	2	24	98
5	Ph–C(=O), OMs	indole (H)	Ph–C(=O) product	1	24	88
6	MeO–C(=O), OMs	indole (H)	MeO–C(=O) product	1	24	79

[a] Reaction condtions: ArOMs (1.0 mmol), *N*-heterocycle (1.5 mmol), K$_2$CO$_3$ (2.5 mmol), Pd(OAc)$_2$:**CM-phos** = 1:4 (mol% as indicated), PhB(OH)$_2$ (0.04 mmol), *t*-BuOH (4.0 mL), at 110 °C under N$_2$ for indicated period of time. [b] Yields of isolated product. [c] ArOMs (1.0 mmol), carbazole (1.0 mmol) were used.

Table 5. Palladium-catalyzed Suzuki-Miyaura coupling of aryl mesylates with arylboronic acids[a]

Entry	ArOMs	Ar'B(OH)$_2$	Product	Pd (mol%)	Time (h)	Yield (%)[b]
1				1	19	91
2				0.5	24	88
3				4	8	70
4				2	3	92
5				2	3	81
6				2	3	95
7				2	3	97
8				2	3	89
9				4	8	89

[a] Reaction conditions: ArOMs (1.0 mmol), Ar'B(OH)$_2$ (2.0 mmol), K$_3$PO$_4$ (3.0 mmol), Pd(OAc)$_2$:**CM-phos** = 1:4 (mol% as indicated), t-BuOH (3.0 mL), at 110 °C under N$_2$ for indicated period of time. [b] Yields of isolated product.

Table 6. Palladium-catalyzed Suzuki-Miyaura coupling of heteroaryl mesylates with arylboronic acids[a]

Entry	Het-OMs	Ar'B(OH)₂	Product	Pd (mol%)	Time (h)	Yield (%)[b]
1				2	3	91
2				2	3	77
3				2	3	84
4				2	3	84
5				2	3	85

[a] Reaction conditions: Het-OMs (1.0 mmol), Ar'B(OH)₂ (2.0 mmol), K₃PO₄ (3.0 mmol), Pd(OAc)₂:**CM-phos** = 1:4 (mol% as indicated), t-BuOH (3.0 mL), at 110 °C under N₂ for indicated period of time. [b] Yields of isolated product. Het-OMs = heteroaryl mesylate

References

1. State Key Laboratory of Chirosciences and Department of Applied Biology and Chemical Technology, The Hong Kong Polytechnic University, Hung Hom, Kowloon, Hong Kong. E-mail: chau.ming.so@polyu.edu.hk, fuk-yee.kwong@polyu.edu.hk; Fax: +852-2364-9932. We thank the Research Grants Council of Hong Kong (PolyU153008/14P) for financial support.

2. (a) de Meijere, A.; Diederich, F., Eds. *Metal-Catalyzed Cross-Coupling Reactions*, 2nd ed.; Wiley-VCH: Weinheim, Germany, 2004; Vols. 1-2. (b) Beller, M.; Bolm, C. *Transition Metals for Organic Synthesis : Building Blocks and Fine Chemicals*, 2nd ed.; Wiley-VCH: Weinheim, Germany, 2004; Vols. 1-2. (c) Negishi, E., Ed. *Handbook of Organopalladium for Organic Synthesis*; Wiley-Interscience: Chichester, UK, 2002; Vols. 1-2. (d) Hassan, J.; Sévignon, M.; Gozzi, C.; Schulz, E.; Lemaire, M. *Chem. Rev.* **2002**, *102*, 1359. (e) Tsuji, J. *Palladium Reagents and Catalysts*, 2nd ed.; Wiley-Interscience: Chichester, UK, 2004 (f) Yin, L.; Liebscher, J., *Chem. Rev.* **2007**, *107*, 133. (g) Corbet, J.-P.; Mignani, G. *Chem. Rev.* **2006**, *106*, 2651. (h) Roglans, A.; Pla-Quintana, A.; Moreno-Mañas, M. *Chem. Rev.* **2006**, *106*, 4622.

3. (a) King, A. O.; Yasuda, N. In *Organometallics in process chemistry*; Larsen, R. D., Ed.; Springer-Verlag: Berlin Heidelberg, 2004; pp. 205-245. (b) Suzuki, A. In *Modern arene chemistry*; Astruc, D., Ed.; Wiley-VCH: Weinheim, Germany, 2002; pp. 53-106. (c) Miyaura, N. *J. Organomet. Chem.* **2002**, *653*, 54. (d) Miyaura, N. *Top. Curr. Chem.* **2002**, *219*, 11. (e) Muci, A. R.; Buchwald, S. L. *Top. Curr. Chem.* **2002**, *219*, 211.

4. (a) So, C. M.; Zhou, Z.; Lau, C. P.; Kwong, F. Y. *Angew. Chem. Int. Ed.* **2008**, *47*, 6402. (b) So, C. M.; Lau, C. P.; Kwong, F. Y. *Angew. Chem. Int. Ed.* **2008**, *47*, 8059. (c) So, C. M.; Kwong, F. Y. *Chem. Soc. Rev.* **2011**, *40*, 4963.

5. Trost, B. M. *Angew. Chem., Int. Ed. Engl.* **1995**, *34*, 259.

Appendix
Chemical Abstracts Nomenclature (Registry Number)

4-*tert*-Butylphenol; (98-54-4)
Triethylamine; (121-44-8)
Methanesulfonyl chloride; (124-63-0)
Palladium(II) acetate; (3375-31-3)
CM-phos: 2-(2-(Dicyclohexylphosphino)phenyl)-1-methyl-1*H*-indole, (1067883-58-2)
Phenylboronic acid; (98-80-6)
N-Methylaniline; (100-61-8)
tert-Butanol; (75-65-0)
Potassium phosphate; (7778-77-0)

Organic
Syntheses

Shun Man Wong received his B.Sc. in Chemical Technology from The Hong Kong Polytechnic University in 2010. He pursued his postgraduate study at the same university and obtained his Ph.D. degree in 2014. He is currently a research associate under the supervision of Prof. Fuk Yee Kwong, researching the synthesis of new heterocyclic phosphine ligands and their potential applications.

Pui Ying Choy received her B.Sc. in Chemical Technology in The Hong Kong Polytechnic University in 2010. She pursued her postgraduate study at the same university and obtained her Ph.D. degree in 2014. She is currently a research associate under the supervision of Prof. Fuk Yee Kwong, researching the synthesis of new heterocyclic phosphine ligands and their potential applications in transition-metal catalysis.

On Ying Yuen received her B.Sc. (1st class honors) in Chemical Technology from the Hong Kong Polytechnic University in 2011. Currently, she is pursuing her Ph.D. under the guidance of Prof. Fuk Yee Kwong. Her main research focuses on palladium-catalyzed direct functionalization of aromatics: process and catalyst design.

Organic
Syntheses

Chau Ming So is currently a Visiting Assistant Professor in the Department of Applied Biology and Chemical Technology of The Hong Kong Polytechnic University. He received his B.Sc. (1st class honor) from PolyU in 2006. He pursued his postgraduate study at the same university and obtained his Ph.D. degree in 2010. He received the Hong Kong Young Scientist Award in the same year. Moreover, he was the winner of Eli Lilly the Best Thesis Award (1st Prize). In 2012-2013, he moved to Institute of Materials Research and Engineering (IMRE) as postdoctoral fellow in Prof. Tamio Hayashi's research group.

Fuk Yee (Michael) Kwong is currently a Professor of the Department of Applied Biology and Chemical Technology at The Hong Kong Polytechnic University. He received his B.Sc. in 1996, and completed his Ph.D. at The Chinese University of Hong Kong in 2000 under the supervision of Professor Kin Shing Chan. In 2001–2003, he was at the Massachusetts Institute of Technology (MIT), USA, as a Croucher Foundation postdoctoral fellow in Professor Stephen L. Buchwald's research group. Kwong's research interests are new cross-coupling methodologies, carbon–hydrogen bond functionalization, and catalytic enantioselective transformations.

Kyohei Matsushita was born in 1989 in Chiba, Japan. He received his B.Sc. degree in 2012 at Chuo University under the supervision of Prof. Shin-ichi Fukuzawa. In the same year, he joined the research group of Prof. Keisuke Suzuki at Tokyo Institute of Technology. In 2014, he received his M.Sc., and currently, is pursuing his Ph.D.

Synthesis of Optically Active 1,2,3,4-Tetrahydroquinolines *via* Asymmetric Hydrogenation Using Iridium-Diamine Catalyst

Fei Chen, Zi-Yuan Ding, Yan-Mei He, and Qing-Hua Fan[*1]

Beijing National Laboratory for Molecular Sciences, CAS Key Laboratory of Molecular Recognition and Function, Institute of Chemistry, Chinese Academy of Sciences (CAS), Beijing 100190, P. R. China.

Checked by Douglass C. Duquette and Brian Stoltz

Procedure

A. *(1S,2S)-(–)-N-4-(Trifluoromethyl)benzenesulfonylated-DPEN* *((S,S)-1)*. An oven-dried 100-mL three-necked, round-bottomed flask equipped with an oven-dried 50-mL dropping funnel in the middle neck, an argon line attached to a glass gas adaptor in one of the side necks, a glass stopper in the other side neck, and an octagonal magnetic stir bar (6 mm x 25 mm) is

Org. Synth. **2015**, *92*, 213-226

213

DOI: 10.15227/orgsyn.092.0213

Published on the Web 7/8/2015

© 2015 Organic Syntheses, Inc.

assembled hot under an atmosphere of argon and cooled to room temperature. The glass stopper is removed under positive pressure of argon, and the flask is charged with (1S,2S)-(−)-1,2-diphenyl-1,2-ethanediamine (DPEN) (Note 1) (1.06 g, 5.00 mmol, 1.00 equiv), triethylamine (Note 2) (1.30 mL, 10.00 mmol, 2.00 equiv) and dichloromethane (Note 3) (40 mL), after which point the glass stopper is replaced with a rubber septum pierced with a thermometer. The resulting solution is cooled to 1 °C (internal temperature) with an ice bath. A solution of 4-(trifluoromethyl)benzenesulfonyl chloride (Note 4) (1.22 g, 5.00 mmol, 1.00 equiv) in dichloromethane (20 mL) is added dropwise from the dropping funnel over 30 min into the reaction mixture, reaching a maximum internal temperature of 3 °C. After the completion of addition, the resulting mixture is warmed to room temperature (20 °C) and stirred for additional 6 h, at which point the reaction is white and heterogeneous. The reaction mixture is transferred to a 125-mL separatory funnel, washed with water (20 mL) and saturated aqueous sodium chloride (Note 5) solution (20 mL), and then dried over anhydrous sodium sulfate (Note 6) (10 g) for 30 min. After filtration through a medium-porosity fritted funnel, the organic solvent is removed under reduced pressure (40 °C, 30 mmHg) by rotary evaporation to give a white solid (1.90–1.99 g). The resulting crude product is dissolved in ethyl acetate (Note 7) (22 mL) under refluxing conditions, and then petroleum ether (Note 8) (6 mL) is added dropwise until the solution becomes slightly turbid. The mixture is cooled initially to room temperature to provide white crystals, and then left standing at −20 °C in refrigerator for another 4 h. The crystalline product is isolated by filtration through a Büchner funnel (Φ 60 mm), washed with cooled ethyl acetate/petroleum ether solution (1/1, v/v, 6 mL; −20 °C), and then dried in vacuo (50 °C at 6 mmHg for 24 h) to provide 1.71–1.75 g (81–84% yield) of (S,S)-1 (Note 9) as white crystals.

    B.   *[IrOTf(Cp*)((S,S)-N-4-(Trifluoromethyl)benzenesulfonylated-DPEN)]* *((S,S)-2)*. An oven-dried 50-mL round-bottomed flask is equipped with an octagonal magnetic stir bar (6 mm x 25 mm) and a rubber septum fitted with an argon inlet needle. The flask is flushed with argon and charged with (pentamethylcyclopentadienyl)iridium(III) dichloride dimer ([IrCp*Cl₂]₂) (Note 10) (0.796 g, 1.00 mmol, 1.00 equiv) and (S,S)-1 (0.840 g, 2.00 mmol, 2.00 equiv). A solution of triethylamine (0.58 mL, 4.00 mmol, 4.00 equiv) in dichloromethane (20 mL) is added to the flask through the septum by syringe. The resulting deep red-orange mixture is stirred at 23 °C for 12 h, and the reaction is monitored by TLC analysis (Note 11).[3] After the reaction is complete, the organic solvent and the excess triethylamine are removed under reduced pressure (40 °C, 30 mmHg) by rotary evaporation. The resulting residue (1.85 g yellow solid) is dissolved in dichloromethane (6 mL), and subjected to flash column chromatography over silica gel (20 g, dichloromethane/methanol, 100/1, v/v, 300 mL) (Notes 12 and 13) to remove the triethylamine hydrochloride salt, providing the crude product (1.66–1.71 g) as a light yellow solid.

An oven-dried 50-mL round-bottomed flask is equipped with an octagonal magnetic stir bar (6 mm x 25 mm) and a rubber septum fitted with an argon inlet needle. The flask is flushed with argon and charged with the crude product (1.66–1.71 g) and silver trifluoromethanesulfonate (Note 14) (0.514 g, 2.00 mmol, 1.00 equiv). Dichloromethane (20 mL) is added to the flask through the septum by syringe under argon atmosphere. The resulting mixture is stirred at 25 °C for 2 h. After separation of the silver chloride precipitate by filtration through a Büchner funnel (Φ 60 mm) packed with Celite (Note 15) (3 g), the organic solvent is removed under reduced pressure (40 °C, 30 mmHg) by rotary evaporation to give 1.78 g of dark red solid. The resulting crude product is dissolved in ethyl acetate (6 mL) at 25 °C, and then petroleum ether (3 mL) is added dropwise until the solution becomes slightly turbid. The mixture is left standing at room temperature for 2 h, allowing slow evaporation of the solvent. When little crystal seeds appear, shaking the flask achieves complete crystallization. The crystalline product is isolated by filtration through a Büchner funnel (Φ 60 mm), washed with ethyl acetate/petroleum ether solution (2/1, v/v, 3 mL), and dried in *vacuo* (50 °C and 5 mmHg for 3 days) (Note 16) to provide 1.55–1.61 g (83–86% total yield) of (*S,S*)-**2** (Note 17) as a dark red solid.

C. *2-Methyl-1,2,3,4-tetrahydroquinoline* ((*S*)-**4a**). A 50-mL glass tube equipped with an octagonal magnetic stir bar (6 mm x 25 mm) is charged with 2-methylquinoline (Note 18) (10.00 g, 69.90 mmol, 1.00 equiv) and (*S,S*)-**2** (0.125 g, 0.133 mmol, 0.002 equiv). A solution of trifluoroacetic acid (Note 19) (0.797 g, 6.99 mmol, 0.10 equiv) in undegassed methanol (Note 20) (7 mL) is added to the tube under air atmosphere. The glass tube is then placed into a stainless steel autoclave. The autoclave is closed and connected to a hydrogen source from a cylinder (Note 21). After the autoclave is purged with hydrogen three times *via* pressurization and depressurization, the autoclave is pressurized with 50 atm of hydrogen (Note 22). The mixture is stirred at 18 °C (room temperature) for 15 h. During this period of time, the hydrogen pressure is kept at 50 atm by occasional introduction of hydrogen from the cylinder. When the consumption of hydrogen ceases, the gas-inlet tube is disconnected. After the excess hydrogen gas is carefully released by opening the valve, the autoclave is opened. The reddish reaction mixture is transferred into a 125-mL Erlenmeyer flask and diluted with dichloromethane (30 mL) and saturated aqueous sodium carbonate (Note 23) solution (30 mL). The resulting solution is stirred for 20 min and then transferred to a 125-mL separatory funnel. The organic layer is separated, and the aqueous layer is extracted with dichloromethane (30 mL). The combined organic layer is dried over anhydrous sodium sulfate (10.0 g) for 30 min and concentrated under reduced pressure (40 °C, 40 mmHg) by rotary evaporation to afford the crude product. Purification is performed by column chromatography over silica (Note 24) to give 9.84–9.90 g of nearly pure product as a yellow oil. This oil is transferred to a 25 mL round-bottomed flask and distilled using a Kugelrohr apparatus (oven temperature 95 °C, 5–7 mmHg) to afford 9.19–9.28 (91–92% yield) of (*S*)-**4a** (Note 25) as a pale yellow oil. The enantiomeric excess of (*S*)-**4a** is 94% determined by chiral HPLC with a chiral OJ-H column (Notes 26 and 27).

Notes

1. (1*S*,2*S*)-(-)-1,2-Diphenyl-1,2-ethanediamine (98+%) was purchased from Alfa Aesar Chemical Company, Inc., and used without further purification.

2. Triethylamine (99%) was purchased from Alfa Aesar Chemical Company, Inc., and used without further purification.

3. Dichloromethane (99.7+%) was purchased from Alfa Aesar Chemical Company, Inc., and used without further purification.

4. 4-(Trifluoromethyl)benzenesulfonyl chloride (98%) was purchased from Alfa Aesar Chemical Company, Inc., and used without further purification.

5. Sodium chloride (99%) was purchased from Alfa Aesar Chemical Company, Inc., and used without further purification.

6. Sodium sulfate (99%) was purchased from Alfa Aesar Chemical Company, Inc., and used without further purification.

7. Ethyl acetate (99.5+%) was purchased from Alfa Aesar Chemical Company, Inc., and used without further purification.

8. Petroleum ether 40/60 was purchased from Alfa Aesar Chemical Company, Inc., and used without further purification.

9. (S,S)-**1**: White crystals; mp 208–211 °C; $[\alpha]_D^{20}$ = +21.4 (c 0.5, chloroform), [Lit.[2] $[\alpha]_D^{26}$ = +22.4 (c 0.5, chloroform), 100% ee for (1S,2S) enantiomer]; IR (thin film, NaCl) 3337, 3289, 3085, 2854, 1597, 1454, 1403, 1330, 1162, 1150, 1127, 1108, 1098, 1055 cm^{-1}; ^{1}H NMR (500 MHz, DMSO-d_6) δ: 3.98 (d, J = 7.1 Hz, 1 H), 4.16 (br, s, 3 H), 4.37 (d, J = 7.1 Hz, 1 H), 6.94 (d, J = 4.1 Hz, 5 H), 7.00–7.21 (m, 5 H), 7.57 (s, 4 H); ^{13}C NMR (126 MHz, DMSO-d_6) δ: 60.4, 65.0, 122.4, 124.5, 125.6, 125.6, 125.6, 125.7, 126.4, 126.6, 126.7, 127.0, 127.2, 127.3, 127.4, 127.6, 127.8, 130.9, 131.1, 131.4, 131.7, 139.2, 142.4, 144.9; Anal. Calcd. for $C_{21}H_{19}F_3N_2O_2S$ C, 59.99; H, 4.55; N, 6.66; Found C, 60.23; H, 4.73; N, 6.75.

10. (Pentamethylcyclopentadienyl)iridium(III) dichloride dimer (99%) was purchased from Alfa Aesar Chemical Company, Inc., and used without further purification.

11. Checkers performed thin layer chromatography (TLC) using E. Merck silica gel 60 F254 precoated plates (0.25 mm) eluting with dichloromethane/methanol (20/1, v/v), and visualized by a 254-nm UV lamp. The observed R_f value is 0.48 for (S,S)-**1**. Submitters report thin layer chromatography (TLC) performed on precoated silica gel plates (SGF254, 0.2 mm±0.03 mm) purchased from Yantai Chemical Industry Research Institute eluting with dichloromethane/methanol (20/1, v/v), and visualized by a 254-nm UV lamp. The observed R_f value is 0.53 for (S,S)-**1**.

12. Checkers used Silicycle SiliaFlash® P60 Academic Silica gel (particle size 40–63 nm). Submitters used silica gel 60 (zcx-3 II, 200–300 mesh) purchased from Qingdao Haiyang Chemical Company, Ltd.

13. Flash column chromatography was performed on a silica gel column (3.5 cm width x 10.0 cm length) using 20 g of 40–63 nm particle size silica with 500 mL dichloromethane/methanol (100/1, v/v) as eluent. The crude product was collected in fractions 3–9 (50 mL each), which were combined and concentrated by rotary evaporation under reduced pressure (40 °C, 30 mmHg) to provide an orange solid.

14. Silver trifluoromethanesulfonate (99%) was purchased from Alfa Aesar Chemical Company, Inc., and used without further purification.

15. Celite was purchased from Alfa Aesar Chemical Company, Inc., and used without further purification.

16. The organometallic complex was isolated as a 2:1 complex with ethyl acetate, even after drying under the described conditions. In a separate run, the sample was dried on a diffusion pump at 23 °C, 8 x 10^{-3} mmHg for 5 d without any change in the amount of EtOAc present by ^{1}H NMR. A spectrum free of EtOAc could be obtained by dissolving 20 mg of (S,S)-2 in 5 mL CDCl$_3$ and removing the solvent under reduced pressure (40 °C, 30 mmHg) by rotary evaporation, repeating this process a total of four times.

17. (S,S)-2: Dark red solid; mp 152–157 °C (dec); $[\alpha]_D^{20}$ = +59.9 (c 1.9, chloroform); IR (thin film, NaCl) 3207, 3101, 2922, 1719, 1577, 1496, 1451, 1404, 1323, 1276, 1245, 1157, 1106, 1089, 1062, 1030 cm^{-1}; ^{1}H NMR (500 MHz, CDCl$_3$) δ: 1.26 (t, J = 7.1 Hz, 1.5 H), 1.87 (s, 15 H), 2.04 (s, 1.5 H), 4.12 (q, J = 7.1 Hz, 1 H), 4.25 (s, 1 H), 4.79 (s, 1 H), 5.33 (d, J = 13.5 Hz, 1 H), 6.21 (d, J = 13.1 Hz, 1 H), 6.86–6.97 (m, 2 H), 7.07–7.23 (m, 8 H), 7.29–7.40 (m, 4 H); ^{13}C NMR (126 MHz, CDCl$_3$) δ: 10.5, 14.3, 21.2, 60.5, 91.6, 110.1, 119.0, 121.6, 122.2, 124.3, 125.6, 126.6, 126.7, 127.8, 128.5, 128.6, 128.8, 128.9, 133.5, 133.8, 136.3, 137.8, 142.1, 171.3; Anal. Calcd. for C$_{32}$H$_{33}$F$_6$IrN$_2$O$_5$S$_2$•1/2(C$_4$H$_8$O$_2$) C, 43.44; H, 3.97; N, 2.98; Found C, 43.22; H, 4.18; N, 2.81.

18. 2-Methyl quinoline (97+%) was purchased from Alfa Aesar Chemical Company, Inc., and used without further purification.

19. Trifluoroacetic acid (99%) was purchased from Alfa Aesar Chemical Company, Inc., and used without further purification.

20. Methanol (99.8+%) was purchased from Alfa Aesar Chemical Company, Inc., and used without further purification.

21. The purity of hydrogen gas used by checkers was 99.999%. The purity of hydrogen gas used by submitters was 99.99%.
22. The gas-inlet tube was attached to the autoclave, with a three-way valve. After pressurization of the autoclave, the valve was opened to release extra hydrogen pressure and then turned to repressurize. This procedure was repeated three times.
23. Sodium carbonate (99%) was purchased from Alfa Aesar Chemical Company, Inc., and used without further purification.
24. Column chromatography was performed on a silica gel column (3.5 cm width x 15.0 cm length) using 30 g of 40–63 nm particle size silica gel with 300 mL petroleum ether/triethylamine (95/5, v/v) as eluent. The product was collected in fractions 2-6 (50 mL each), which were combined and concentrated by rotary evaporation under reduced pressure (40 °C, 30 mmHg) to provide a light yellow oil. The submitters report that the resulting product was dried in a 40 °C oil bath at 5 mmHg.
25. (S)-4a: Light yellow oil; bp 89 °C (5 mm Hg); IR (thin film, NaCl) 3393, 2961, 2924, 2843, 1727, 1608, 1583, 1491, 1309, 1277 cm^{-1}; 94% ee, $[\alpha]_D^{25}$ = –80.4 (c 0.20, chloroform); ^{1}H NMR (500 MHz, CDCl$_3$) δ: 1.22 (d, J = 6.3 Hz, 3 H), 1.50–1.70 (m, 1 H), 1.90–1.97 (m, 1 H), 2.70–2.77 (m, 1 H), 2.81–2.90 (m, 1 H), 3.34–3.45 (m, 1 H), 3.80 (br, s, 1 H), 6.49 (d, J = 7.8 Hz, 1H), 6.62 (t, J = 7.4 Hz, 1 H), 6.97 (t, J = 6.5 Hz, 2 H); ^{13}C NMR (126 MHz, CDCl$_3$) δ: 22.7, 26.7, 30.3, 47.4, 114.2, 117.2, 121.3, 126.8, 129.4, 144.8; Anal. Calcd. for C$_{10}$H$_{13}$N C, 81.59; H, 8.90; N, 9.51; Found C, 81.19; H, 8.98; N, 9.39.
26. The checkers performed HPLC analysis on an Agilent 1100 series liquid chromatograph with a chiral column (OJ-H, eluent: Hexane/i-PrOH = 95/5, v/v, detector: 254 nm, flow rate: 1.2 mL/min), major isomer (S): t_{R1} = 12.8 min, minor isomer (R): t_{R2} = 14.5 min. The submitters report HPLC analysis performed on a Varian Prostar 210 liquid chromatograph with a chiral column (OJ-H, eluent: Hexane/i-PrOH = 90/10, v/v, detector: 254 nm, flow rate: 1.0 mL/min), major isomer (S): t_{R1} = 10.1 min, minor isomer (R): t_{R2} = 11.3 min.
27. The racemic product was prepared according to the published method.[4]

Working with Hazardous Chemicals

The procedures in *Organic Syntheses* are intended for use only by persons with proper training in experimental organic chemistry. All hazardous materials should be handled using the standard procedures for work with chemicals described in references such as "Prudent Practices in the Laboratory" (The National Academies Press, Washington, D.C., 2011; the full text can be accessed free of charge at http://www.nap.edu/catalog.php?record_id=12654). All chemical waste should be disposed of in accordance with local regulations. For general guidelines for the management of chemical waste, see Chapter 8 of Prudent Practices.

In some articles in *Organic Syntheses*, chemical-specific hazards are highlighted in red "Caution Notes" within a procedure. It is important to recognize that the absence of a caution note does not imply that no significant hazards are associated with the chemicals involved in that procedure. Prior to performing a reaction, a thorough risk assessment should be carried out that includes a review of the potential hazards associated with each chemical and experimental operation on the scale that is planned for the procedure. Guidelines for carrying out a risk assessment and for analyzing the hazards associated with chemicals can be found in Chapter 4 of Prudent Practices.

The procedures described in *Organic Syntheses* are provided as published and are conducted at one's own risk. *Organic Syntheses, Inc.*, its Editors, and its Board of Directors do not warrant or guarantee the safety of individuals using these procedures and hereby disclaim any liability for any injuries or damages claimed to have resulted from or related in any way to the procedures herein.

Discussion

Optically pure tetrahydroquinoline derivatives are important organic synthetic intermediates and building blocks for the stereoselective synthesis of biologically active compounds.[5] The preparation of these chiral tetrahydroquinoline derivatives by the transition metal-catalyzed asymmetric hydrogenation of quinolines is one of the most straightforward and convenient methods. Recently, a number of iridium-phosphine complexes have been reported to be effective in the asymmetric hydrogenation of quinolines since the first example reported by Zhou and co-workers.[6] However, a frequently encountered problem associated with the use of phosphine-containing catalysts is the air-sensitivity. In addition, most of these reported catalytic systems suffered from low catalyst efficiency.

In comparison with chiral phosphorus ligands, chiral diamine ligands are more readily available and air-stable. Their transition metal (Ru, Rh, and Ir) complexes have been extensively studied in the asymmetric transfer hydrogenation of aromatic ketones and imines.[7] However, they are long neglected in the hydrogenation of unsaturated compounds.[8,9] The procedure described herein provides a highly efficient method for the preparation of the desired chiral 2-substituted 1,2,3,4-tetrahydroquinolines in high enantiomeric excess *via* asymmetric hydrogenation using readily available and air stable chiral cationic Cp*Ir(OTf)(*N*-sulfonylated diamine) complexes[10] as catalysts. In addition, the hydrogenation can be carried out in undegassed methanol and with no need for inert gas protection throughout the entire operation.

Hydrogenation of 2-methylquinoline (10.00 g, 69.90 mmol) catalyzed by Ir-catalyst (*S,S*)-**2** proceeds smoothly with a substrate-to-catalyst molar ratio as high as 500:1 in undegassed methanol, affording 2-methyl-1,2,3,4-tetrahydroquinoline in 96% isolated yield with 96% ee (Entry 1). In addition, the present method has been successfully applied to the asymmetric hydrogenation of a series of 2-substituted and 2,6-disubstituted quinoline derivatives with catalyst (*S,S*)-**2**. Excellent enantioselectivities and reactivities have been obtained in all cases except for 2-phenyl quinoline (see Table 1).

Table 1. Asymmetric Hydrogenation of 2-Substituted Quinoline Derivatives Catalyzed by (S,S)-2[a]

R^1 ⟶ (quinoline, 3a-n) + H$_2$ (50 atm) $\xrightarrow[\text{undegassed methanol}]{0.2 \text{ mol\% } (S,S)\text{-}2}$ R^1 ⟶ (tetrahydroquinoline, 4a-n)

Entry	Compound	R^1	R^2	Yield (%)[c]	Ee (%)[d]
1[b]	4a	H	Me	96	96 (S)[e]
2	4b	H	Et	98	98
3	4c	H	n-Pr	98	96
4	4d	H	n-Bu	97	97
5	4e	H	n-Pentyl	98	96
6	4f	H	(phenethyl group)	97	97
7	4g	H	(methylenedioxyphenyl ethyl group)	97	97
8	4h	H	(dimethoxyphenyl ethyl group, MeO, MeO)	96	96
9	4i	H	(hydroxyl group, OH, Me, Me)	97	99
10	4j	H	(cyclohexanol group, OH)	97	99
11	4k	MeO	Me	97	97
12	4l	Me	Me	95	97
13	4m	F	Me	98	94
14	4n	H	Ph	90	79

[a] Reaction conditions: 0.75 mmol substrate in 1 mL undegassed MeOH, 0.2 mol% (S,S)-2, 10 mol% TFA, 50 atm H$_2$, 15 °C, 24-48 h. All manipulations were conducted in air, and the autoclave was purged with H$_2$ for three times before reaction. [b] Reaction conditions: (10.0 g, 69.9 mmol) 2-methyl quinoline in 7 mL undegassed MeOH, 0.2 mol% (S,S)-2, 10 mol% TFA, 50 atm H$_2$, 15 °C, 15 h. [c] Isolated yield (without distillation). [d] Determined by chiral HPLC analysis. [e] Absolute configuration was determined by comparison of optical rotation with literature data.[6]

References

1. fanqh@iccas.ac.cn; Beijing National Laboratory for Molecular Sciences, CAS Key Laboratory of Molecular Recognition and Function, Institute of Chemistry, Chinese Academy of Sciences (CAS), Beijing 100190, P. R. China. We thank the financial supports from the National Natural Science Foundation of China (21232008) and the National Basic Research Program of China (2010CB833300).
2. Martins, J. E. D.; Wills, M. *Tetrahedron* **2009**, *65*, 5782–5786.
3. Murata, K.; Ikariya, T. *J. Org. Chem.* **1999**, *64*, 2186–2187.
4. Nose, A.; Kudo, T. *Chem. Pharm. Bull.* **1984**, *6*, 2421–2425.
5. (a) Keay, J. G. in Comprehensive Organic Synthesis; ed. Trost, B. M. and Fleming, I., Pergamon: Oxford, **1991**, vol. *8*, p 579. (b) Kartritzky, A. R.; Rachwal, S.; Rachwal, B. *Tetrahedron* **1996**, *52*, 1503–1507. (c) Comprehensive Natural Products Chemistry, Barton, D. H.; Nakanishi, K.; Meth-Cohn, O., Elsevier, Oxford, **1999**, vol. *1–9*. (d) Sridharan, V.; Suryavanshi, P. A.; Menéndez, J. C. *Chem. Rev.* **2011**, *111*, 7157–7259.
6. (a) Wang, W. B.; Lu, S. M.; Yang, P. Y.; Han, X. W.; Zhou, Y. G. *J. Am. Chem. Soc.* **2003**, *125*, 10536–10537. (b) Wang, D. S.; Chen, Q. A.; Lu, S. M.; Zhou, Y. G. *Chem. Rev.* **2012**, *112*, 2557–2590.
7. (a) Noyori, R.; Hashiguchi, S. *Acc. Chem. Res.* **1997**, *30*, 97–102. (b) Hashiguchi, S.; Fujii, A.; Takehara, J.; Ikariya, T.; Noyori, R. *J. Am. Chem. Soc.* **1995**, *117*, 7562–7563. (c) Uematsu, N.; Fujii, A.; Hashiguchi, S.; Ikariya, T.; Noyori, R. *J. Am. Chem. Soc.* **1996**, *118*, 4916–4917.
8. (a) Ohkuma, T.; Utsumi, N.; Tsutsumi, K.; Murata, K.; Sandoval, C. A.; Noyori, R. *J. Am. Chem. Soc.* **2006**, *128*, 8724–8725. (b) Ohkuma, T.; Utsumi, N.; Watanabe, M.; Tsutsumi, K.; Arai, N.; Murata, K. *Org. Lett.* **2007**, *9*, 2565–2567. (c) Zhou, H. F.; Li, Z. W.; Wang, Z. J.; Wang, T. L.; Xu, L. J.; He, Y. M.; Fan, Q. H.; Pan, J.; Gu, L. Q.; Chan, A. S. C. *Angew. Chem. Int. Ed.* **2008**, *47*, 8464–8467. (d) Wang, Z. J.; Zhou, H. F.; Wang, T. L.; He, Y. M.; Fan, Q. H. *Green Chem.* **2009**, *11*, 767–769. (e) Wang, T. L.; Zhuo, L. G.; Li, Z. W.; Chen, F.; Ding, Z. Y.; He, Y. M.; Fan, Q. H.; Xiang, J. F.; Yu, Z. X.; Chan, A. S. C. *J. Am. Chem. Soc.* **2011**, *133*, 9878–9891.
9. Li, Z. W.; Wang, T. L.; He, Y. M.; Wang, Z. J.; Fan, Q. H.; Pan, J.; Xu, L. J. *Org. Lett.* **2008**, *10*, 5265–5268.
10. (a) Heiden, Z. M.; Rauchfuss, T. B. *J. Am. Chem. Soc.* **2007**, *129*, 14303–14304. (b) Arita, S.; Koike, T.; Kayaki, Y.; Ikariya, T. *Angew. Chem. Int. Ed.* **2008**, *47*, 2447–2449.

Appendix
Chemical Abstracts Nomenclature (Registry Number)

4-(Trifluoromethyl)benzenesulfonyl chloride; (2991-42-6)
(1S,2S)-(–)-1,2-Diphenyl-1,2-ethanediamine ; (29841-69-8)
Triethylamine; (121-44-8)
(Pentamethylcyclopentadienyl)iridium(III) Dichloride Dimer; (12354-84-6)
Silver trifluoromethanesulfonate; (2923-28-6)
2-Methyl quinoline: Quinaldine; (91-63-4)
Trifluoroacetic acid; (76-05-1)

Fei Chen was born in 1984 in Jiangxi Province, China. He received his B. S. degree in Chemistry in 2006 from Northeast Normal University, Changchun. He obtained a Ph. D. degree in 2012 from Institute of Chemistry of the Chinese Academy of Sciences under the supervision of Professor Qing-Hua Fan. Now he is an assistant professor working in the group of Professor Qing-Hua Fan in the same institute. His current research interest focuses on transition metal-catalyzed asymmetric hydrogenation of imines and its application in the synthesis of biologically active N-containing compounds..

Zi-Yuan Ding was born in 1984 in Anhui Province, China. He received his B. S. degree in 2006 and his M. S. degree in 2010 from the University of Science and Technology of China. He then began his Ph. D. study at Institute of Chemistry of the Chinese Academy of Sciences under the supervision of Professor Qing-Hua Fan. His current research focuses on Rudiamine complex-catalyzed enantioselective hydrogenation of benzodiazepines.

Yan-Mei He was born in 1968 in Beijing, China. She obtained her B. S. degree in 1990 and M. S. degree in 1993 from the Department of Chemistry in Peking University. After working in Institute of Materia Medica CAMS for five years on developing new drugs, she moved to the United States. In 2003, she joined the research group of Professor Qing-Hua Fan in Institute of Chemistry of the Chinese Academy of Sciences. Now she is an associate professor level senior engineer, and her research interests include development of advanced functional materials and asymmetric catalysis.

Organic Syntheses

Qing-Hua Fan was born in 1966 in Hunan Province. China. He received his M. S. degree in 1992 from Institute of Chemistry of the Chinese Academy of Sciences (ICCAS) and Ph. D. degree in 1998 from The Hong Kong Polytechnic University under the supervision of Professor Albert S. C. Chan. He then came back to ICCAS as associate professor and research group leader. Since 2003 he has been a full professor of Organic Chemistry in the same institute. His research interests include asymmetric catalysis and synthesis of biologically active heterocyclic compounds, environmentally benign catalytic organic reactions, and molecular design and self-assembly of functional dendrimers.

Douglas C. Duquette was born in Springfield, Massachusetts in 1987. In 2009 he received his B. A. and M. A. in chemistry from Harvard University, where he did undergraduate research in the laboratory of Professor David A. Evans. Douglas is pursuing his graduate studies in the research group of Professor Brian M. Stoltz.

Zirconium (IV) chloride catalyzed amide formation from carboxylic acid and amine: (S)-tert-butyl 2-(benzylcarbamoyl)pyrrolidine-1-carboxylate

Fredrik Tinnis,[1] Helena Lundberg,[1] Tove Kivijärvi,[1] and Hans Adolfsson[1]*

Department of Organic Chemistry, Stockholm University, SE-10691 Stockholm, Sweden.

Checked by Yuxing Wang and Huw M. L. Davies

A.

$ZrCl_4$ (10 mol %)
MS 4Å
———————————→
THF, 0.2M
reflux, 48 h

Procedure

(S)-tert-Butyl 2-(benzylcarbamoyl)pyrrolidine-1-carboxylate (2). Molecular sieves (Note 1) are activated (Note 2) and 18 g are transferred to a 250-mL single-necked, round-bottomed flask equipped with a gas adaptor with stopcock. The flask is filled with argon, charged with Boc-L-proline (4.0 g, 18.6 mmol, 1.0 equiv) (Note 3), a magnetic stir bar (Note 4), and $ZrCl_4$ (0.432 g, 1.86 mmol, 0.1 equiv) (Note 5). Once the addition is complete, the adaptor is replaced with an oven-dried condenser fitted with a rubber septum, and the atmosphere is evacuated and replaced with argon (Note 6). Dry THF (93 mL) (Note 7) is added through the septum of the condenser with a syringe equipped with a long needle and the reaction mixture is then heated to reflux in an oil bath under vigorous stirring (Figure 1). At reflux, benzylamine (2.44 mL, 22.3 mmol, 1.2 equiv) (Note 8) is added over a period of five minutes into the reaction mixture through the septum using a

Org. Synth. **2015**, *92*, 227-236
DOI: 10.15227/orgsyn.092.0227

Published on the Web 7/11/2015
© 2015 Organic Syntheses, Inc.

syringe equipped with a long needle. The rubber septum is replaced with an adaptor connected to an argon line. The mixture is refluxed for 48 h, allowed to cool to room temperature, and filtered through a pad of silica in a sintered glass filter funnel (Note 9), directly into a round-bottomed flask (250 mL) with the aid of water suction (ca. 300 mmHg). The original round-bottomed flask is washed three times with a total of 500 mL EtOAc:Et$_3$N (200:1), which is subsequently filtered through the silica pad. The solvent is concentrated under reduced pressure (18 mmHg, 35 °C) and the resulting white solid is dissolved in 100 mL EtOAc and washed with 150 mL saturated aqueous NaCl. The aqueous layer is then extracted with EtOAc (2 x 100 mL). The organic layer is combined and dried over MgSO$_4$. The solvent is concentrated under reduced pressure, and the resulting white solid is dissolved in a minimal amount (~ 40 mL) of hot (56 °C) acetone (Note 10), and then allowed to cool to room temperature. A minimal

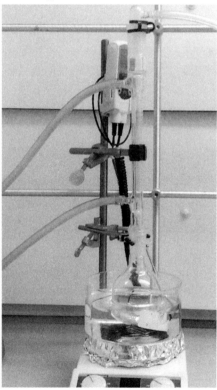

amount (~ 35 mL) of water is added until tiny precipitates begin to appear. The flask is left to stand in a refrigerator at –4 °C for 24 h. The precipitate is collected by vacuum filtration and the filtrate is transferred to another flask. The filtered material is washed with 100 mL deionized water. A portion of the water washings (~ 50 mL) is added to the original filtrate until tiny precipitates begin to appear. The flask is left to stand in a refrigerator for 48 h at –4 °C and the resulting precipitate is collected by vacuum filtration. The combined white solid is dried on a rotary evaporator in a water bath at 65 °C, and then under vacuum (1-2 mmHg) connected to a manifold for 6 h to yield the title compound (**2**) (3.78 g, 67%) (Note 11) as a white solid (Note 12).

Figure 1. Reaction Set-up

)rganic
Syntheses ———————————————————————

Notes

1. Molecular sieves (pellets, diameter 1.6 mm, pore size 4 Å) were purchased from Sigma Aldrich Co.
2. Molecular sieves (30 g) were oven dried at 140 °C for 2 days, flame-dried (Bunsen-burner) in a 250 mL round-bottomed flask fitted with a stopcock under vacuum (1-2 mmHg) for 7 min. The flask was allowed to cool to room temperature under vacuum and then filled with nitrogen.
3. Boc-L-Pro-OH 99.5% was purchased from Sigma Aldrich Co. and used as received.
4. An oven-dried, Teflon· coated, 35 x 8 mm stir bar was used.
5. Zirconium (IV) chloride ≥ 99.5% trace metals basis was purchased from Sigma Aldrich Co. and used as received. The Zirconium (IV) chloride was stored under a nitrogen atmosphere but weighed under normal atmosphere and quickly transferred to the reaction flask.
6. The atmosphere was exchanged for Ar by means of a needle in the septum, connected to a vacuum manifold. The procedure was repeated three times.
7. THF (≥ 99.9%) was purchased from Sigma Aldrich Co., dried in a Glass Contour solvent purifying system.
8. Benzylamine ≥ 98% was purchased from Sigma Aldrich Co, and was used as received.
9. The sintered glass filter funnel (45 mm h x 55 mm diameter) was filled with 24 g of silica.
10. Acetone (≥ 99.8%) was purchased from E. Merck Co. and used as received.
11. When the reaction was performed at half-scale, a yield of 69% was obtained.
12. IR (neat) 3310, 2976, 2871, 1682, 1652, 1526, 1393, 1370, 1157, 1127, 766, 724, 693 cm^{-1}. mp 124–125 °C. ^{1}H NMR (600 MHz, d-6 DMSO) Major rotamer δ: 1.28 (s, 6 H), 1.41 (s, 3 H), 1.67 – 1.90 (m, 3 H), 2.04 – 2.18 (m, 1 H), 3.25 – 3.32 (m, 1 H), 3.36 – 3.43 (m, 1 H), 4.08 (dd, J = 8.6, 3.1 Hz, 1 H), 4.21 (dd, J = 14.0, 5.9 Hz, 1 H), 4.34 (dd, J = 14.0, 5.9 Hz, 1 H), 7.20–7.34 (m, 5 H), 8.39 (t, J = 5.9 Hz, 1 H). Minor rotamer δ: 1.28 (s, 6 H), 1.41 (s, 3 H), 1.67 – 1.90 (m, 3 H), 2.04 – 2.18 (m, 1H), 3.25 – 3.32 (m, 1 H), 3.36 – 3.43 (m, 1 H), 4.12 (dd, J = 8.6, 3.1 Hz, 1 H), 4.19 (dd, J = 14.0, 5.9 Hz, 1 H), 4.34 (dd, J = 14.0, 5.9 Hz, 1 H), 7.20-7.34 (m, 5 H), 8.34 (t, J =

5.9 Hz, 1 H). ¹H NMR (600 MHz, DMSO, 60 °C) δ: 1.29 (s, 9 H), 1.67 –
1.91 (m, 2 H), 1.99 – 2.20 (m, 1 H), 3.29 (dt, *J* = 10.5, 6.8 Hz, 1 H),
3.38 (ddd, *J* = 10.5, 7.5, 5.1 Hz, 1 H), 4.10 (s, 1 H), 4.21 (dd, *J* = 14.9,
6.0 Hz, 1 H), 4.31 (dd, *J* = 14.9, 6.0 Hz, 1 H), 7.20 (t, *J* = 6.8 Hz, 1 H), 7.23 –
7.31 (m, 4 H), 8.16 (s, 1 H). ¹³C NMR (150 MHz, *d-6* DMSO) Major
rotamer δ: 23.6, 28.4, 31.6, 42.5, 47.0, 60.4, 78.9, 127.8, 128.6, 140.1, 153.8,
172.9; Minor rotamer δ: 24.5, 28.66, 30.5, 42.3, 47.2, 60.3, 79.1, 127.0,
127.19, 127.3, 140.1, 154.2, 172.7. HRMS *m/z* calcd for $C_{17}H_{25}N_2O_3$ [M +
H]⁺: 305.1860; Found: 305.1856. Anal. calcd for C₁₇H₂₄N₂O₃: C, 67.08; H,
7.95; N, 9.20; Found: C, 67.02; H, 7.89; N, 9.04. Chiral HPLC AD-H
column, 90:10 isohexane:2-PrOH, flow 0.7 mL/min, 30 °C, retention time
18.54 min (Retention times for racemic sample = 7.16 min and 19.93
min).

Working with Hazardous Chemicals

The procedures in *Organic Syntheses* are intended for use only by
persons with proper training in experimental organic chemistry. All
hazardous materials should be handled using the standard procedures for
work with chemicals described in references such as "Prudent Practices in
the Laboratory" (The National Academies Press, Washington, D.C., 2011;
the full text can be accessed free of charge at
http://www.nap.edu/catalog.php?record_id=12654). All chemical waste
should be disposed of in accordance with local regulations. For general
guidelines for the management of chemical waste, see Chapter 8 of Prudent
Practices.

In some articles in *Organic Syntheses*, chemical-specific hazards are
highlighted in red "Caution Notes" within a procedure. It is important to
recognize that the absence of a caution note does not imply that no
significant hazards are associated with the chemicals involved in that
procedure. Prior to performing a reaction, a thorough risk assessment
should be carried out that includes a review of the potential hazards
associated with each chemical and experimental operation on the scale that
is planned for the procedure. Guidelines for carrying out a risk assessment
and for analyzing the hazards associated with chemicals can be found in
Chapter 4 of Prudent Practices.

The procedures described in *Organic Syntheses* are provided as published and are conducted at one's own risk. *Organic Syntheses, Inc.*, its Editors, and its Board of Directors do not warrant or guarantee the safety of individuals using these procedures and hereby disclaim any liability for any injuries or damages claimed to have resulted from or related in any way to the procedures herein.

Discussion

The procedure described herein is a direct amide coupling of non-activated carboxylic acids and amines catalyzed by zirconium (IV) chloride. The amide functionality is of major importance in many scientific areas, for example in material science, chemistry, and biology. Furthermore, the amide functionality is one of the most synthesized within the pharmaceutical industry, and it has been estimated that 25% of the pharmaceutical compounds available on the market contain at least one amide bond.[2] The importance of the amide bond is further emphasized by the fact that 2/3 of 128 drug candidates covered in a survey from Process Chemistry R&D departments of GlaxoSmithKline, AstraZeneca and Pfizer in 2006 made use of amide couplings.[3] The most common method for amide coupling is the reaction of an activated carboxylic acid with an amine, mainly by the use of acid chlorides, coupling reagents or mixed anhydride methods, to avoid the salt formation between the carboxylic acid and amine.[3] These methodologies all rely on stoichiometric amounts of additional reagents and equally produces one equivalent of waste along with the desired product. For this reason, catalytic amidations have been considered as greener alternatives.[4] Only a limited number of high-yielding catalytic methods are currently known which employs non-activated carboxylic acids and amines as starting material. The most well-known protocol utilizes boronic acids as catalysts, most commonly performed at high reaction temperatures.[5] In contrast, the method described herein is a mild and selective metal-catalyzed protocol, giving rise to high yields of the desired products at moderate reaction temperatures, and without racemizing chiral amino acids (Table 1). The method stems from our recent work where we have shown that early transition metal complexes such as

Table 1. Examples of amides formed by ZrCl$_4$-catalysis from carboxylic acids and amines[a]

Entry	Amide	Isoated yield (%)
1		85[b,c]
2		>99[d]
3		63[b]
4		62[b,c]
5		97[g]
6		89[d]
7		65[b,c]
8		93[d]
9		93
10		82[b,f]

[a]Results taken from reference 6. [b]10 mol% ZrCl$_4$. [c]Reaction temperature 100°C. [d]5 mol% ZrCl$_4$. [e]No racemization was detected with chiral HPLC. [f]Carboxylic acid (1 mmol), amine (1.5 mmol). [g]Carboxylic acid concentration 0.2M

$ZrCl_4$ or $Ti(OPr)_4$ are excellent catalysts for a range of substrates, including the formation of both secondary and tertiary amides, as well as chiral amides.[6,7] $ZrCl_4$ was recently also shown by Williams and co-workers to catalyze the same reaction at higher reaction temperatures and higher catalyst loadings, however, without molecular sieves.[8]

This journal has previously published a reliable catalytic method for direct amidation of carboxylic acids and amines, employing 3,4,5-(trifluorophenyl)boronic acid as catalyst.[9] This particular protocol requires higher reaction temperatures and higher catalyst loading compared to the method described herein, and the boronic acid catalyst has either to be synthesized, or purchased at a price more than fifteen times higher than that of $ZrCl_4$.[10] A different protocol using boric acid as catalyst has also been published in this journal.[11] The method we present above using $ZrCl_4$ as catalyst has advantages over both protocols based on boron catalysts. The reaction conditions are significantly milder, with lower reaction temperature and no observed racemization of enantiopure starting materials. Furthermore, boric acid is considerably more toxic in comparison to the zirconium-catalyst.[12]

References

1. Department of Organic Chemistry, Stockholm University, SE-10691 Stockholm, Sweden. hans.adolfsson@su.se This work was supported by the Swedish Research Council, and the K&A Wallenberg foundation.
2. Ghose, A.K.; Viswanadhan, V. N.; Wendoloski, J. J. *J. Comb. Chem.* **1999**, *1*, 55–68.
3. Carey, J. S.; Laffan, D.; Thomson, C.; Williams, M. T. *Org. Biomol. Chem.* **2006**, *4*, 2337–2347.
4. See the following reviews and references therein: (a) Allen, C. L.; Williams, J. M. J. *Chem. Soc. Rev.* **2011**, *40*, 3405–3415. (b) Roy, S.; Roy, S.; Gribble, G. W. *Tetrahedron* **2012**, *68*, 9867–9923. (c) Lundberg, H.; Tinnis, F.; Selander, N.; Adolfsson, H. *Chem. Soc. Rev.* **2014**, *43*, 2714–2742.
5. (a) Ishihara, K.; Ohara, S.; Yamamoto, H. *J. Org. Chem.* **1996**, *61*, 4196–4197. (b) Maki, T.; Ishihara, K.; Yamamoto, H. *Org. Lett.* **2006**, *8*, 1431–1434. (c) Al-Zoubi, R. M. *Angew. Chem. Int. Ed.* **2008**, *47*, 2876–2879. (d) Arnold, K.; Davies, B.; Herault, D.; Whiting, A. *Angew. Chem. Int. Ed.* **2008**, *47*, 2673–2676.

6. Lundberg, H.; Tinnis, F.; Adolfsson, H. *Chem. Eur. J.* **2012**, *18*, 3822–3826.
7. Lundberg, H.; Tinnis, F.; Adolfsson, H. *Synlett* **2012**, *23*, 2201–2204.
8. Allen, C. L.; Chhatwal, A. R.; Williams, J. M. J. *Chem. Commun.* **2012**, *48*, 666–668.
9. K. Ishihara, S. Ohara, H. Yamamoto, *Org. Synth.* **2002**, *79*, 176–182.
10. 3,4,5-(Trifluorophenyl)boronic acid (CAS 143418-49-9) is sold from Sigma Aldrich Co. for approximately 745 Euro (July 2015) per five grams, and ZrCl₄ (CAS 10026-11-6) has a retail price of approximately 45 Euro for the same amount from the same company during the same period.
11. Tang, P. *Org. Synth.* **2005**, *81*, 262–272.
12. According to the EC Regulation No 1272/2008 [EU-GHS/CLP], and the EU Directives 67/548/EEC or 1999/45/EC, boric acid shows reproductive toxicity (Category 1B), may impair fertility, and may cause harm to the unborn child. Zirconium(IV) chloride has the properties of an acid and should be treated as such, however, there are no long-term biological problems reported upon exposure to the compound.

Appendix
Chemical Abstracts Nomenclature (Registry Number)

(Phenylthio)acetic acid: (Phenylmercapto)acetic acid, S-Phenylthioglycolic acid, Thiophenoxyacetic acid; (103-04-8)

ZrCl₄: Zirconium(IV) Chloride, Tetrachlorozirconium, Zirconium tetrachloride; (10026-11-6)

Benzylamine: α-Aminotoluene; (100-46-9)

Et₃N: Triethylamine; (121-44-8)

Boc-L-Proline: Boc-Pro-OH, N-(*tert*-Butoxycarbonyl)-L-proline; (15761-39-4)

Fredrik Tinnis obtained his Ph.D. degree from Stockholm University (2014) under the supervision of Prof. Hans Adolfsson, where he focused on the development of catalytic procedures for the formation of amides. Fredrik is now a post-doctoral fellow in the same research group and is currently working on chemoselective reduction of amides.

.

Helena Lundberg obtained her Ph. D. degree in June 2015 under the supervision of Professor Hans Adolfsson at the Department of Organic Chemistry, Stockholm University. She is currently pursuing research in the same research group in collaboration with Prof. Fahmi Himo, primarily focusing on the mechanistic aspects of direct catalytic amide formation using group (IV) metal complexes.

.

Tove Kivijärvi received her B. Sc. in Environmental Chemistry in June 2013 and her M. Sc. in Organic Chemistry in June 2015 from Stockholm University. Her graduate research under the supervision of Prof. Hans Adolfsson focused on ruthenium-catalyzed asymmetric transfer hydrogenation of sterically demanding ketones.

Hans Adolfsson, professor in organometallic chemistry since 2007, is currently the Pro Vice-Chancellor of Stockholm University. He conducted his undergraduate studies at Stockholm University and graduated in 1989. He moved to the Royal Institute of Technology (KTH) in Stockholm for Ph. D. studies under the guidance of Professor Christina Moberg. After his graduation in November 1995 he continued as a post-doctoral fellow at KTH until August 1996 when he moved to the Scripps Research Institute for a two-year post-doctoral stay with Prof. K. Barry Sharpless. In November 1998 he moved back to Sweden and started as an assistant professor at Stockholm University. In 2002 he was promoted to associate professor, and in 2007 to full professor. His research interests are in the fields of practical and selective catalysis comprising oxidations, reductions and addition reactions.

Yuxing Wang was born in Xi'an, China. He earned his B. Sc. in Chemistry in 2009 and M.Sc. in Organic Chemistry under the guidance of Professor Chanjuan Xi in 2011 from Tsinghua University. The same year he started his Ph. D. studies at Emory University under the guidance of Professor Huw M. L. Davies. His research mainly focuses on the development of transition metal-catalyzed donor-acceptor carbene reactions.

Organic Syntheses

Aminocarbonylation Using Electron-rich Di-*tert*-butylphosphinoferrocene

Carl A. Busacca, Magnus C. Eriksson,[1*] Bo Qu,[1*] Heewon Lee, Zhibin Li, and Chris H. Senanayake

Chemical Development, Boehringer Ingelheim Pharmaceuticals, Inc., P. O. Box 368, Ridgefield, CT 06877

Checked by Beau P. Pritchett, Nicholas R. O'Connor, and Brian M. Stoltz

Procedure

A. *Pyrimidine-5-carboxylic acid p-tolylamide methanesulfonate (2).* A 475 mL Parr vessel containing a cylindrical Teflon-coated magnetic stir bar (5 cm in length, 1 cm in diameter) is charged with *p*-tolylamine (4.01 g, 37.1 mmol, 1.0 equiv) (Note 1), 5-bromo-pyrimidine (7.34 g, 44.8 mmol, 1.2 equiv) (Note 2), palladium acetate (257 mg, 1.23 mmol, 0.03 equiv) (Note 3), and ligand **1** (970 mg, 2.25 mmol, 0.06 equiv) in open air (Note 4). Acetonitrile (80.0 mL) (Note 5) and *N,N*-diisopropylethylamine (26.1 mL,

149 mmol, 4.0 equiv) (Note 6) are added to the vessel. The vessel is sealed and purged with carbon monoxide three times. The vessel is then pressurized with carbon monoxide to 100 psi, and placed in a room-temperature oil bath. The oil bath is heated to 100 °C. The reaction is stirred at 100 °C (bath temperature) and 100 psi CO for 4 h (Notes 7 and 8). The reaction vessel is removed from the oil bath and cooled to ambient temperature while stirring. After releasing the pressure, HPLC analysis shows that *p*-tolylamine is fully consumed. The reaction mixture is transferred to a 250 mL one-necked round-bottomed flask and concentrated on a rotary evaporator (Note 9) to fully remove acetonitrile and the tertiary amine. To the resulting dark semi-solid is added 2-MeTHF (100 mL) (Note 10) and 1 M aqueous NaOH (50 mL), and the resulting dark mixture is stirred vigorously for 1 h (Note 11). The two-phase mixture is then filtered through a pad of Celite (1 cm) in a 150 mL fritted funnel (medium porosity, Note 12) into a one-necked 500 mL round-bottomed flask using house vacuum. The Celite pad is washed with 2-MeTHF (2 x 100 mL). The combined filtrates are transferred to a 1 L separatory funnel and allowed to stand for 15 min. The lower aqueous layer is discarded. The upper organic layer is washed with water (2 x 100 mL) and then transferred to a one-

necked 500 mL round-bottomed flask (Note 13). Dry silica gel (16 g) is added to the flask, and the resultant mixture is dried on a rotary evaporator until a brown powder is obtained (Note 14). This material is then placed atop a silica gel column (250 g, 8 cm diameter) that had been conditioned with 450 mL 20% EtOAc/hexanes. Sand is added on top of the silica gel. The column is eluted first with 1300 mL 20% EtOAc/hexanes to remove non-polar impurities. Pure EtOAc is then used to elute the product. The fractions are then combined and concentrated in vacuo to give 6.1 g of the crude free base product as a light brown solid (Note 15). A final purification is then carried out (Note 16). The solids are transferred to a three-necked 250 mL flask (the necks contain an inert gas valve and two septa, one containing a thermocouple), followed by addition of 2-MeTHF (85 mL). The mixture is stirred for ca 5 min at 65 °C under N_2 to give a brown solution. Methanesulfonic acid (1.95 mL, 30 mmol, 0.8 equiv) (Notes 17, 18 and 19) is then added at once via syringe, causing the immediate formation of a thick slurry of the MsOH salt. The suspension is stirred at 65 °C for 30 min, then heat is turned off and the mixture is allowed to cool slowly to ambient temperature (approximately three hours). The slurry is then filtered through a medium-fritted filter funnel using house vacuum, washing the cake with 2-MeTHF (1 x 20 mL). After 30 min, the dry solids are collected to give 8.7 g (76%) of the product salt as a tan powder (Notes 20, 21, and 22).

Notes

1. *p*-Tolylamine (99.0%) was purchased from Aldrich and used as received.
2. 5-Bromo-pyrimidine (97%) was purchased from Aldrich and used as received.
3. Palladium (II) acetate (98%) was purchased from Aldrich and used as received.
4. Ligand L-1 was prepared as described in *Org. Synth.* **2013**, *90*, 316–326.
5. Anhydrous acetonitrile was purchased from Aldrich and used as received.
6. *N,N*-Diisopropylethylamine (99.5%) was purchased from Aldrich and used as received.

7. The checkers observed an increase in pressure to 110 psi upon heating, followed by a decrease in pressure to 88 psi after approximately 90 min reaction time. The reaction pressure remained at 88 psi for the remainder of the reaction.

8. The reaction mixture becomes darker at extended reaction times, making visualization of the subsequent phase separation difficult. The reaction is completed in approximately 4 h.

9. Vacuum (25-50 mmHg) and water bath (60 °C) were used. Full removal of the tertiary amine is critical to the success of the salt formation. Successful removal of tertiary amine can be monitored by ^{1}H NMR of the crude in d_6-DMSO.

10. 2-MeTHF (>99%) was purchased from Aldrich and used as received.

11. The carbonylation generates significant amounts (~ 10-20%) of an impurity derived from the desired product: it is the imide in which *two* acyl-pyrimidine fragments are on the aniline nitrogen atom ($C_{17}H_{13}N_5O_2$, HRMS [M+H]$^+$ calc 320.1142, found: 320.1141). Treating the reaction mixture with 1M NaOH converts this material to additional product (plus pyrimidine carboxylic acid), increasing the isolated yield of **2**.

12. The filtration removes dark materials that obscure the phase separation.

13. If an emulsion is observed at this stage, longer periods of settling may be required.

14. If necessary, a spatula can be used to scrape some of the material off the inside wall of the flask.

15. Fractions of 65 mL were collected. The desired free-base intermediate was collected in fractions 11–26. R$_f$ = 0.4 (75% EtOAc in hexanes) visualized by UV irradiation.

16. Some mixed fractions can be included as the product is further purified during salt formation.

17. Methanesulfonic acid (99.5%) was purchased from Aldrich and used as received.

18. The amount of methanesulfonic acid used (0.8 equiv) was determined based on the crude yield of the free-base. A second crop of product was obtained by a subsequent salt formation reaction using 0.2 equiv of MsOH. This process only produced an additional 2% of **2•MsOH** (in decreased purity).

19. The checkers observed an increase in internal temperature to 73 °C upon addition of MsOH.

20. Compound **2•MsOH** exhibits the following analytical data: ^{1}H NMR and ^{13}C NMR spectra are reported relative to d_6-DMSO (δ 2.50 ppm and δ 39.52 ppm, respectively). ^{1}H NMR (20 mg solid in 0.6 mL d_6-DMSO, 500 MHz) δ: 2.28 (s, 3 H), 2.47 (s, 3 H), 7.17 (d, J = 8 Hz, 2 H), 7.63 (d, J = 8 Hz, 2 H), 9.25 (s, 2 H), 9.34 (s, 1 H), 10.52 (s, 1 H), 10.94 (br s, 1 H); ^{13}C NMR (d_6-DMSO, 126 MHz) δ: 20.6, 39.7, 120.4, 128.7, 129.2, 133.4, 136.1, 156.2, 160.1, 162.0; IR (neat film, NaCl): 3411, 3278, 3111, 3034, 2930, 1648, 1620, 1601, 1540, 1514, 1414, 1191, 1150, 1042, 1023.6, 918, 823, 811, 697 cm^{-1}; mp 209.6–210.1 °C; HRMS (FAB+) for free base $C_{12}H_{12}N_3O$ [M+H]$^+$: calcd, 214.0980; found, 214.0987; Elem. Anal. calcd. for $C_{13}H_{15}N_3O_4S$: C, 50.48; H, 4.89; N, 13.58; found: C, 50.66; H, 4.87; N, 13.44.

21. The checkers observed that the chemical shift of the broad singlet at 10.94 ppm is dependent on the ^{1}H NMR sample concentration (increases with concentration).

22. On a half-scale run, the checkers isolated compound **2•MsOH** in 76% yield.

Working with Hazardous Chemicals

The procedures in *Organic Syntheses* are intended for use only by persons with proper training in experimental organic chemistry. All hazardous materials should be handled using the standard procedures for work with chemicals described in references such as "Prudent Practices in the Laboratory" (The National Academies Press, Washington, D.C., 2011; the full text can be accessed free of charge at http://www.nap.edu/catalog.php?record_id=12654). All chemical waste should be disposed of in accordance with local regulations. For general guidelines for the management of chemical waste, see Chapter 8 of Prudent Practices.

In some articles in *Organic Syntheses*, chemical-specific hazards are highlighted in red "Caution Notes" within a procedure. It is important to recognize that the absence of a caution note does not imply that no significant hazards are associated with the chemicals involved in that procedure. Prior to performing a reaction, a thorough risk assessment should be carried out that includes a review of the potential hazards associated with each chemical and experimental operation on the scale that is planned for the procedure. Guidelines for carrying out a risk assessment and for analyzing the hazards associated with chemicals can be found in Chapter 4 of Prudent Practices.

The procedures described in *Organic Syntheses* are provided as published and are conducted at one's own risk. *Organic Syntheses, Inc.*, its Editors, and its Board of Directors do not warrant or guarantee the safety of individuals using these procedures and hereby disclaim any liability for any injuries or damages claimed to have resulted from or related in any way to the procedures herein.

Discussion

The direct palladium-catalyzed carbonylation of aryl halides in the presence of alcohols or amines is an efficient way to synthesize esters and amides, respectively.[2] The synthesis of esters is relatively well precedented, whereas the amino carbonylation, which is of great importance for the synthesis of pharmaceuticals, remains a bigger challenge. An early report by Heck[3] used triphenylphosphine as ligand with aryl and vinyl bromides and later Milstein[4] introduced di-isopropylphosphinopropane (dippp) for aryl chlorides. Buchwald reported more recently[5] on the use of Xantphos for aryl bromides as well as for two examples of hetero aryl bromides. We demonstrated[6] the amino carbonylation of hetero aryl bromides and iodides using the electron-rich ligand di-*tert*-butyl-phosphinoferrocene. The synthesis of this ligand is the subject of a preceding *Organic Syntheses* procedure.[7] The ligand is isolated as the HBF$_4$-salt (L-**1**). To our knowledge, this is the first example of gram scale aminocarbonylation that employs base hydrolysis of an imide by-product to increase the isolated yield.

References

1. Chemical Development, Boehringer Ingelheim Pharmaceuticals, Inc., P. O. Box 368, Ridgefield, CT 06877. E-mail:magnus.eriksson@boehringer-ingelheim.com; bo.qu@boehringer-ingelheim.com
2. Barnard, C. F. *Organometallics* **2008**, *27*, 5402–5422.
3. Schoenberg, A.; Heck, R. F. *J. Org. Chem.* **1974**, *39*, 3327–3331.

4. Ben-David, Y.; Portnoy, M.; Milstein, D. *J. Am. Chem. Soc.* **1989**, *111*, 8742–8744.
5. Martinelli, J. R.; Watson, D. A.; Freckmann, D. M. M.; Barder, T. E.; Buchwald, S. L. *J. Org. Chem.* **2008**, *73*, 7102–7107.
6. Qu, B.; Haddad, N.; Han, Z. S.; Rodriguez, S.; Lorenz, J. C.; Grinberg, N.; Lee, H.; Busacca, C. A.; Krishnamurthy, D.; Senanayake, C. H. *Tetrahedron Lett.* **2009**, *50*, 6126–6129.
7. Busacca, C. A., Eriksson, M. C., Haddad, N., Han, Z. S., Lorenz, J. C., Qu, B., Zeng, X., Senanayake, C. H. *Org. Synth.* **2013**, *90*, 316–326.

Appendix
Chemical Abstracts Nomenclature (Registry Number)

Di-*tert*-butylphosphinoferrocene: Ferrocene, [bis(1,1-dimethylethyl)phosphino]-; (223655-16-1)
Palladium acetate: Acetic acid, palladium(2+) salt (2:1); (3375-31-3)
p-Tolylamine: Benzenamine, 4-methyl-; (106-49-0)
5-Bromo-pyrimidine: Pyrimidine, 5-bromo-; (4595-59-9)
N,N-Diisopropylethylamine: *N*-Ethyl-*N,N*-diisopropylamine; (7087-68-5)

Dr. Carl Busacca received his B.S. in Chemistry from North Carolina State University, and did undergraduate research in Raman spectroscopy and ^{60}Co radiolyses. After three years with Union Carbide, he moved to the labs of A.I. Meyers at Colorado State University, earning his Ph.D. in 1989 studying asymmetric cycloadditions. He worked first for Sterling Winthrop before joining Boehringer-Ingelheim in 1994. He has worked extensively with anti-virals, and done research in organopalladium chemistry, ligand design, organophosphorus chemistry, asymmetric catalysis, NMR spectroscopy, and the design of efficient chemical processes. He is deeply interested in the nucleosynthesis of transition metals in supernovae.

Organic
Syntheses

Dr. Magnus Eriksson was born in Stockholm, Sweden. He received his undergraduate degree in Chemical Engineering and his Ph.D. in Organic Chemistry from Chalmers University of Technology in Gothenburg in 1995 under the guidance of Professor Martin Nilsson working on copper-promoted 1,4-additions to carbonyl compounds. After post-doctoral work at Boehringer Ingelheim Pharmaceuticals and at MIT with Professor Stephen Buchwald, he joined Boehringer Ingelheim Pharmaceuticals in 2000 where he is currently a Principal Scientist. His research interests include Process Research, catalytic transformations and synthetic methodology.

Dr. Bo Qu was born in China, where she received a B.S. degree in chemistry. She then completed her M.S. at University of Science and Technology of China. She obtained her Ph.D. from the University of South Carolina in 2002 under the guidance of Prof. Richard Adams. After 3 years of postdoctoral studies at Cornell University with Prof. David Collum, she joined the Department of Chemical Development at Boehringer Ingelheim Pharmaceuticals in Ridgefield, CT, where she is currently a Senior Scientist. Dr. Qu's research interests focus on development of new catalytic transformations for efficient chemical processes, organometallic chemistry, and automated parallel syntheses.

Dr. Heewon Lee obtained her B.S. in Chemistry and MS in Physical Chemistry from Seoul National University in Seoul, South Korea. She earned her Ph.D. in Analytical Chemistry at the University of Michigan in Ann Arbor. After postdoctoral positions, she worked at ArQule for two years, and joined Boehringer Ingelheim Pharmaceuticals in 2000. Currently, she is Senior Associate Director in Chemical Development department and leads Analytical Research Group. She is responsible for analytical method development, in-process control for Process R&D, and quality control of outsourced materials. She is also involved in the Genotoxic Impurity Council and Process Analytical Technology (PAT).

Zhibin Li joined Boehringer-Ingelheim Pharmaceutical (Ridgefield, US) in 2006, where he has been working in the areas of manual and high through-put polymorph / salt / cocrystal screen, physical form characterization, chiral separation by crystallization, and crystallization development of active pharmaceutical ingredients. Zhibin has a Ph.D. in Chemistry.

Dr. Chris H. Senanayake obtained his Ph.D. with Professor James H. Rigby at Wayne State University followed by postdoctoral fellow with Professor Carl R. Johnson. In 1989, he joined Process Development at Dow Chemical Co. In 1990, he joined the Merck Process Research Group. After Merck, he accepted a position at Sepracor, Inc. in 1996 where he was appointed to Executive Director of Chemical Process Research. In 2002, he joined Boehringer Ingelheim Pharmaceuticals. Currently, he is the Vice President of Chemical Development. He is the co-author more than 340 papers and patents in many areas of synthetic organic chemistry.

Beau P. Pritchett received his B.S. degree in Chemistry and B.S.E. degree in Chemical Engineering from Tulane University in 2012. In the fall of 2012, he joined the laboratories of Professor Brian M. Stoltz at Caltech where he has pursued his Ph.D. as an NSF predoctoral fellow. His research interests include chemical synthesis, reaction design, and their applications in natural product synthesis and human medicine.

O rganic
S yntheses

Nicholas R. O'Connor received a B.A. in chemistry from Macalester College in 2011, conducting research with Professor Rebecca C. Hoye. He then moved to the California Institute of Technology and began his doctoral studies under the direction of Professor Brian M. Stoltz. His graduate research focuses on cycloadditions of strained rings and the application of these reactions to natural product synthesis.

Organic
Syntheses

Preparation of (S)-tert-ButylPyOx and Palladium-Catalyzed Asymmetric Conjugate Addition of Arylboronic Acids

Jeffrey C. Holder, Samantha E. Shockley, Mario P. Wiesenfeldt,
Hideki Shimizu, and Brian M. Stoltz[*1]

The Warren and Katharine Schlinger Laboratory for Chemistry and
Chemical Engineering, Division of Chemistry and Chemical Engineering,
California Institute of Technology, 1200 East California Boulevard, MC 101-
20, Pasadena, California, 91125, United States

Checked by Brian M. Cochran and Margaret Faul

A.

1. *i*-BuOCOCl, NMM, CH$_2$Cl$_2$, 0 °C

2. (*S*)-*tert*-leucinol, NMM, CH$_2$Cl$_2$
0 °C → 23 °C

1 → **2**

B.

1. SOCl$_2$, toluene, 60 °C

2. NaOMe, MeOH, 55 °C

2 → **3**

C.

Pd(OCOCF$_3$)$_2$, (*S*)-*tert*-BuPyOx (**3**),
NH$_4$PF$_6$, H$_2$O

ClCH$_2$CH$_2$Cl, 40 °C

4 + **5** → **6**

Procedure

A. *(S)-N-(1-Hydroxy-3,3-dimethylbutan-2-yl)picolinamide* (**2**). A 1 L one-necked round-bottomed flask equipped with a 3.0 cm x 1.4 cm, egg-shaped, Teflon-coated magnetic stirring bar is sealed with a septum and connected via needle adapter to a two-tap Schlenk adapter attached to an oil bubbler and a nitrogen/vacuum manifold (Note 1). The flask is dried with a heat gun under vacuum and cooled under a stream of nitrogen. The flask is charged with 2-picolinic acid (**1**) (6.15 g, 50.0 mmol, 1.00 equiv) (Note 2), evacuated and back-filled with nitrogen three times, then charged with dichloromethane (300 mL, 0.17 M) (Note 3) and *N*-methylmorpholine (7.59 g, 8.25 mL, 75.0 mmol, 1.50 equiv). The flask is cooled in an ice/water bath and *iso*-butylchloroformate (6.86 mL, 7.17 g, 52.5 mmol, 1.05 equiv) is added dropwise over 30 min by syringe pump. The reaction mixture is stirred for an additional 30 min while remaining submerged in the ice/water bath. A separate 100 mL one-necked round-bottomed flask is sealed with a septum and connected via needle adapter to the two-tap Schlenk adapter and manifold, dried with a heat gun under vacuum, and allowed to cool under a stream of nitrogen. This flask is charged with (*S*)-*tert*-leucinol (6.45 g, 55.0 mmol, 1.10 equiv), dichloromethane (40 mL), and *N*-methylmorpholine (6.07 mL, 5.56 g, 55.0 mmol, 1.10 equiv). The resulting clear solution is taken up in a syringe and transferred dropwise using a syringe pump over the course of 1 h to the stirring reaction mixture in the ice/water bath. The cooling bath is removed, and the pale gold colored reaction mixture is stirred for an additional 6 h at 23 °C. Upon consumption of starting material (Note 4), the mixture is quenched at ambient temperature with a single addition of an aqueous solution of saturated NH$_4$Cl (50 mL), diluted with additional H$_2$O (25 mL), and transferred into a 1 L separatory funnel. The phases are separated, and the aqueous phase is extracted with CH$_2$Cl$_2$ (3 x 100 mL). The combined organic phases are washed with an aqueous solution of saturated NaHCO$_3$ (1 x 50 mL) and brine (1 x 50 mL). The combined organic phases are dried over Na$_2$SO$_4$ (10 g, 15 min while agitating), filtered through a M pore glass frit, and concentrated by rotary evaporation (28 °C, 15 mmHg). Excess *N*-methylmorpholine is further removed by placing the crude residue under high vacuum (< 12 mmHg, 12 h) to provide a pale red solid (Note 5). The crude residue is dissolved in 10 mL of acetone and purified via silica gel flash chromatography (Note 6). The combined product-containing fractions are concentrated by rotary evaporation (40 °C, 15 mmHg) to yield a solid, which is dried under high vacuum (< 12 mmHg, 12 h) to afford (*S*)-*N*-(1-hydroxy-3,3-dimethylbutan-2-yl)picolinamide (**2**) as a white amorphous solid (9.88–9.95 g, 44.4–44.8 mmol, 89–90% yield) (Note 7).

B. *(S)-4-(tert-Butyl)-2-(pyridin-2-yl)-4,5-dihydrooxazole* **(3)**. A 500 mL three-necked round-bottomed flask is equipped with a 3.0 cm x 1.4 cm, egg-shaped, Teflon-coated magnetic stirring bar. All necks are sealed with rubber septa and the center neck is attached via needle adapter to a two-tap Schlenk adapter connected to a bubbler and a nitrogen/vacuum manifold (Note 1). One side neck is fitted with a thermometer pierced through a rubber septum. The flask is dried with a heat gun under vacuum, cooled under a stream of nitrogen and charged with (*S*)-*N*-(1-hydroxy-3,3-dimethylbutan-2-yl)picolinamide (**2**) (8.89 g, 40.0 mmol, 1.00 equiv) and toluene (140 mL, 0.285 M) (Note 3). The resulting clear solution is heated at 60 °C in a preheated bath of armor beads (Note 8). In a separate 50 mL flask, SOCl₂ (5.64 mL, 80.0 mmol, 2.00 equiv, Note 2) is dissolved in toluene (20 mL) and taken up with a syringe. The SOCl₂ solution is added dropwise over 70 min to the vigorously stirring reaction mixture using a syringe

pump (Note 9). Upon complete addition, the mixture is stirred for an additional 4 h. The resulting slurry is allowed to cool to ambient temperature, concentrated by rotary evaporation (55 °C, 15 mmHg), crushed with a spatula, and further dried under high vacuum (< 12 mmHg, 12 h) to afford (*S*)-*N*-(1-chloro-3,3-dimethylbutan-2-yl)picolinamide hydrochloride salt as a tan powder (10.68–10.86 g, 38.5–39.2 mmol, 96–98% yield) (Note 10). This material is used without purification.

A 500 mL one-necked round-bottomed flask equipped with a 3.0 cm x 1.4 cm, egg-shaped, Teflon-coated magnetic stirring bar is charged with (S)-N-(1-chloro-3,3-dimethylbutan-2-yl)picolinamide   hydrochloride   salt (10.26 g, 37.0 mmol, 1.00 equiv) and methanol (100 mL, 0.37 M) (Note 3). To the clear, stirring solution is added powdered sodium methoxide (9.99 g, 185.0 mmol, 5.00 equiv) (Notes 2 and 11) in one portion, and the resulting mixture is lowered into an armor bead bath preheated at 55 °C. The slurry

is stirred for 18 h (Note 12). Afterwards, the armor bead bath is removed, the reaction is allowed to cool to ambient temperature, and toluene (100 mL) is added (Note 13). The reaction mixture is partially concentrated by rotary evaporation to remove the methanol (40 °C, 60 mmHg), at which time the resulting slurry is transferred to a 250 mL separatory funnel and washed with water (100 mL) (Note 14). The layers are separated, and the aqueous layer is back extracted with toluene (3 x 40 mL). The combined organic extracts are dried over Na_2SO_4 (20 g) with stirring for 20 min, filtered through a M pore glass frit, and concentrated by rotary evaporation (40 °C, 15 mmHg). The crude residue is purified by silica gel flash chromatography (Note 15). The combined product-containing fractions are concentrated by rotary evaporation (28 °C, 15 mmHg) to yield a solid, which is further dried under high vacuum (< 12 mmHg, 12 h) to afford (S)-tert-ButylPyOx (3) as a white solid (5.64–5.91 g, 27.6–28.9 mmol, 75–78% yield; 72–76% yield from 2) (Note 16).

C. (R)-3-(4-Chlorophenyl)-3-methylcyclohexanone (6). A 500 mL, one-necked round-bottomed flask is equipped with a 4.2 x 2.0 cm, egg-shaped, Teflon-coated magnetic stirring bar and charged sequentially with 1,2-

dichloroethane (150 mL, 0.23 M), (*S*)-*tert*-ButylPyOx (**3**) (0.429 g, 2.10 mmol, 0.06 equiv), palladium(II) trifluoroacetate (0.582 g, 1.75 mmol, 0.05 equiv), ammonium hexafluorophosphate (1.71 g, 10.5 mmol, 0.30 equiv), and 4-chlorophenylboronic acid (**5**) (6.57 g, 42 mmol, 1.2 equiv) (Note 2). The suspension is stirred at ambient temperature for 5 min, at which time a pale yellow color is observed. Not all solids are dissolved at this time. 3-Methyl-2-cyclohexenone (**4**) (3.97 mL, 35.0 mmol, 1 equiv) is transferred via syringe to the stirring suspension at ambient temperature. Water (3.15 mL, 175 mmol, 5 equiv) (Note 3) is added in one portion via syringe to the stirring mixture and the flask is placed in an armor bead bath preheated at 40 °C. The reaction is vigorously stirred for 12–24 h (Note 17).

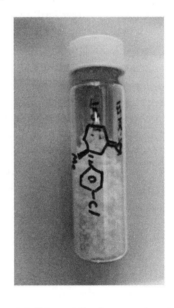

Upon consumption of the starting material (Note 18), the reaction is cooled to ambient temperature and stirring is halted, allowing the solids to settle. The reaction mixture, including the solid matter, is filtered over a pad of silica gel (Note 19). The filter cake is washed with additional 1,2-dichloroethane (150 mL). The filtrate is concentrated by rotary evaporation (28 °C, 15 mmHg) and the crude residue is purified by silica gel flash chromatography (Note 20). The combined product-containing fractions are concentrated by rotary evaporation (28 °C, 15 mmHg) to yield a solid, which is further dried under high vacuum (< 12 mmHg, 12 h) to afford (*R*)-

3-(4-chlorophenyl)-3-methylcyclohexanone (**6**) (6.80–7.07 g, 30.5–31.7 mmol, 87–91% yield, 93% ee) as a colorless crystalline solid (Note 21).

Notes

1. A two-tap Schlenk adapter connected to a bubbler and an argon/vacuum manifold is illustrated in Yu, J.; Truc, V.; Riebel, P.; Hierl, E.; Mudryk, B. *Org. Synth.* **2008**, *85*, 64–71.

2. *Iso*-butylchloroformate (98%) and 2-picolinic acid (99%) were purchased from Acros Organics and used without further purification. *N*-Methylmorpholine (redistilled, >99.5%), thionyl chloride (>99%), sodium methoxide (95%), ammonium hexafluorophosphate (99.9%), and (*S*)-*tert*-leucinol (98%) were purchased from Sigma Aldrich and used as received. The submitters prepared (*S*)-*tert*-leucinol as described by Krout, M. R.; Mohr, J. T.; Stoltz, B. M. *Org. Synth.* **2009**, *86*, 181–193. 4-Chlorophenylboronic acid (98%) and 3-methylcyclohexen-1-one (98%) were purchased from Combi-Blocks and used as received. The submitters noted a variation in ee and yield with different commercial sources and batches of 4-chlorophenylboronic acid, presumably due to the presence of variable amounts of boroxine. Palladium(II) trifluoroacetate (97%) was purchased from Strem Chemicals and 3-methylcyclohexen-1-one (98%) was purchased from Combi-Blocks, and each was used as received.

3. Methylene chloride (> 99.8%, anhydrous), toluene (99.8%, anhydrous), methanol (99.8%, anhydrous) and reagent grade 1,2-dichloroethane (99.8%, anhydrous) were purchased from Sigma Aldrich and used as received. In house deionized water was used without alteration.

4. The reaction can be monitored by TLC analysis using 3:2 hexanes/acetone as the eluent (E. Merck Silica gel 60 F254 precoated plates, 250 nm), visualizing with UV fluorescence quenching and *p*-anisaldehyde staining or iodine staining. Product amide (**2**) R_f = 0.42, impurity R_f = 0.38 (consistent with 3,3-dimethyl-2-(picolinamido)butyl picolinate), *N*-methylmorpholine R_f = 0.06 (iodine stain). Picolinic acid (**1**) and *tert*-leucinol remain at the baseline (R_f = 0), as determined by TLC.

5. If *N*-methylmorpholine is not removed prior to chromatography, it will overload the silica gel column and will contaminate product fractions.

Excess *N*-methylmorpholine complicates product solidification post chromatography.

6. Silica gel column dimensions: 5 cm diameter x 20 cm height, ca. 220 g silica gel (RediSep RF column from Teledyne Isco, catalog 69-2203-422), eluting with 4:1 hexanes/acetone. Pre-run of 200 mL is collected before fractions are taken. Fractions are collected in 18 mm x 150 mm test tubes. Fraction purity can be assayed by TLC analysis using 3:2 hexanes/acetone with UV visualization. Product (**2**) R_f = 0.42, impurity R_f = 0.38 (consistent with 3,3-dimethyl-2-(picolinamido)butyl picolinate), *N*-methylmorpholine R_f = 0.06 (iodine stain). Both starting materials remain at the baseline (R_f= 0).

7. (*S*)-*N*-(1-Hydroxy-3,3-dimethylbutan-2-yl)picolinamide (**2**) exhibited the following characterization data: R_f = 0.42 with 3:2 hexanes/acetone; mp 78.7–79.8 °C; ^{1}H NMR (400 MHz, CDCl$_3$) δ: 1.05 (s, 9 H), 2.60 (br s, 1 H), 3.70 (dd, *J* = 8.4, 11.2 Hz, 1 H), 3.96–4.03 (m, 2 H), 7.44 (dd, *J* = 5.2, 6.8 Hz, 1 H), 7.85 (dt, *J* = 1.2, 8.0 Hz, 1 H), 8.19 (d, *J* = 8.0 Hz, 1 H), 8.32 (br d, *J* = 8.0 Hz, 1 H), 8.56 (d, *J* = 4.8 Hz, 1 H); ^{13}C NMR (100 MHz, CDCl$_3$) δ: 27.0, 33.7, 60.5, 63.6, 122.4, 126.2, 137.4, 148.0, 149.6, 165.5; IR (film): 3390, 3241, 2966, 1646, 1535, 1464, 1435, 1368, 1293, 1241, 1083, 1054, 1022, 999, 863, 750, 697 cm^{-1}; HRMS (Multimode ESI/FTMS) *m/z* calcd for C$_{12}$H$_{19}$N$_2$O$_2$ [M + H]$^+$: 223.1441, found: 223.1439; $[\alpha]_D^{24}$ = –9.09 (*c* 1.728, CHCl$_3$); Anal calcd for C$_{12}$H$_{18}$N$_2$O$_2$: C, 64.84; H, 8.17; N, 12.61. Found: C, 64.65; H, 8.19; N, 12.60.

8. The submitters used an oil bath heated to 60 °C.

9. Care should be taken to ensure the addition rate is initially very slow to avoid formation of a brown, tar-like byproduct. By addition of the first 3.5 mL over 30 min and the remaining solution (~ 23 mL) over 40 min by syringe pump, no byproduct was seen.

10. (*S*)-*N*-(1-Chloro-3,3-dimethylbutan-2-yl)picolinamide hydrochloride salt exhibited the following characterization data: ^{1}H NMR (400 MHz, DMSO-*d*$_6$) δ: 0.93 (s, 9 H), 3.88–3.97 (m, 2 H), 4.08 (dt, *J* = 3.6, 9.6 Hz, 1 H), 7.68 (ddd, *J* = 1.2, 4.8, 7.2 Hz, 1 H), 8.09 (dt, *J* = 1.2, 7.6 Hz, 1 H), 8.15 (d, *J* = 7.6 Hz, 1 H), 8.68–8.71 (m, 2 H), 13.24 (br s, 1 H); ^{13}C NMR (100 MHz, DMSO-*d*$_6$) δ: 26.6, 35.3, 45.1, 59.4, 122.5, 126.9, 138.8, 147.9, 149.0, 163.7; IR (Neat): 3192, 3023, 2953, 2922, 1680, 1601, 1561, 1518, 1472, 1345, 1291, 1211, 1181, 1036, 1002, 971 cm^{-1}; HRMS (MultiMode ESI/FTMS) *m/z* calcd for C$_{12}$H$_{18}$ClN$_2$O [M+H]$^+$: 241.1102, found 241.1103; $[\alpha]_D^{22}$ = +25.09 (*c* 1.132, MeOH). Anal calcd for C$_{12}$H$_{18}$Cl$_2$N$_2$O:

C, 52.00; H, 6.55; N, 10.11; Cl 25.58. Found: C, 52.29; H, 6.59; N, 10.10; Cl, 25.57. A broad resonance at ~ 5 ppm in the ^{1}H NMR spectrum is found with incomplete drying of the product; by D_2O quenching we determined this resonance to be the hydrate.

11. Sodium methoxide is used from a freshly opened bottle or retrieved from storage in a nitrogen-atmosphere glovebox, free of adventitious moisture. The addition to the reaction mixture is exothermic.

12. The reaction can be monitored by TLC analysis using 3:2/hexanes:acetone as the eluent, visualizing with UV fluorescence quenching and p-anisaldehyde staining. Product R_f = 0.44, starting material R_f = 0.32.

13. Toluene is added to prevent the crude ligand from exposure to concentrated sodium methoxide. The ligand is unstable to concentrated acids or bases.

14. After toluene was added, approx 140–160 mL of solvent were removed on the rotovap. During the work-up and phase separation, a persistent rag and emulsions formed. The rag is to be kept with the top organic layer.

15. Silica gel column dimensions: 5 cm diameter x 20 cm height, ca. 200 g silica gel (Silica Gel ZEOprep® 60 ECO 40-63 Micron from American International Chemical, Inc. It is necessary to use this specific silica gel to suppress reversion to (S)-N-(1-hydroxy-3,3-dimethylbutan-2-yl)picolinamide during chromatography, which was observed when employing silica gel from other commercial sources. The submitters report the reaction's yield is decreased by 5–10% when another silica gel is used.) (Note: The checker used this brand of Silica Gel as received from Prof. Stoltz's lab. No other silica brands were examined) eluting with 4:1 hexanes/acetone. Fractions are collected in 18 mm x 150 mm test tubes. Fraction purity can be assayed by TLC analysis using 3:2 hexanes/acetone with UV visualization. Product (**3**) R_f = 0.44. This slow eluent is used for chromatography to remove a yellow impurity of similar polarity that contaminates the white solid product if the column is eluted with higher polarity eluent. This impurity appears to be an indiscrete decomposition product not readily identified. This impurity is not easily removed by recrystallization, and contaminated material should be resubmitted to flash chromatography.

16. (S)-tert-ButylPyOx (**3**) exhibited the following characterization data R_f = 0.44 with 3:2 hexanes/acetone; mp 70.4–71.0 °C; ^{1}H NMR (400 MHz, CDCl$_3$) δ: 0.99 (s, 9 H), 4.13 (dd, J = 8.8, 10.4 Hz, 1 H), 4.32 (t, J = 8.4 Hz,

1 H), 4.46 (t, J = 8.8 Hz, 1 H), 7.39 (ddd, J = 1.2, 4.4, 7.2 Hz, 1 H), 7.77 (dt, J = 1.2, 8.0 Hz, 1 H), 8.10 (d, J = 8.0 Hz, 1 H), 8.72 (d, J = 4.0 Hz, 1 H); ^{13}C NMR (100 MHz, CDCl$_3$) δ: 26.0, 34.0, 69.3, 76.5, 124.0, 125.4, 136.5, 147.0, 149.7, 162.5; IR (film): 2954, 2903, 2867, 1640, 1565, 1466, 1358, 1345, 1272, 1244, 1211, 1095, 1037, 967 cm^{-1}; HRMS (MultiMode ESI/APCI) m/z calcd for C$_{12}$H$_{17}$N$_2$O [M+H]$^+$: 205.1335, found 205.1336; $[\alpha]^{22}_D$ = -91.85 (c 5.09, CHCl$_3$); Anal calcd for C$_{12}$H$_{16}$N$_2$O, C 70.56, H 7.90, N 13.71, found C 71.09, H 8.12, N 13.29. Purity of **3** was assessed at 99.3 wt% by quantitative ^{1}H NMR in CDCl$_3$ using methyl phenyl sulfone as a standard. If desired, the white solid can be recrystallized from minimal hot n-heptane, with crystals collected at –20 °C (after cooling to room temperature) as white needles. However, the ligand is more difficult to weigh in this state due to its tendency to cling with static electricity. Enantiomeric excess can be determined via analytical chiral SFC (Jasco SFC utilizing a Chiralcel OB-H column (4.6 mm x 25 cm) obtained from Daicel Chemical Industries, Ltd) with visualization at 210 nm and flow rate of 5 mL/min eluting with 10% MeOH/CO$_2$. Major enantiomer retention time: 2.51 min, minor enantiomer retention time: 2.20 min. (In addition to the high optical purity as seen by polarimetry, the material was analysed by SFC using the method above and the ligand was determined to be 99.7% ee.) Attempts were made to separate enantiomers via analytical chiral HPLC using various columns (Chiralcel OD-H, Chiralcel OJ-H, Chiralpak AD, Chiralpak AS, Chiralcel OB-H) and solvent systems, however adequate separation of peaks was not achieved. In the absence of an analytical chiral SFC, optical rotation measurements can be utilized to give the enantiomeric excess by way of optical purity calculations (optical purity (%) = $\frac{[\alpha]_{observed}}{[\alpha]_{maximal}} \times 100$). The optical rotation listed above may be used as $\alpha_{maximal}$ and a sufficiently large concentration (e.g. c 5.00) as well as multiple trials should be used in order to minimize error. It is imperative to store the ligand in a desiccator to minimize hydrolysis, which results in diminished enantiocontrol in the conjugate addition reaction. For long-term storage, the ligand should be stored frozen under an inert atmosphere; we recommend a nitrogen atmosphere glovebox freezer.

17. Vigorous stirring that results in a visible vortex is essential for reaction conversion, as dispersion of water throughout the 1,2-dichloroethane is necessary. After water is added to the reaction, a viscous gel is formed

which prevented the magnetic stir bar from spinning. The reaction was manually agitated until the stir bar became free.

18. The reaction progress can be monitored by TLC analysis with 4:1 hexanes/ethyl acetate, using a *p*-anisaldehyde stain; product (**6**) R_f = 0.38 (stains yellow/orange with fresh *p*-anisaldehyde stain), 3-methyl-2-cyclohexenone (**4**) R_f = 0.19 (stains tan/brown), 4,4'-dichloro-1,1'-biphenyl R_f = 0.65 (does not stain). Reaction times typically are 12–24 h, with conversion slowing considerably after the first several hours.

19. Silica gel plug dimensions: 3 cm diameter x 4 cm height, ca. 12 g silica gel.

20. The product is very crystalline. When loading onto the column, the product may crystallize. Silica gel column dimensions: 5 cm diameter x 20 cm height, ca. 220 g silica gel (RediSep RF column from Teledyne Isco, catalog 69-2203-422). The column is eluted with 92:8 hexanes/ethyl acetate until the product is collected. Fractions are collected in 18 mm x 150 mm test tubes. Fraction purity can be assayed by TLC analysis using 4:1 hexanes/ethyl acetate with UV fluorescence quenching visualization and heating with *p*-anisaldehyde; product (**6**) R_f = 0.38, (stains yellow/orange with fresh *p*-anisaldehyde stain), 3-methyl-2-cyclohexenone (**4**) R_f = 0.19 (stains tan/brown), 4,4'-dichloro-1,1'-biphenyl R_f = 0.65 (does not stain).

21. (*R*)-3-(4-Chlorophenyl)-3-methylcyclohexanone (**6**) exhibited the following characterization data R_f = 0.38 (hexanes/ethyl acetate 4:1), ^{1}H NMR (400 MHz, CDCl₃) δ: 1.30 (s, 3 H), 1.60-1.69 (m, 1 H), 1.85-1.94 (m, 2 H), 2.12-2.19 (m, 1 H), 2.31 (t, *J* = 6.8 Hz, 2 H), 2.42 (d, *J* = 14.4 Hz, 1 H), 2.83 (d, *J* = 14.4 Hz, 1 H), 7.23-7.25 (m, 2 H), 7.27-7.30 (m, 2 H); ^{13}C NMR (100 MHz, CDCl₃) δ: 21.9, 29.9, 37.9, 40.7, 42.6, 52.9, 127.1, 128.6, 132.0, 145.8, 210.9; IR (thin film): 2944, 2865, 1693, 1589, 1492, 1450, 1432, 1352, 1265, 1229, 1092, 1010, 952, 824 759, 737, 721 cm^{-1}; HRMS (MultiMode ESI/APCI) *m/z* calcd for $C_{13}H_{14}ClO$ [M+H]$^+$: 223.0884, found 223.0885; [α]$_D^{22}$ = -65.8 (*c* 1.912, CDCl₃); Anal calcd $C_{13}H_{15}ClO$: C 70.11, H 6.79, O 7.18, Cl 15.92 found C 69.73, H 7.02, O 7.49, Cl 15.88. Enantiomeric excess of 93% is determined via analytical chiral HPLC (Agilent 1100 Series HPLC utilizing a Chiralcel OB-H column (4.6 mm x 25 cm) obtained from Daicel Chemical Industries, Ltd) with visualization at 254 nm and flow rate of 1 mL/min eluting with 1% *iso*-propanol/hexanes. The sample was prepared in a

1:1 v/v solution of *iso*-propanol/hexanes. Major enantiomer retention time 16.47 min, minor enantiomer retention time 14.55 min.

Working with Hazardous Chemicals

The procedures in *Organic Syntheses* are intended for use only by persons with proper training in experimental organic chemistry. All hazardous materials should be handled using the standard procedures for work with chemicals described in references such as "Prudent Practices in the Laboratory" (The National Academies Press, Washington, D.C., 2011; the full text can be accessed free of charge at http://www.nap.edu/catalog.php?record_id=12654). All chemical waste should be disposed of in accordance with local regulations. For general guidelines for the management of chemical waste, see Chapter 8 of Prudent Practices.

In some articles in *Organic Syntheses*, chemical-specific hazards are highlighted in red "Caution Notes" within a procedure. It is important to recognize that the absence of a caution note does not imply that no significant hazards are associated with the chemicals involved in that procedure. Prior to performing a reaction, a thorough risk assessment should be carried out that includes a review of the potential hazards associated with each chemical and experimental operation on the scale that is planned for the procedure. Guidelines for carrying out a risk assessment and for analyzing the hazards associated with chemicals can be found in Chapter 4 of Prudent Practices.

The procedures described in *Organic Syntheses* are provided as published and are conducted at one's own risk. *Organic Syntheses, Inc.*, its Editors, and its Board of Directors do not warrant or guarantee the safety of individuals using these procedures and hereby disclaim any liability for any injuries or damages claimed to have resulted from or related in any way to the procedures herein.

Discussion

Pyridinooxazoline (PyOx) ligands represent a growing class of bidentate, dinitrogen ligands used in asymmetric catalysis.[2] Our laboratory has reported the palladium-catalyzed asymmetric conjugate addition of arylboronic acids to cyclic, β,β-disubstituted enones utilizing (S)-tert-ButylPyOx as the chiral ligand.[3] This robust reaction does not require an inert atmosphere, is highly tolerant of water,[4] and provides cyclic ketones bearing β-benzylic quaternary stereocenters in high yields and enantioselectivities. While the reaction itself proved to be amenable to multi-gram scale, no reliable method for the large-scale synthesis of (S)-tert-ButylPyOx was known.[5] We sought to address this shortcoming by developing an efficient route starting from a cheap, commercially available precursor to pyridinooxazoline ligands. In our initial experiments, (S)-tert-ButylPyOx was synthesized by methanolysis of 2-cyanopyridine, and subsequent acid-catalyzed cyclization with (S)-tert-leucinol to afford the (S)-tert-ButylPyOx ligand.[6] We found the yields of this reaction sequence to be highly variable, and the purification by silica gel chromatography to be tedious due to the presence of numerous impurities.

The above reported procedure begins with activation of 2-picolinic acid by treatment with iso-butylchloroformate and N-methylmorpholine, facilitating the desired amidation with (S)-tert-leucinol in high isolated yield, albeit requiring column chromatography.[7] Many alternative coupling methods were screened, but mixed anhydride activation gave the highest yield, and the chromatographic purification proved simple even on multi-gram scale. Furthermore, unlike the methoxyimidate intermediate from the initial synthesis, (S)-N-(1-hydroxy-3,3-dimethylbutan-2-yl)picolinamide is bench stable at room temperature, allowing long-term storage of material. The cyclization of the amide alcohol to (S)–tert-ButylPyOx proved more challenging than anticipated. Activation as mesylate and tosylate followed by in situ cyclization gave the desired product in low yield and incomplete conversion. This could be a result of ligand hydrolysis under the reaction conditions.[8] As an alternative to in situ cyclization of an activated intermediate, the amide alcohol was treated with thionyl chloride to yield the hydrochloride salt of (S)-N-(1-chloro-3,3-dimethylbutan-2-yl)picolinamide. After drying under vacuum, this compound proved to be bench stable and was spectroscopically unchanged after exposure to ambient atmosphere and adventitious moisture for more than one week. Furthermore, this chloride salt proved to be a competent cyclization substrate. A series of bases were screened, and sodium methoxide was found to be optimal, as slower rates of hydrolysis to (S)-N-(1-hydroxy-3,3-dimethylbutan-2-yl)picolinamide were observed when compared to the use of other bases.

Our asymmetric conjugate addition method is noteworthy for its simple procedure, owing to its tolerance of both water and atmospheric oxygen. As the catalytic, enantioselective construction of all-carbon quaternary stereocenters remains a challenging problem in synthetic chemistry,[9] asymmetric conjugate addition of carbon-based nucleophiles to suitable α,β-unsaturated carbonyl acceptors has garnered much attention as a reliable method for the formation of quaternary stereocenters.[10] However, the majority of well-established asymmetric conjugate additions involve the use of highly reactive and water-sensitive organometallic reagents (e.g., diorganozinc,[11] triorganoaluminum,[12] and organomagnesium reagents[13]) and, therefore, require rigorously anhydrous reaction conditions. The notable exception is the Hayashi and Shintani rhodium/diene system,[14] one of few quaternary-center forging reactions that avoid use of organozinc or organoaluminum reagents.[15] Furthermore, our system advantageously uses commercially available arylboronic acids, as opposed to sodium tetraarylborates or boroxins, and palladium as the transition metal catalyst, as opposed to relatively rare and expensive rhodium.[16] We believe these advantages uniquely dispose our system toward use in large-scale asymmetric conjugate addition reactions. Furthermore, a large number of 1,4-addition products have been prepared utilizing this methodology featuring a wide array of functional groups (Table 1).

Table 1. β–Quaternary Ketones Prepared via Asymmetric Conjugate Addition[17]

84% yield
92% ee

99% yield
87% ee

58% yield
69% ee

91% yield
95% ee

99% yield
91% ee

40% yield
92% ee

99% yield
96% ee

99% yield
96% ee

91% yield
93% ee

84% yield
91% ee

85% yield
93% ee

96% yield
92% ee

86% yield
79% ee

68% yield
88% ee

95% yield
91% ee

74% yield
91% ee

86% yield
85% ee

References

1. The Warren and Katharine Schlinger Laboratory for Chemistry and Chemical Engineering, Division of Chemistry and Chemical

Engineering, California Institute of Technology, 1200 East California Boulevard, MC 101-20, Pasadena, California, 91125, United States. E-mail: stoltz@caltech.edu. This publication is based on work supported by NIH-NIGMS (R01GM080269-01), and the authors additionally thank Amgen, Abbott, Boehringer Ingelheim and Caltech for financial support. J.C.H. wishes to thank the American Chemical Society Division of Organic Chemistry for a graduate fellowship. We also thank Shionogi & Co., Ltd. for a research grant and a fellowship to H.S.

2.   (a) Podhajsky, S. M.; Iwai, Y.; Cook-Sneathen, A.; Sigman, M. S.
     Tetrahedron **2011**, *67*, 4435–4441.   (b) Aranda, C.; Cornejo, A.; Fraile, J.
     M.; García-Verdugo, E.; Gil, M. J.; Luis, S. V.; Mayoral, J. A.; Martínez-
     Merino, V.; Ochoa, Z. *Green Chem.* **2011**, *13*, 983–990.   (c) Pathak, T. P.;
     Gligorich, K. M.; Welm, B. E.; Sigman, M. S. *J. Am. Chem. Soc.* **2010**, *132*,
     7870–7871.   (d) Jiang, F.; Wu, Z.; Zhang, W. *Tetrahedron Lett.* **2010**, *51*,
     5124–5126.   (e) Jensen, K. H.; Pathak, T. P.; Zhang, Y.; Sigman, M. S. *J.
     Am. Chem. Soc.* **2009**, *131*, 17074–17075.   (f) He, W.; Yip, K.-T.; Zhu, N.-Y.;
     Yang, D. *Org. Lett.* **2009**, *11*, 5626–5628.   (g) Dai, H.; Lu, X. *Tetrahedron
     Lett.* **2009**, *50*, 3478-3481.   (h) Linder, D.; Buron, F.; Constant, S.; Lacour,
     J. *Eur. J. Org. Chem.* **2008**, 5778–5785.   (i) Schiffner, J. A.; Machotta, A. B.;
     Oestreich, M. *Synlett* **2008**, 2271–2274.   (j) Koskinen, A. M. P.; Oila, M. J.;
     Tois, J. E. *Lett. Org. Chem.* **2008**, *5*, 11–16.   (k) Zhang, Y.; Sigman, M. S. *J.
     Am. Chem. Soc.* **2007**, *129*, 3076–3077.   (l) Yoo, K. S.; Park, C. P.; Yoon, C.
     H.; Sakaguchi, S.; O'Neill, J.; Jung, K. W. *Org. Lett.* **2007**, *9*, 3933–3935.
     (m) Dhawan, R.; Dghaym, R. D.; St. Cyr, D. J.; Arndtsen, B. A. *Org. Lett.*
     2006, *8*, 3927–3930.   (n) Xu, W.; Kong, A.; Lu, X. *J. Org. Chem.* **2006**, *71*,
     3854–3858.   (o) Malkov, A. V.; Stewart Liddon, A. J. P.; Ramírez-López,
     P.; Bendová, L.; Haigh, D.; Kocovsky, P. *Angew. Chem., Int. Ed.* **2006**, *45*,
     1432–1435.   (p) Abrunhosa, I.; Delain-Bioton, L.; Gaumont, A.-C.; Gulea,
     M.; Masson, S. *Tetrahedron* **2004**, *60*, 9263–9272.   (q) Brunner, H.; Kagan,
     H. B.; Kreutzer, G. *Tetrahedron: Asymmetry* **2003**, *14*, 2177–2187.   (r)
     Cornejo, A.; Fraile, J. M.; García, J. I.; Gil, M. J.; Herrerías, C. I.;
     Legarreta, G.; Martínez-Merino, V.; Mayoral, J. A. *J. Mol. Catal. A: Chem.*
     2003, *196*, 101–108.   (s) Zhang, Q.; Lu, X.; Han, X. *J. Org. Chem.* **2001**, *66*,
     7676–7684.   (t) Zhang, Q.; Lu, X. *J. Am. Chem. Soc.* **2000**, *122*, 7604–7605.
     (u) Perch, N. S.; Pei, T.; Widenhoefer, R. A. *J. Org. Chem.* **2000**, *65*, 3836–
     3845.   (v) Bremberg, U.; Rahm, F.; Moberg, C. *Tetrahedron: Asymmetry*
     1998, *9*, 3437–3443.   (w) Brunner, H.; Obermann, U.; Wimmer, P.
     Organometallics **1989**, *8*, 821–826.

3. (a) Kikushima, K.; Holder, J. C.; Gatti, M.; Stoltz, B. M. *J. Am. Chem. Soc.* **2011**, *133*, 6902-6905. (b) Holder, J. C.; Zou, L.; Marziale, A. N.; Liu, P.; Lan, Y.; Gatti, M.; Kikushima, K.; Houk, K. N.; Stoltz, B. M. *J. Am. Chem. Soc.* **2013**, *135*, 14996–15007; (c) Holder, J. C.; Marziale, A. N.; Gatti, M.; Mao, B.; Stoltz, B. M. *Chem. Eur. J.* **2013**, *19*, 74–77; (d) Holder, J. C.; Goodman, E. D.; Kikushima, K.; Gatti, M.; Marziale, A. N.; Stoltz, B. M. *Tetrahedron* **2015**, *71*, 5781–5792. (e) Shockley, S. E.; Holder, J. C.; Stoltz, B. M. *Org. Process Res. Dev.* **2015**, http://pubs.acs.org/doi/pdf/10.1021/acs.oprd.5b00169.

4. The reaction requires a small amount of water to run efficiently. Typically 5 equiv of water are added to small-scale reactions.

5. A number of syntheses are known, including: Brunner. H.; Obermann. U. *Chem. Ber.* **1989**, *122*, 499-507, and ref 3b.

6. This route is adapted from the synthesis reported in ref 5

7. Jensen, K. H.; Webb, J. D.; Sigman, M. S. *J. Am. Chem. Soc.* **2010**, *132*, 17471–17482. While Sigman and coworkers utilize these conditions to make derivatives of PyOx ligands, such conditions are not reported for the synthesis of *tert*-ButylPyOx.

8. Degradation experiments demonstrate that (*S*)-*tert*-ButylPyOx is susceptible to hydrolysis. Exposure of *tert*-ButylPyOx to 3N HCl results in complete hydrolysis to (*S*)-*N*-(1-hydroxy-3,3-dimethylbutan-2-yl)picolinamide as observed by ¹H NMR.

9. For reviews on the synthesis of quaternary stereocenters, see: (a) Denissova, I.; Barriault, L. *Tetrahedron* **2003**, *59*, 10105–10146. (b) Douglas, C. J.; Overman, L. E. *Proc. Natl. Acad. Sci. U.S.A.* **2004**, *101*, 5363–5367. (c) Christoffers, J.; Baro, A. *Adv. Synth. Catal.* **2005**, *347*, 1473–1482. (d) Trost, B. M.; Jiang, C. *Synthesis* **2006**, 369–396. (e) Mohr, J. T.; Stoltz, B. M. *Chem.–Asian J.* **2007**, *2*, 1476–1491. (f) Cozzi, P. G.; Hilgraf, R.; Zimmermann, N. *Eur. J. Org. Chem.* **2007**, *36*, 5969–5994.

10. For an excellent comprehensive review, see: Hawner, C.; Alexakis, A. *Chem. Commun.* **2010**, *46*, 7295–7306.

11. (a) Wu, J.; Mampreian D. M.; Hoveyda, A. H. *J. Am. Chem. Soc.* **2005**, *127*, 4584–4585. (b) Hird A. W.; Hoveyda, A. H. *J. Am. Chem. Soc.* **2005**, *127*, 14988–14989. (c) Lee, K.; Brown, M. K.; Hird, A. W.; Hoveyda, A. H. *J. Am. Chem. Soc.* **2006**, *128*, 7182–7184. (d) Brown, M. K.; May, T. L.; Baxter, C. A.; Hoveyda, A. H. *Angew. Chem., Int. Ed.* **2007**, *46*, 1097–1100. (e) Wilsily, A.; Fillion, E. *J. Am. Chem. Soc.* **2006**, *128*, 2774–2775. (f) Wilsily, A.; Fillion, E. *J. Org. Chem.* **2009**, *74*, 8583–8594. (g) Dumas, A. M.; Fillion, E. *Acc. Chem. Res.* **2010**, *43*, 440–454. (h) Feringa, B. L. *Acc.*

Chem. Res. **2000**, *33*, 346–353. (i) Wilsily, A.; Fillion, E. *Org. Lett.* **2008**, *10*, 2801–2804.

12. (a) d'Augustin, M.; Palais, L.; Alexakis A. *Angew. Chem., Int. Ed.* **2005**, *44*, 1376–1378. (b) Vuagnoux-d'Augustin, M.; Alexakis, A. *Chem. Eur. J.* **2007**, *13*, 9647–9662. (c) Palais, L.; Alexakis A. *Chem. Eur. J.* **2009**, *15*, 10473–10485. (d) Fuchs, N.; d'Augustin, M.; Humam, M.; Alexakis, A.; Taras, R.; Gladiali, S. *Tetrahedron: Asymmetry*, **2005**, *16*, 3143–3146. (e) Vuagnoux-d'Augustin, M.; Kehrli, S.; Alexakis, A. *Synlett*, **2007**, 2057–2060. (f) May, T. L.; Brown, M. K.; Hoveyda, A. H. *Angew. Chem., Int. Ed.* **2008**, *47*, 7358–7362. (g) Ladjel, C.; Fuchs, N.; Zhao, J.; Bernardinelli, G. Alexakis, A. *J. Org. Chem.* **2009**, 4949–4955. (h) Palais, L.; Mikhel, I. S.; Bournaud, C.; Micouin, L.; Falciola, C. A.; Vuagnoux-d'Augustin, M.; Rosset, S.; Bernardinelli, G.; Alexakis, A. *Angew. Chem., Int. Ed.* **2007**, *46*, 7462–7465. (i) Hawner, C.; Li, K.; Cirriez, V.; Alexakis, A. *Angew. Chem., Int. Ed.* **2008**, *47*, 8211–8214. (j) Müller, D.; Hawner, C.; Tissot, M.; Palais, L.; Alexakis, A. *Synlett*, **2010**, 1694–1698. (k) Hawner, C.; Müller, D.; Gremaud, L.; Felouat, A.; Woodward, S.; Alexakis, A. *Angew. Chem., Int. Ed.* **2010**, *49*, 7769–7772.

13. (a) Martin, D.; Kehrli, S.; d'Augustin, M.; Clavier, H.; Mauduit, M.; Alexakis, A. *J. Am. Chem. Soc.* **2006**, *128*, 8416–8417. (b) Kehrli, S.; Martin, D.; Rix, D.; Mauduit, M.; Alexakis, A. *Chem. Eur. J.* **2010**, *16*, 9890–9904. (c) Hénon, H.; Mauduit, M.; Alexakis, A. *Angew. Chem., Int. Ed.* **2008**, *47*, 9122–9124. (d) Matsumoto, Y.; Yamada, K.-i.; Tomioka, K. *J. Org. Chem.* **2008**, *73*, 4578–4581.

14. (a) Shintani, R.; Tsutsumi, Y.; Nagaosa, M.; Nishimura, T.; Hayashi, T. *J. Am. Chem. Soc.* **2009**, *131*, 13588–13589. (b) Shintani, R.; Takeda, M.; Nishimura, T.; Hayashi, T. *Angew. Chem., Int. Ed.* **2010**, *49*, 3969–3971.

15. (a) Mauleón, P.; Carretero, J. C. *Chem. Commun.* **2005**, 4961–4963. (b) Shintani, R. Duan, W.-L.; Hayashi, T. *J. Am. Chem. Soc.* **2006**, *128*, 5628–5629.

16. A paper describing the use of a Rh•OlefOx (olefin-oxazoline) complex provided a single example of a phenyl boronic acid addition to 3-methylcyclohexenone. Unfortunately, the product was isolated in only 36% yield and 85% ee, see: Hahn, B. T.; Tewes, F.; Fröhlich, R.; Glorius, F. *Angew Chem., Int. Ed.* **2010**, *49*, 1143–1146. For more recent examples of palladium-catalyzed conjugate addition of arylboronic acids, see: Gottumukkala, A. L.; (a) Matcha, K.; Lutz, M.; de Vries, J. G.; Minnaard, A. J. *Eur. J. Chem.* **2012**, *18*, 6907–6914. (b) Buter, J.; Moezelaar, R.; Minnaard, A. J. *Org. Biomol. Chem.* **2014**, *12*, 5883–5890.

17. Yields in Table 1 are without ammonium hexafluorophosphate as an additive, which can greatly improve isolated yield of many arylboronic acid additions.

Appendix
Chemical Abstracts Nomenclature (Registry Number)

2-Picolinic Acid: Pyridine-2-carboxylic acid; (98-98-6)
iso-Butylchloroformate; (543-27-1)
N-Methylmorpholine: NMM; (109-02-4)
(S)-tert-Leucinol: (S)-2-Amino-3,3-dimethyl-1-butanol; (112245-13-3)
(S)-N-(1-Hydroxy-3,3-dimethylbutan-2-yl)picolinamide; (1476785-62-2)
thionyl chloride; (7719-09-7)
Sodium methoxide: sodium methylate; (124-41-4)
(S)-tert-ButylPyOx: (S)-4-(tert-Butyl)-2-(pyridin-2-yl)-4,5-dihydrooxazole (117408-98-7)
Palladium(II) trifluoroacetate (42196-31-6)
Ammonium hexafluorophosphate (16941-11-0)
3-Methyl-2-cyclohexenone (1193-18-6)
4-Chlorophenylboronic acid (1679-18-1)
(R)-3-(4-Chlorophenyl)-3-methylcyclohexanone (1235989-03-3)

Jeffrey C. Holder graduated from Harvard University in 2009 where he conducted research with Professors Daniel Kahne and E. J. Corey. In 2014, he completed his Ph.D. in the laboratory of Professor Brian M. Stoltz at the California Institute of Technology, where he studied palladium-catalyzed asymmetric conjugate addition reactions and their application in total synthesis. Currently, he is a National Institutes of Health postdoctoral fellow in the laboratory of Professor John F. Hartwig at the University of California, Berkeley. His research interests include the development and study of transition-metal catalyzed reactions, and their applications in natural product synthesis.

Organic
Syntheses

Samantha E. Shockley was born in Birmingham, AL in 1990. She received her B.S. degree in Chemistry from the University of Chicago in 2012 where she conducted research for Professor Richard F. Jordan. After graduating from the College, she worked with Professor Martin Banwell at the Australian National University as a U.S. Fulbright scholar. She is now pursuing her graduate studies at the California Institute of Technology under the guidance of Professor Brian M. Stoltz. Her graduate research focuses on the total synthesis of novel natural products.

Mario P. Wiesenfeldt was born in 1989 in Ludwigshafen am Rhein, Germany. He received his B.S. degree in Chemistry from Ruprecht-Karls-Universität Heidelberg in 2012 where he conducted research with Professor Lutz H. Gade. He started his Masters studies in Heidelberg in the same year with an exchange semester at the University of York, United Kingdom. As a visiting student in the laboratory of Professor Brian M. Stoltz, he has conducted research on palladium-catalyzed enantioselective conjugate addition reactions towards his Master's Thesis. After successfully completing his master degree in early 2015, he joined the group of Professor Frank Glorius at the Westfälische Wilhelms-Universität Münster as a graduate student. His research interests include organometallic chemistry and the development of new synthetic methodology.

Hideki Shimizu was born in 1976 in Osaka, Japan. He received his M. S. with Professor Munehiro Nakatani from Kagoshima University in 2001 and Ph. D. in Organic Chemistry from Kyushu University in 2004 under the supervision of Professors Tsutomu Katsuki. He joined the Department of Process Chemistry at Shionogi & Co., Ltd. in 2004. During his time at Shionogi, he worked as a visiting postdoctoral fellow under the direction of Professor Brian M. Stoltz at the California Institute of Technology from 2010 to 2011. He is currently an Associate Director at Shionogi, working on production planning management in the Global Supply Chain Management Division.

Organic **S**yntheses

Brian M. Stoltz was born in Philadelphia, PA in 1970 and obtained his B.S. degree from the Indiana University of Pennsylvania in Indiana, PA. After graduate work at Yale University in the labs of John L. Wood and an NIH postdoctoral fellowship at Harvard in the Corey labs he took a position at the California Institute of Technology. A member of the Caltech faculty since 2000, he currently is a Professor of Chemistry. His research interests lie in the development of new methodology for general applications in synthetic chemistry.

Brian Cochran obtained his Bachelor's degree in chemistry in 1999 at Grinnell College. His organic chemistry career started as a medicinal chemist while working for Albany Molecular Research, Inc. followed by Genzyme, Inc. Brian returned to academia and received his Ph. D. in 2009 developing and studying new amination methods at the University of Washington under Professor Forrest Michael. He moved to Colorado to pursue his postdoctoral studies with Professor Tomislav Rovis developing new cyclic anhydride desymmetrization methods with application towards the total synthesis of Ionomycin. In 2012, Brian joined the process chemistry group at Amgen in Thousand Oaks, California.

Phosphorus(III)-Mediated Reductive Condensation of α-Keto Esters and Protic Pronucleophiles

Wei Zhao and Alexander T. Radosevich[*1]

Department of Chemistry, The Pennsylvania State University, University Park, Pennsylvania 16802, United States

Checked by Jared T. Moore and Brian M. Stoltz

Procedure

A. *Methyl 2-(N-Benzyl-4-Methylphenylsulfonamido)-2-Phenylacetate* (*3*). 4-Methyl-*N*-(phenylmethyl)benzenesulfonamide (**2**) (8.49 g, 32.5 mmol, 1.1 equiv) is placed in an oven-dried, 1-L 3-necked round-bottomed flask equipped with a magnetic stir bar (1.5 × 3.5 cm, Teflon-coated, egg-shaped). The center neck of the reaction flask is fitted with a rubber septum-capped 250-mL pressure-equalizing addition funnel. One of the side necks is fitted with a rubber septum and the other is fitted with a nitrogen inlet (Figure 1). The reaction vessel is then purged with nitrogen atmosphere by three evacuation-backfill cycles (Note 1). Dry dichloromethane (296 mL, 0.1 M) is added to the round-bottomed flask via cannula transfer from an oven-dried 1-L Schlenk flask (Note 2). Methyl benzoylformate (**1**) (4.20 mL, 4.85 g, 29.6 mmol, 1.0 equiv) is added to the solution through the side neck (Note 3). The addition funnel is then charged with tris(dimethylamino)phosphine (5.90 mL, 5.30 g, 32.5 mmol, 1.1 equiv) through the septum on top (Note 4). Dichloromethane (33 mL) is added to the addition funnel in the same way using a 60-mL syringe and an oven-dried steel needle. The reaction flask is cooled to –78 °C with a dry ice-acetone bath in an appropriately sized cooling vessel (Note 5). The tris(dimethylamino)phosphine solution is added dropwise to the flask over 20–30 min while stirring (Notes 6 and 7). Upon complete addition of the tris(dimethylamino)phosphine solution, the cooling bath is removed and the reaction mixture is stirred for 2 h, during which time it warms to ambient temperature (Note 8).

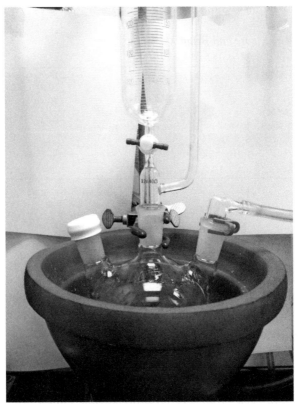

Figure 1. Reaction Set-up

When the reaction reaches completion as determined by TLC (Note 9), the dropping funnel is removed and distilled water (300 mL) is added to the reaction mixture in one portion. The biphasic mixture is then transferred to a 1-L separatory funnel (Note 10). The organic layer is separated, washed with saturated aqueous sodium chloride solution (3 × 400 mL), dried over anhydrous sodium sulfate (150 g) for 15 min (Note 11). The sodium sulfate is filtered and the solution is concentrated in vacuo using a rotary evaporator (ca. 100 mmHg, water bath temperature 30 °C). The crude residue is purified by silica gel flash column chromatography using EtOAc and hexanes as eluent (Notes 12, 13, 14 and 15), yielding the title compound as an air and moisture stable white amorphous solid (10.7 g, 88%) (Notes 16 and 17) (Figure 2).

Figure 2. Amorphous white product

Notes

1. 4-Methyl-*N*-(phenylmethyl)benzenesulfonamide was synthesized according to the known procedure: Coste, A.; Couty, F.; Evano, G. *Org. Synth.* **2010**, *87*, 231.
2. Dichloromethane was purchased from Fischer Scientific (D138-4) and degassed by bubbling argon for 1.5 h before passing through an activated alumina column using a Glass Contour solvent drying system. The solvent was collected in an oven-dried 1-L Schlenk flask , which had been purged with argon atmosphere by three evacuation-backfill cycles.
3. Methyl benzoylformate (>97.0%) was purchased from TCI and used as received.

4. Tris(dimethylamino)phosphine (97%) was purchased from Alfa Aesar and used as received.

5. 4-Methyl-N-(phenylmethyl)benzenesulfonamide may not be completely dissolved at this temperature.

6. The rate of the addition was held at approximately 2 drops per second.

7. The mixture was stirred at 600 rpm throughout the reaction.

8. The reaction should remain colorless to faint yellow. A bright yellow color usually indicates the formation of side products.

9. Thin layer chromatograph was performed on silica gel 60 F_{254} TLC plate (EMD Millipore TLC Silica Gel 60 Glass Plates, purchased from Fischer Scientific) with 1:6 EtOAc:hexanes as eluent. The following R_f values were observed (visualized under 254 nm UV light): R_f (methyl benzoylformate): 0.40; R_f (4-methyl-N-(phenylmethyl)benzene-sulfonamide): 0.14; R_f (product): 0.22. Picture of TLC plate is shown below (left lane: methyl benzoylformate; middle lane: reaction mixture; right lane: 4-methyl-N-(phenylmethyl)benzenesulfonamide) (Figure 3).

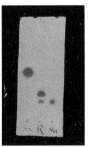

Figure 3. Image of TLC Analysis

10. Extractions should be performed carefully, since HMPA is produced as a byproduct of the reaction.

11. Sodium chloride and sodium sulfate were purchased from VWR and used as received.

12. Silica gel (230-400 mesh) was purchased from SiliCycle and used as received.

13. Ethyl acetate and hexanes were purchased from Fischer Scientific and used as received.

14. Dichloromethane (5 mL) was added to the crude mixture. The crude product was loaded onto a column packed with silica gel slurry in 1:10 EtOAc:hexanes. Column is 8 cm in diameter, height of silica gel is

27 cm. After 1 L of initial elution with 1:10 EtOAc:hexanes, 1:8 EtOAc:hexanes (1,800 mL), 1:7 EtOAc:hexanes (1,600 mL) and 1:6 EtOAc:hexanes (2,500 mL) were collected in 90 mL fractions.

15. The product was isolated as a white amorphous solid and had the following spectral characteristics: ^{1}H NMR (500 MHz, CDCl$_3$) δ: 2.44 (s, 3 H), 3.60 (s, 3 H), 4.42 (d, J = 16.2 Hz, 1 H), 4.65 (d, J = 16.2 Hz, 1 H), 5.80 (s, 1 H), 6.87 (dd, J = 7.7, 1.8 Hz, 2 H), 7.01–7.06 (m, 3 H), 7.10–7.15 (m, 2 H), 7.17–7.23 (m, 3 H), 7.25–7.28 (m, 2 H), 7.61–7.65 (m, 2 H). ^{13}C NMR (125 MHz, CDCl$_3$) δ: 21.7, 49.4, 52.3, 63.2, 126.8, 127.5, 127.8, 128.0, 128.8, 128.9, 129.4, 129.6, 133.5, 136.9, 137.5, 143.6, 170.6. IR (neat film, NaCl): 3031, 2951, 1747, 1598, 1496, 1454, 1437, 1342, 1206, 1162, 1091, 1029, 932, 814, 747, 696, 661 cm^{-1}. HRMS (MM: ESI/APCI): m/z calcd for C$_{23}$H$_{24}$NO$_4$S [M+H]$^+$: 410.1426. Found: 410.1421. Anal. Calcd for C$_{23}$H$_{23}$NO$_4$S: C, 67.46; H, 5.66; N, 3.42; O, 15.63; S, 7.83. Found: C, 67.40; H, 5.70; N, 3.39; O, 15.73; S, 7.85.

16. The melting point of the solid was determined to be 84–86 °C. The submitters report a mp of 96–98 °C. All other characterization data was identical for products produced by the checkers and the submitters.

17. Reaction run on one-half scale resulted in a white amorphous solid (5.22 g, 85% yield).

Working with Hazardous Chemicals

The procedures in *Organic Syntheses* are intended for use only by persons with proper training in experimental organic chemistry. All hazardous materials should be handled using the standard procedures for work with chemicals described in references such as "Prudent Practices in the Laboratory" (The National Academies Press, Washington, D.C., 2011; the full text can be accessed free of charge at http://www.nap.edu/catalog.php?record_id=12654). All chemical waste should be disposed of in accordance with local regulations. For general guidelines for the management of chemical waste, see Chapter 8 of Prudent Practices.

In some articles in *Organic Syntheses*, chemical-specific hazards are highlighted in red "Caution Notes" within a procedure. It is important to recognize that the absence of a caution note does not imply that no significant hazards are associated with the chemicals involved in that procedure. Prior to performing a reaction, a thorough risk assessment should be carried out that includes a review of the potential hazards associated with each chemical and experimental operation on the scale that is planned for the procedure. Guidelines for carrying out a risk assessment and for analyzing the hazards associated with chemicals can be found in Chapter 4 of Prudent Practices.

The procedures described in *Organic Syntheses* are provided as published and are conducted at one's own risk. *Organic Syntheses, Inc.*, its Editors, and its Board of Directors do not warrant or guarantee the safety of individuals using these procedures and hereby disclaim any liability for any injuries or damages claimed to have resulted from or related in any way to the procedures herein.

Discussion

Phosphorus(III) reagents are known to undergo reaction with 1,2-dicarbonyl compounds to give adducts of formal *P*-addition to the carbonyl oxygen (Kukhtin-Ramirez reaction).[2] We have shown that these Kukhtin-Ramirez adducts further react to incorporate a range of *N*-, *O*-, and C-based protic pronucleophiles with expulsion of a phosphine oxide by-product.[3,4] The process likely proceeds in stepwise fashion, initiated by proton transfer from the protic pronucleophile to the Kukhtin-Ramirez adduct, followed by Arbuzov-like displacement of the phosphine oxide leaving group. This reaction sequence therefore represents a convenient one pot process for access to a range of α–functionalized carbonyl compounds from readily available reagents and precursors.[3,4]

The synthetic method is exemplified in the above procedure, which demonstrates the synthesis of methyl 2-(*N*-benzyl-4-methylphenylsulfonamido)-2-phenylacetate (**3**), an α–amino ester derivative.[5] By direct reductive construction of the α-C-N bond, our approach takes advantage of the wide available α–keto esters as starting materials, and provides an operationally simple and chemoselective alternative to transamination and reductive amination strategies.[6,7] Furthermore, since this method does not involve the intermediacy of imine equivalents, useful C-N bonds from N-pronucleophiles that do not form imines (e.g. azoles) can be successfully synthesized using this method (Table 1).[7]

Table 1. Scope of the phosphorus(III)-mediated α-amino ester synthesis[a]

8aa, R' = H	70%	**8b** 60%	**8c** 60%
8ab, R' = Ph	89%		
8ac, R' = Bn	82%		
8ad, R' = allyl	85%		
8ae, R' = (CH₂)₂OTBS	85%	**8d** 83%	**8e** 83%

[a] Yields under literature reported conditions[2]

The reaction is tolerant of a range of solvents (CH$_2$Cl$_2$, THF, PhMe), but the use of dichloromethane reliably provides the highest yields and streamlines the aqueous workup on the laboratory scale. The use of commercially available P(NMe$_2$)$_3$ results in the formation of O=P(NMe$_2$)$_3$ (i.e. HMPA), a water soluble byproduct that is readily eliminated by aqueous extraction. In view of potential handling concerns resulting from the toxicity of HMPA, we note that the use of alternative phosphorous triamide reagents, specifically tris(1-pyrrolidinyl)phosphine (which generates a less toxic phosphorus(V) oxide by-product), provide similarly satisfactory results (Scheme 1, 1.8 mmol scale).[8]

Scheme 1. Phosphorus(III)-mediated synthesis of α-amino esters.

As noted in our previous studies, the scope of the reaction includes a diverse of O-based (phenols, carboxylic acids and some alcohols) and C-based (readily enolizable 1,3-dicarbonyls and related derivatives) protic pronucleophiles. The selection of protic pronucleophile is bracketed by pK_a, with only those species capable of proton transfer to the Kukhtin-Ramirez adduct (pK_a ca. 25-27 in DMSO) being reactive under these conditions (Table 2).

Table 2. Additional examples of phosphorus(III)-mediate carbonyl funct-ionalization with O- and C- pronucleophiles

selected examples using O-pronucleophiles

11a 99%   **11b** 89%   **11c** 60%   **11d** 60%

selected examples using C-pronucleophiles

11e 55%   **11f** 88%, 4.8:1 dr   **11g** 93%, 2.5:1 dr   **11h** 90%, 1.4:1 dr

References

1. Department of Chemistry, Pennsylvania State University, University Park, PA 16802, radosevich@psu.edu. We thank NIGMS (GM114547), Alfred P. Sloan Foundation, and the Pennsylvania State University for funding.
2. Osman, F. H.; El-Samahy, F. A. *Chem. Rev.* **2002**, *102*, 629-678.
3. Miller, E. J.; Zhao, W.; Herr, J. D.; Radosevich, A. T. *Angew. Chem. Int. Ed.* **2012**, *51*, 10605-10609.
4. Zhao, W.; Fink, D. M.; Labutta, C. A.; Radosevich, A. T. *Org. Lett.* **2013**, *15*, 3090-3093.
5. Williams, R. M. *Synthesis of Optically Active α-Amino Acids*, Pergamon, Oxford, **1989**.
6. Genet, J. P.; Greck, C.; Lavergne, D. In *Modern Amination Methods*; Ricci, A., Ed.; Wiley-VCH: Weinheim, Germany, **2000**; Chapter 3.
7. (a) Abel-Magid, A. F.; Mehrman, S. J. *Org. Process. Res. Dev.* **2006**, *10*, 971-1031. (b) Gomez, S.; Peters, J. A.; Maschmeyer, T. *Adv. Synth. Catal.* **2002**, *344*, 1037-1057. (c) Baxter, E. W.; Reitz, A. B. *Org. React.* **2002**, *59*, 1.
8. (a) Coste, J.; Le-Nguyen, D.; Castro, B. *Tetrahedron Lett.* **1990**, *31*, 205-208. (b) Coste, J.; Frerot, E.; Jouin, P. *Tetrahedron Lett.* **1991**, *32*, 1967-1970. (c) Kang, F.-A.; Sui, Z.; Murray, W. V. *J. Am. Chem. Soc.* **2008**, *130*, 11300-11302.

Appendix
Chemical Abstracts Nomenclature (Registry Number)

4-Methyl-*N*-(phenylmethyl)benzenesulfonamide: Benzenesulfonamide, 4-methyl-*N*-(phenylmethyl)-; (1576-37-0)

Dichloromethane: Methane, dichloro-; (75-09-2)

Methyl benzoylformate: Benzeneacetic acid, α-oxo, methyl ester; (15206-55-0)

Tris(dimethylamino)phosphine: Phosphorus triamide, *N,N,N,N',N',N'*-hexamethyl-; (1608-26-0)

Sodium Chloride: sodium chloride; (7647-14-5)

Sodium Sulfate: sulfuric acid sodium salt (1:2); (7757-82-6)

Organic
Syntheses

Wei Zhao is from Jinan, Shandong Province, P. R. China. He completed his BSc at Xiamen University, working with Prof. Pei-Qiang Huang and Prof. Xiao Zheng. In fall 2010 he joined Penn State Chemistry working with Prof. Alexander Radosevich. He is now a senior graduate student with research focused on redox catalysis at geometrically constrained organophosphorus compounds.

Alex Radosevich is from Waukegan, IL and received his B.S. from Notre Dame (2002). He obtained a Ph.D. from UC Berkeley (2007) working with Prof. Dean Toste. Following postdoctoral research at MIT with Prof. Dan Nocera, he joined the department of chemistry at Penn State in 2010 as an assistant professor, where his research has focused on the design, development, and implementation of new synthetic methodology.

Jared Moore is from Libertyville, IL and attended Cal Poly, San Luis Obispo where he received his B.S. (2009). He obtained a Ph.D. from UC Davis in 2014 in the laboratory of Prof. Jared Shaw, where his research was focused on the development of new synthetic methods. Jared is currently an NIH postdoctoral research fellow at Caltech where he investigates the total synthesis of meroterpene natural products with Prof. Brian M. Stoltz.

One-pot Hydrozirconation/Copper-catalyzed Conjugate Addition of Alkylzirconocenes to Enones

David Arnold, Tanja Krainz, and Peter Wipf[*1]

Department of Chemistry, University of Pittsburgh, 219 Parkman Avenue, Pittsburgh, PA

Checked by Koichi Fujiwara and John Wood

A.

$$\text{(4-penten-1-ol)} \xrightarrow[\substack{\text{toluene, reflux} \\ \text{2 h}}]{0.3 \text{ equiv B(OH)}_3} \left[\text{O}\right]_3 \text{B} \quad \mathbf{1}$$

B.

$$\left[\text{O}\right]_3 \text{B} \xrightarrow[\substack{1:1 \text{ toluene:THF, rt} \\ \text{2 h}}]{3.6 \text{ equiv Cp}_2\text{ZrHCl}} \left[\text{Cp}_2\text{ClZr} \cdots \text{O}\right]_3 \text{B} \quad \mathbf{2}$$

C.

$$\left[\text{ZrClCp}_2 \cdots \text{O}\right]_3 \text{B} \; + \; \text{(cyclohexenone)} \xrightarrow[\text{rt, 2 h}]{0.3 \text{ equiv CuBr·Me}_2\text{S}} \text{(product)} \quad \mathbf{3}$$

Procedure

A. *Tripent-4-enyl borate (1)*. A 300-mL, three-necked, flame-dried, round-bottomed flask equipped with a Teflon-coated stir bar (3 cm), two septa (Necks 1 and 3) and a Dean-Stark trap (20 mL, Neck 2) wrapped in aluminum foil and fitted with a reflux condenser (20 cm), and a nitrogen gas inlet adaptor (Note 1) is charged with boric acid (886 mg, 14.3 mmol, 1.0 equiv) (Note 2), toluene (70 mL) (Note 3), and 4-penten-1-ol (4.48 mL,

43.4 mmol, 3.0 equiv) (Note 2) at room temperature (Figure 1). The resulting suspension is slowly heated in an oil bath (135–136 °C) to reflux over 30 min and heating is continued at reflux for 1.5 h (Note 4). The reaction mixture turns homogeneous after 50 min of heating.

Figure 1. Apparatus Assembly for Step A

After heating at reflux for 1.5 h, the flask is removed from the oil bath and stirred at room temperature for 10 min before the septum (Neck 1) is replaced with a thermometer adaptor/thermometer (Note 5) and the Dean-Stark trap/reflux condenser (Neck 2) is replaced with a dry reflux condenser (20 cm) equipped with a nitrogen gas inlet adaptor (Figure 2). The resulting apparatus is further cooled for 20 min with the aid of an ice/water bath (Note 4) to 23-24 °C (internal temperature). The cooling bath is removed and the *in situ* generated tripent-4-enyl borate used directly for the next step (Note 6).

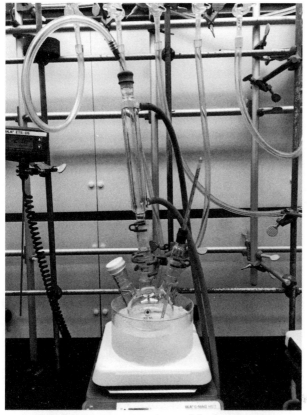

Figure 2. Apparatus Assembly for Step B

B. *Tris[5-(bis(cyclopentadienyl)zirconium(IV)chloride)pentyl] borate (2).* To the clear tripent-4-enyl borate solution is quickly added THF (50 mL) by syringe (Note 3), followed by Cp₂ZrHCl (14.7 g, 57.0 mmol, 4.0 equiv) (Note 7) in one portion via a powder funnel (Neck 3) at 24 °C (internal temperature) (Note 8). The resulting white suspension (Figure 3) bubbles gently for approximately 5 s, turns yellow/orange in color (Figure 4), and then clear (Figure 5) after approximately 15 min (Note 9). The hydrozirconation reaction is complete after 2 h (Note 10), and the solution of the alkyl zirconocene product formed is used without purification for the next step.

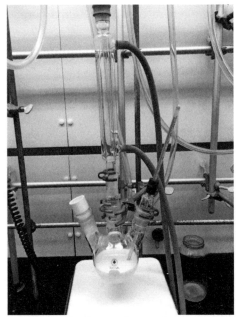

Figure 3. Initial white solution formed in Step B

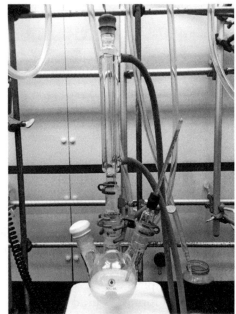

Figure 4. Yellow suspension formed in Step B

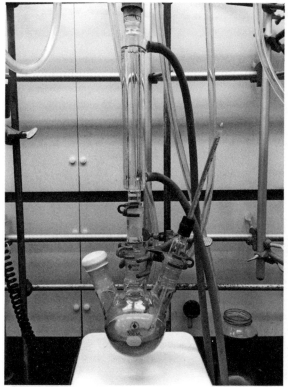

Figure 5. Clear solution formed in Step B

C. *3-(5-Hydroxypentyl)cyclohexan-1-one* *(3)*. To the crude tris[5-(bis(cyclopentadienyl)zirconium(IV)chloride)pentyl] borate solution is added 2-cyclohexen-1-one (4.58 mL, 47.3 mmol, 3.3 equiv) by syringe (Notes 8 and 11) followed by CuBr•Me₂S (884 mg, 4.30 mmol, 0.3 equiv) (Note 8) via a powder funnel (Neck 3) in one portion at 24 °C. A strongly exothermic reaction ensues upon addition of CuBr•Me₂S, with the reaction temperature rising from 24 °C to 38 °C within 10 min (Note 12). The heterogeneous reaction mixture turns black (Figure 6) (Note 13) and the conjugate addition reaction is found to be complete after 2 h (Note 14).

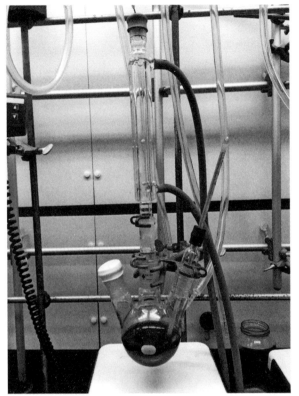

Figure 6. Black solution formed in Step C

The reaction is quenched by pouring the reaction mixture into a 500-mL Erlenmeyer flask that contains a concentrated NH_4OH solution (150 mL) (Note 3) and a Teflon-coated stir bar (5.0 cm). The remaining residue in the three-necked, round-bottomed flask is transferred to the Erlenmeyer flask with an ether wash (100 mL) (Note 3) and the resulting heterogeneous, biphasic, grey mixture is vigorously stirred at room temperature for 2 h.

After 2 h, the mixture turns light blue in color (Figure 7) and is suction filtered (37 mmHg) through a 350-mL glass filter (porosity: 40-60 μm) into a 1000-mL filtration flask. The resulting pale blue solids and the 500-mL Erlenmeyer flask are washed with deionized water (3 x 40 mL) followed by ether (3 x 40 mL), and the filtrate is transferred to a 500-mL separatory funnel (Figure 8).

Figure 7. Light blue solution formed in Step C

The dark yellow ether layer is removed and the dark blue aqueous layer is extracted with ether (2 x 100 mL). The ether extracts are combined in a 1000-mL Erlenmeyer flask, dried over Na₂SO₄ (20.3 g, 15 min) (Note 15) and filtered (slight over pressure of air) through a plug of SiO_2 (25.0 g, height = 10.0 cm, diameter = 2.5 cm) into a 1000-mL round-bottomed flask with excess ether (100 mL), which is also used to rinse the Erlenmeyer flask. The yellow filtrate is concentrated by rotary evaporation (20 to 50 °C/50 mmHg) and the resulting oil (Note 16) is loaded with a 1:1 hexanes:EtOAc solution (10.0 mL) onto a 4.5-cm diameter flash column, dry packed (pre-flushed with 1:1 hexanes:EtOAc solution until uniformly solvated) with SiO_2 (162 g) (Note 17). Following the sample addition, the column is eluted with a 1:1 hexanes:EtOAc solution (1.6 L total volume, 33 mL/min. average flow rate) and the first 325 mL of eluent are collected as one fraction. At this time, a

Figure 8. Biphasic solution in separatory funnel

yellow band (Note 18) begins to elute from the column and fractions 60 mL in volume are collected. The desired product elutes over fractions 13-52 and a sample of the desired product enriched with the regioisomeric byproduct 3-(5-hydroxypentan-2-yl)cyclohexanone (**4**) is isolated over fractions 53-60 (Notes 19 and 20). Fractions 13-52 are combined (each test tube is rinsed with ether (2 mL)), concentrated by rotary evaporation (20 to 40 °C/50 mmHg), transferred to a pre-weighed 50-mL round-bottomed flask with ether (10 mL), re-concentrated by rotary evaporation (20 to 40 °C/50 mmHg) and dried under high vacuum (14 mmHg, 12 h) to afford 3-(5-hydroxypentyl)cyclohexan-1-one (**3**) as a pale yellow oil (5.20 g, 66%) (Notes 21 and 22).

Notes

1. All glassware contained 24/40 joints and was either flame-dried or oven-dried (110 °C) overnight prior to use, if not otherwise noted. All reaction steps and reagents were performed under a partial positive

nitrogen gas atmosphere using a nitrogen gas line connected to an external mineral oil bubbler.

2. Copper(I)bromide dimethylsulfide (99%, Acros Organics), 4-penten-1-ol (99%, Acros Organics), and boric acid (>95%, Fisher Scientific, Reagent grade) were used as received. 2-Cyclohexene-1-one (>95%, Sigma-Aldrich) was freshly distilled (76 °C, 37 mmHg) prior to use. The submitters mentioned boric acid was either used as received or recrystallized two times from boiling deionized water and dried under high vacuum.[2] No appreciable difference in overall yield or reactivity was observed by the checkers using recrystallized vs. non-recrystallized boric acid.

3. Toluene and tetrahydrofuran were dried using a solvent purification system manufactured by SG Water U.S.A., LLC. Column chromatography solvents EtOAc and hexanes were used as received. Diethyl ether (ACS grade) and NH_4OH (ACS grade) were used as received.

4. A Pyrex® crystallization dish with a diameter of 15.0 cm and a height of 7.0 cm was used.

5. A mercury thermometer with a range of –20 to 110 °C was used.

6. The submitter followed the reaction progress by removing an aliquot (0.15 mL) of the reaction mixture after cooling to 22–23 °C and directly transferring it to an NMR tube for ^{1}H NMR analysis. The analysis showed ~85% conversion of the reagents to a single borate species: ^{1}H NMR (300 MHz, dry $CDCl_3$) δ: 1.77–1.84 (m, 2 H), 2.26–2.29 (m, 2 H), 3.94–3.99 (m, 2 H), 5.11–5.22 (m, 2 H), 5.92–6.06 (m, 1 H). This reaction has been performed under the same reaction conditions on a 1.0-mmol scale in both benzene and toluene at reflux furnishing the crude tripent-4-enyl borate, after removal of the reaction solvent via distillation and high vacuum, with 95-98% mass recoveries and in nearly identical purity to that of this reaction as analyzed by ^{1}H NMR.

7. Cp_2ZrHCl was prepared employing a method reported by Buchwald and co-workers.[3] The activity of this reagent was measured by a convenient ^{1}H NMR assay: To a suspension of Cp_2ZrHCl (0.05 mmol, 12.9 mg, 1.0 equiv) in dry CH_2Cl_2 (0.50 mL) was immediately added *tert*-butyldiphenyl(prop-2-ynyloxy)silane (0.05 mmol, 14.7 mg, 1.0 equiv). The resulting homogeneous pale yellow reaction mixture was stirred at room temperature for 15 min, quenched by the addition of saturated NH_4Cl solution (0.50 mL), and stirred for 5 min. The organic solvent

layer was separated, filtered through a 1" plug of Celite contained in a disposable 5 ¾" Pasteur pipette, washed with additional CH_2Cl_2 (1.5 mL) and concentrated by rotary evaporation (40 °C). The resulting white solids were dried under high vacuum for 30 min. This procedure was repeated three times on three different samples for each new batch of Cp_2ZrHCl. The percent conversion for each reaction was then determined via a ^1H NMR (300 MHz, $CDCl_3$) analysis by comparison of the integrated areas for the alkyne CH_2O peak at δ 4.32 (d, $J = 2.4$ Hz, 2 H) to the alkene CH_2O peak at δ 4.22 (dt, $J = 4.5$, 2.0 Hz, 2 H). Typical values for the average % conversion among the three experiments were found to be 73-83%. The numerical value for the average % conversion for a particular batch of Cp_2ZrHCl was used to calculate the number of equivalents of Cp_2ZrHCl (by mass) used in these hydrozirconations, ensuring that only a slight excess of Cp_2ZrHCl (3.1 equiv) was used in each experiment. Sample calculation: average % conversion = 78%; 3.1 active equiv of Cp_2ZrHCl = (X equiv of Cp_2ZrHCl) x 0.78 => X = 4.0 equiv of Cp_2ZrHCl.

$$\text{OTBDPS} \xrightarrow[\text{2. NH}_4\text{Cl (aq.)}]{\substack{\text{1. 1.0 equiv Cp}_2\text{ZrHCl} \\ \text{CH}_2\text{Cl}_2,\ \text{rt, 15 min}}} \text{OTBDPS}$$

8. The septum in Neck 3 was removed with the apparatus under a positive pressure of nitrogen gas atmosphere and then immediately replaced after the addition of reagent.

9. No exotherm was noted upon the addition of Cp_2ZrHCl to **1**. The submitter mentioned that this observation stands in contrast to the reaction of 4-penten-1-ol (2.07 mL, 20.0 mmol, 1.0 equiv) with Cp_2ZrHCl (11.1 g, 43.0 mmol, 2.15 equiv) in THF (50 mL) at 25 °C (internal temperature) which was found to be exothermic (reaction temperature increasing to 34 °C) and accompanied by a vigorous (H_2) gas release, which persisted for 2 min, as a result of the deprotonation of 4-penten-1-ol with Cp_2ZrHCl.

10. The submitter followed the reaction progress by ^1H NMR (300 MHz, $CDCl_3$) analysis of aliquots (0.10 mL) of the reaction mixture. An approximate % conversion for the hydrozirconation reaction can be obtained by comparing the integrated areas for the tripent-4-enyl borate olefin proton resonance δ 4.97–5.10 (m, CH_2) vs. the newly formed tris[5-(biscyclopentadienyl)zirconium(IV)chloride)pentyl] borate methylene

proton resonance δ 0.90-0.97 (m, 2H). This analysis gave 90%, 97% and 96% conversions for reactions run on 7%, 50% and 100% scales, respectively. Thin layer chromatography (Whatman®, aluminum backed, 250 μm thickness) analysis using a 1:1 mixture of hexanes:EtOAc as the eluent and a p-anisaldehyde solution (2.5 mL of p-anisaldehyde, 2.0 mL of AcOH, and 3.5 mL of conc. H_2SO_4 in 100 mL of 95% EtOH) for visualization showed a dark green/brown spot with an R_f = 0.03 for the hydrozirconation product.

11. The submitter mentioned that a mild exotherm was noted upon the addition of 2-cyclohexen-1-one to the reaction mixture with the reaction temperature rising from 23 °C to 25 °C. This exothermic reaction is quite possibly the result of the reaction of any excess Cp_2ZrHCl with 2-cyclohexen-1-one, since it was found that an exothermic reaction took place upon the addition of Cp_2ZrHCl (134.0 mg, 0.52 mmol, 1.0 equiv) to a solution of 2-cyclohexen-1-one (47.4 μL, 0.52 mmol, 1.0 equiv) in THF (3.0 mL) with the reaction temperature rising from 25 °C to 32 °C. This reaction provided 2-cyclohexen-1-ol in good conversion by crude ^{1}HNMR (300 MHz, $CDCl_3$) analysis and demonstrates the importance of using a slight excess of 2-cyclohexen-1-one in the above protocol.

12. The reaction mixture slowly re-cooled to 23 °C over 1.5 h.

13. The black color is a result of precipitation of copper(0) from the reaction mixture.

14. The submitter reported that the reaction progress was monitored by thin layer chromatography (Whatman® aluminum backed, 250 μm thickness) using a 1:1 mixture of hexanes:EtOAc as the eluent and a p-anisaldehyde solution for visualization. After 2 h reaction time, the conversion of 2-cyclohexen-1-one (R_f = 0.72) to the product 3-(5-hydroxypentyl)cyclohexan-1-one (R_f = 0.43) ceased after near complete consumption of 2-cyclohexen-1-one.

15. Anhydrous sodium sulfate (EMD, ACS grade, granular powder) was used as received.

16. A 3.0 mg sample of the crude mixture was dissolved in CH_2Cl_2 (1.0 mL) and analyzed by GC (HP-5ms agilent 30 m x 0.25 mm x 0.25 μm, helium flow 1.0 mL/min, temperature gradient 110 °C to 280 °C at 30 °C/min, FID detector). GC analysis showed a >96:4 ratio of the desired product 3-(5-hydroxypentanyl)cyclohexanone (**3**) (retention time of 4.74 min) to the side product 3-(5-hydroxypentan-2-yl)cyclohexanone (**4**).

17. SiO_2 40-63 μm D (Silicycle, Quebec City, Canada) was used.

18. The submitter noted that the yellow band was found to contain a complex mixture of reaction byproducts including 2-cyclohen-1-ol, 4-penten-1-ol and 1-pentanol as determined by ^{1}H NMR (300 MHz, CDCl$_3$) analysis.

19. Column chromatography conditions to completely separate 3-(5-hydroxypentyl)cyclohexan-1-one and the minor regioisomer 3-(5-hydroxypentan-2-yl)cyclohexanone have not been identified. The chromatography conditions reported above allow for a partial separation of the two reaction products in which the last fractions of 3-(5-hydroxypentyl)cyclohexan-1-one are enriched with the byproduct 3-(5-hydroxypentan-2-yl)cyclohexanone. Collection, combination and concentration of column fractions 53-60 led to an enriched sample (90 mg) containing a 94:6 mixture of 3-(5-hydroxypentyl)cyclohexan-1-one (**3**): 3-(5-hydroxypentan-2-yl)cyclohexanone (**4**), as determined by GC (conditions in Note 16) and ^{1}H NMR (400 MHz, CDCl$_3$) analysis. The isomer 3-(5-hydroxypentan-2-yl)cyclohexanone (**4**) can be identified by ^{1}H NMR (400 MHz, CDCl$_3$) as a result of its characteristic methyl peak at δ 0.89 (d, J = 6.8 Hz, 3 H, -CH$_3$).

4

20. The submitters reported that attempts to optimize this reaction included: (A) Reacting 4-penten-1-ol (0.269 mL, 2.58 mmol) with 2.0 equivalents of Cp$_2$ZrHCl (1.33 g, 5.16 mmol) at 5 °C in THF (10 mL) either by the addition of 4-penten-1-ol to a suspension of Cp$_2$ZrHCl (2.0 equiv) in THF at 5 °C followed by warming to 20 °C over 20 min or by sequentially reacting Cp$_2$ZrHCl (1.0 equiv) with 4-penten-1-ol (1.0 equiv) at 5 °C and then warming the mixture to 20 °C over 20 min before adding a second equivalent of Cp$_2$ZrHCl. The optimized procedure was found to give a cleaner hydrozirconation reaction product by ^{1}H NMR (300 MHz, CDCl$_3$) analysis. (B) The reaction mixtures were heated to 40 °C for 15-20 min to effect the hydrozirconation reaction, followed by cooling to 25 °C. (C) To the alkyl zirconocene solutions were added 2-cyclohexen-1-one (0.255 mL, 2.58 mmol, 1.0 equiv) followed by CuBr•Me$_2$S (53.5 mg, 0.258 mmol,

1.0 equiv) at 25 °C and the resulting brown mixtures were heated to 40 °C for 30 min to effect the transmetallation/conjugate addition reaction. Finally, these reactions were quenched with NH₄OH (14.8 M, 50 mL). While these reaction variants successfully produced 3-(5-hydroxypentyl)cyclohexan-1-one in comparable yields (60-70%, 283–337 mg), the isolated products were found to be contaminated with ca. 8-10% of the 3-(5-hydroxypentan-2-yl)cyclohexanone byproduct (**4**).

21. The product isolated in fractions 13-52 was found to be 97.6% pure by GC (conditions in Note 16) and contain 0.8% of 3-(5-hydroxypentan-2-yl)cyclohexanone. The product has the following characteristic physicochemical properties: ^{1}H NMR (400 MHz, CDCl₃) δ: 1.23–1.39 (m, 7 H), 1.49–1.69 (m, 3 H), 1.68–1.80 (m, 2 H), 1.81–2.08 (m, 3 H), 2.18–2.42 (m, 3 H), 3.60 (t, J = 6.6 Hz, 2 H); ^{13}C NMR (100 MHz, CDCl₃) δ: 25.4, 25.9, 26.6, 31.4, 32.7, 36.6, 39.1, 41.6, 48.3, 62.9, 212.4; IR (neat) 3412, 2927, 2856, 1704, 1447, 1421, 1346, 1313, 1279, 1053 cm^{-1}; ESI-MS m/z 185 (24), 207 (100); HRMS (ESI) m/z calcd for C₁₁H₂₁O₂ [M+H]$^{+\bullet}$ 185.1536, found 185.1536. A 2.0 g sample was subjected to bulb-to-bulb distillation (150–151 °C/4.0 mmHg) affording 0.957 g (48% recovery) of the product as a clear oil. Anal Calcd for C₁₁H₂₀O₂: C, 71.70; H, 10.94; O, 17.36; Found: C, 71.53, H, 11.10.

22. A reaction checked at half-scale provided 2.69 g (68%) of the product. The submitters report that reactions run on 7% and 50% scales produced the product under identical reaction/workup conditions in 65% yield for both reactions with measured GC purities of ≥98% (conditions in Note 16).

Working with Hazardous Chemicals

The procedures in *Organic Syntheses* are intended for use only by persons with proper training in experimental organic chemistry. All hazardous materials should be handled using the standard procedures for work with chemicals described in references such as "Prudent Practices in the Laboratory" (The National Academies Press, Washington, D.C., 2011; the full text can be accessed free of charge at http://www.nap.edu/catalog.php?record_id=12654). All chemical waste should be disposed of in accordance with local regulations. For general guidelines for the management of chemical waste, see Chapter 8 of Prudent Practices.

In some articles in *Organic Syntheses*, chemical-specific hazards are highlighted in red "Caution Notes" within a procedure. It is important to recognize that the absence of a caution note does not imply that no significant hazards are associated with the chemicals involved in that procedure. Prior to performing a reaction, a thorough risk assessment should be carried out that includes a review of the potential hazards associated with each chemical and experimental operation on the scale that is planned for the procedure. Guidelines for carrying out a risk assessment and for analyzing the hazards associated with chemicals can be found in Chapter 4 of Prudent Practices.

The procedures described in *Organic Syntheses* are provided as published and are conducted at one's own risk. *Organic Syntheses, Inc.*, its Editors, and its Board of Directors do not warrant or guarantee the safety of individuals using these procedures and hereby disclaim any liability for any injuries or damages claimed to have resulted from or related in any way to the procedures herein.

Discussion

The copper-catalyzed conjugate Michael reaction and related additions of alkylzirconocenes to α,β-unsaturated carbonyl compounds have emerged as synthetically useful C-C bond forming processes.[4-16] Recent examples include catalytic asymmetric additions to α,β-unsaturated lactones and enones.[17,18] The scope of this reaction has been previously investigated and some representative examples are shown in Table 1.[9,11,12] As shown, these reactions are tolerant of cyclic and acyclic alkenes, as well as silyl ether, silyl ester and acetal functionalities. A variety of enones have also been successfully employed, including cyclic, acyclic, sterically hindered and chiral substrates.

Table 1. Copper-Catalyzed Conjugate Addition Reactions of Alkylzirconocenes to Enones

Entry	Alkene	Enone	Product	Yield[a]
1	⌒⌒OTBDMS	(cyclohexenone)	(3-substituted cyclohexanone)OTBDMS	76%[b,c]
2	(dihydropyran)	(cyclohexenone)	(product)OH	85%[b,c,d]
3	(allyl dioxolane)	(cyclohexenone)	(product dioxolane)	78%[b,c]
4	(pentenoate)OTIPS	(3-methylcyclohexenone)	(product)OTIPS	78%[b]
5	(1-hexene)	(dimedone-derived enone)	(alkylated product)	46%[b,e]
6	(3-hexene)	Ph⌒C(O)Ph / Ph	Ph C(O) / Ph	51%[b,c]
7	(cyclohexene)	Ph⌒C(O)N-oxazolidinone / Ph	Ph / Ph (oxazolidinone product)	60%[e,f,g]

[a]Reactions were typically conducted in THF by reacting the alkene with Cp$_2$ZrHCl for 10 min at 40 °C followed by cooling to rt, adding the enone, followed by 10% CuBr•Me$_2$S and heating to 40 °C for 10 min. [b]See reference 11. [c]See reference 12. [d]Two equiv of Cp$_2$ZrHCl were used. [e]One equiv of BF$_3$•OEt$_2$ was used. [f]See reference 9. [g]90%de.

The hydrozirconation/copper-catalyzed conjugate addition reaction of 3,4-dihydro-2*H*-pyran (DHP) with cyclohexen-1-one was shown to produce the title compound 3-(5-hydroxypentyl)cyclohexan-1-one in 85% yield.[11,12] The sequence of steps for this reaction is reported to involve an initial hydrozirconation of DHP followed by β-elimination to form intermediate (**I**), and then a second hydrozirconation of (**I**) generating intermediate (**II**) which undergoes the copper-catalyzed conjugate addition reaction with 2-cyclohexen-1-one (Scheme 1). Analysis of the reaction sequence suggested that this reaction could be generalized to include the conjugate addition of unprotected alkenyl alcohols. To test this hypothesis, reaction conditions were optimized for the reaction of 4-penten-1-ol with 2 equiv of Cp_2ZrHCl in THF (Note 20). This reaction proceeds through the initial deprotonation of 4-penten-1-ol with 1 equiv of Cp_2ZrHCl at 5 °C to generate the common intermediate (**I**), followed by hydrozirconation of (**I**) at 40 °C forming (**II**) which undergoes the copper-catalyzed conjugate addition to 2-cyclohexen-1-one in 60-70% yield (Scheme 1). While the yields for this addition reaction are acceptable, the isolated products are contaminated with 8-10% of the regioisomeric 3-(5-hydroxypentan-2-yl)cyclohexanone.

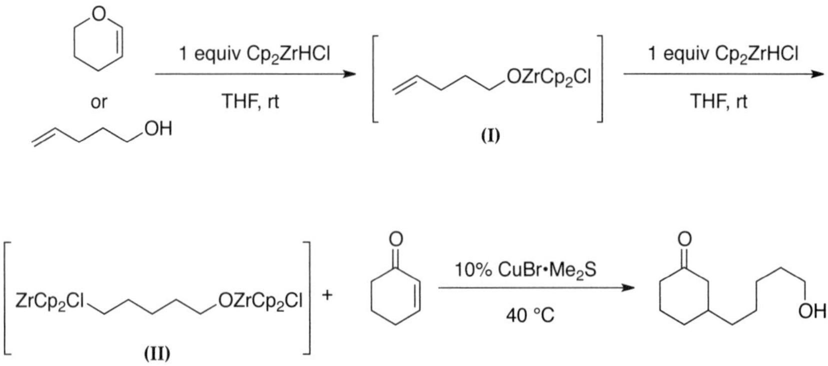

Scheme 1. Hydrozirconation / Copper-Catalyzed Conjugate Addition of 2-Cyclohexene-1-one to DHP / 4-Penten-1-ol

As a result of the cost of Cp_2ZrCl_2 and the high molecular weight of Cp_2ZrHCl, we were interested in developing a more cost effective and atom economical route involving the temporary protection of alkenyl-alcohol substrates to be utilized in this methodology. We found that the condensation of 4-penten-1-ol with boric acid, which changed the temporary organometallic alcohol protecting group from the zirconate RO-ZrCp$_2$Cl to the boronate (RO)$_3$B, was well tolerated in the subsequent hydrozirconation/copper-catalyzed conjugate addition reaction to 2-cyclohexen-1-one, and all steps following the formation of borate could be conducted at room temperature. Significantly, while this reaction led to comparable overall one-pot yields *vs.* the reaction performed with 2 equivalents of Cp$_2$ZrHCl, switching the initially produced metal alkoxide from Cp$_2$ClZr-OR to B-(OR)$_3$ allowed for a relatively fast hydrozirconation/copper-catalyzed conjugate addition reaction sequence at room temperature, which reduced the 3-(5-hydroxypentan-2-yl)cyclohexanone regioisomer formation to ca. ≤ 3.3% as determined by GC analysis of the crude reaction mixtures at all scales.

The procedure reported herein further exemplifies the utility and scalability of the one-pot tandem hydrozirconation/copper-catalyzed conjugate addition of alkylzirconocenes to enones. This reaction sequence allows for the economical use of unprotected alkenyl alcohols by utilizing the relatively benign B(OH)$_3$ as a temporary boron-based alcohol protecting group.

References

1. Department of Chemistry, University of Pittsburgh, Pittsburgh, PA; Email: pwipf@pitt.edu
2. Amarego, W. L. F.; Chai, C. L. L. *Purification of Laboratory Chemicals, 5*ᵗʰ ed.; Elsevier; Butterworth Heinemann; Amsterdam, 2003; p. 403.
3. Buchwald, S. L.; LaMarie, S. J.; Nielsen, R. B.; Watson, B. T.; King, S. M. *Org. Synth.* **1993**, *71*, 77.
4. Wipf, P.; Nunes, R. L. *Tetrahedron* **2004**, *60*, 1269.
5. Wipf, P. *Top. Organomet. Chem.* **2004**, *8*, 1.
6. Wipf, P.; Kendall, C. *Chem. Eur. J.* **2002**, *8*, 1778.
7. Wipf, P.; Takahashi, H.; Zhuang, N. *Pure Appl. Chem.* **1998**, *70*, 1077.
8. Wipf, P.; Xu, W. J.; Takahashi, H.; Jahn, H.; Coish, P. D. G. *Pure Appl. Chem.* **1997**, *69*, 639.
9. Wipf, P.; Takahashi, H. *Chem. Commun.* **1996**, 2675.
10. Wipf, P.; Jahn, H. *Tetrahedron* **1996**, *52*, 12853.
11. Wipf, P.; Xu, W. J.; Smitrovich, J. H.; Lehmann, R.; Venanzi, L. M. *Tetrahedron* **1994**, *50*, 1935.

12. Wipf, P.; Smitrovich, J. H.; Moon, C.-W. *J. Org. Chem.* **1992**, *57*, 3178.
13. Hart, D. W.; Schwartz, J. *J. Am. Chem. Soc.* **1974**, *96*, 8115.
14. Schwartz, J.; Labinger, J. A. *Angew. Chem. Int. Ed. Engl.* **1976**, *15*, 333.
15. Wailes, P. C.; Weigold, H. *J. Organomet. Chem.* **1970**, *24*, 405.
16. Wailes, P. C.; Weigold, H.; Bell, A. P. *J. Organomet. Chem.* **1971**, *27*, 373.
17. Maciver, E. E.; Maksymowicz, R. M.; Wilkinson, N.; Roth, P. M. C.; Fletcher, S. P. *Org. Lett.* **2014**, *16*, 3288.
18. Roth, P. M. C.; Fletcher, S. P. *Org. Lett.* **2015**, *17*, 912.

Appendix
Chemical Abstracts Nomenclature (Registry Number)

Boric acid: trihydrooxidoboron; (10043-35-3)
4-Penten-1-ol: pent-4-en-1-ol; (821-09-0)
Cp_2ZrHCl: bis(cyclopentadienyl)zirconium(IV) chloride hydride; (37342-97-5)
2-Cyclohexen-1-one: (930-69-7)
$CuBr \cdot Me_2S$: bromocopper-methylsulfanylmethane; (54678-23-8)
NH_4OH: ammonium hydroxide; (1336-21-6)

David M. Arnold obtained his B.S. in Chemistry from Kings College in 2005 and graduated with a Master of Science degree from the University of Pittsburgh in 2010. Under the direction of Prof. Peter Wipf, he worked on the synthesis of biologically active heterocycles and the development of new organometallic methods.

Organic
Syntheses

Tanja Krainz received her Dipl. Ing. from the Vienna University of Technology, Austria. In 2010, she moved to the University of Queensland, Australia and obtained her Ph.D. in 2014 in the field of natural product synthesis under the supervision of Associate Professor Craig M. Williams. She is currently a postdoctoral research associate in the Wipf group.

Peter Wipf received his Dipl. Chem. in 1984 and his Ph.D. in 1987 from the University of Zürich under the direction of Professor Heinz Heimgartner. After a Swiss NSF postdoctoral fellowship with Professor Robert E. Ireland at the University of Virginia, Wipf began his appointment at the University of Pittsburgh in the fall of 1990. Since 2004, he is a Distinguished University Professor of Chemistry. He also serves as a co-Leader of the UPCI Cancer Therapeutics Program and he is the editor of Volume 87 of *Organic Syntheses*.

Koichi Fujiwara received a B.S. degree in Pharmaceutical Sciences in 2010 from the Tokyo University of Science. He then moved to Tohoku University where, under the direction of Professor Takayuki Doi, he earned the M.S and Ph.D. degree in Pharmaceutical Sciences. In 2015, Koichi moved to Baylor University for postdoctoral studies under the direction of Professor John L. Wood.

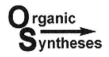

Dipeptide Syntheses via Activated α-Aminoesters

Jean-Simon Suppo, Renata Marcia de Figueiredo,* and Jean-Marc Campagne*[1]

ICGM-UMR 5253 CNRS-UM-ENSCM, Ecole Nationale Supérieure de Chimie, 8 Rue de l'Ecole Normale, 34296 Montpellier Cedex 5, France

Checked by John T. Colyer, Christopher J. Borths and Margaret Faul

A.

$HCl\cdot H_2N$–Ala CO_2Me + [imidazole-CO-imidazole] $\xrightarrow[\text{23 °C, 24 h}]{\text{DIPEA} \atop CH_2Cl_2}$ product **1**

B.

Boc-Phe CO_2H + **1** $\xrightarrow[\text{CH}_2\text{Cl}_2 \text{ (1 M)} \atop \text{25 °C, 20 h}]{\text{CuBr}_2 \text{ (cat)} \atop \text{HOBt (cat)}}$ product **2**

Procedure

A. *Methyl (1H-imidazole-1-carbonyl)-L-alaninate (1).* A 500-mL round-bottomed, single-necked flask, equipped with a 50 x 20 mm, Teflon-coated, oval magnetic stir bar, is charged with N,N'-carbonyldiimidazole (CDI) (11.6 g, 71.6 mmol) (Note 1), which is weighed into the flask under air, followed by dichloromethane (40 mL) (Note 2). Gentle stirring produces a slightly turbid suspension. The flask is sealed with a rubber septum, into which is inserted a digital thermometer probe. The flask is placed in an ice water bath to bring the temperature of the solution to 0 °C (Note 3) (Note 4). A second 250-mL round-bottomed, single-necked flask equipped with a 50 x 20 mm, Teflon-coated, oval magnetic stir bar, is then charged with L-alanine methyl ester hydrochloride (HCl•H₂N-Ala-OMe) (10.0 g, 71.6 mmol) (Note 5), which is weighed into the flask under air, followed by

Published on the Web 10/9/2015
© 2015 Organic Syntheses, Inc.

dichloromethane (80 mL) (Note 2). To the resulting white suspension is added diisopropylethylamine (DIPEA) (9.25 g, 12.5 mL, 71.6 mmol) (Note 6). At this stage, the suspension turns to a clear solution (Note 7), which is then transferred to a 250-mL dropping funnel (20 x 5 cm) (Note 8). The dropping funnel is attached to the 500-mL round-bottomed flask containing the chilled CDI suspension and fitted with a calcium chloride filled drying tube (15 g) (Notes 9 and 10). The amino acid/DIPEA solution is then added dropwise over 40 min (Note 11) to the CDI-suspension at 0 °C (Notes 4 and 12). As the reaction progresses, the suspension transforms into a clear, colorless solution. After the complete addition of the L-alanine methyl ester solution, the final mixture is stirred at room temperature (23 °C) for 20 h (Note 13). Upon completion of the reaction (Note 14), the mixture is concentrated using a rotary evaporator (30 °C water bath, 40 mmHg). Ethyl acetate (EtOAc) (200 mL) is added and the solution is transferred to a 1.0 L separatory funnel (Note 15). Additional EtOAc (50 mL) is used to assist transfer. The mixture is washed once with deionized water (150 mL). The aqueous layer is separated and further extracted with EtOAc (2 x 60 mL). The combined organic layers are dried over anh. magnesium sulfate (50 g) (Note 16), filtered through a glass filter funnel filled with cotton (Note 17), and concentrated using a rotary evaporator (30 °C water bath, 10 mmHg) to provide a viscous oil. The crude product is purified by flash chromatography on silica gel to afford **1** (10.21 g, 70.3%) (Notes 18, 19, and 20) as a white solid (Note 21 and 22).

B. *Methyl (tert-butoxycarbonyl)-L-phenylalanyl-L-alaninate (2)*. A 100-mL round-bottomed, single-necked flask equipped with a 30 x 16 mm, Teflon-coated, oval magnetic stir bar, is charged with methyl (1*H*-imidazole-1-carbonyl)-L-alaninate (8.0 g, 40.6 mmol) (Note 23), which is weighed into the flask under air, followed by CH$_2$Cl$_2$ (41 mL, 1.0 M) (Note 2). Gentle stirring at 20–22 °C (Note 24) gives a clear colorless solution, to which is successively added Boc-L-phenylalanine (Boc-Phe-OH) (16.16 g, 60.9 mmol), 1-hydroxybenzotriazole hydrate (HOBt hydrate) (549 mg, 10 mol%, 4.06 mmol) and copper(II) bromide (CuBr$_2$) (907 mg, 10 mol%, 4.06 mmol) (Note 25). The flask is closed with a rubber septum equipped with a needle (Note 26). As the reaction progresses, CuBr$_2$ becomes more

Figure 1. Reaction appearance

and more soluble and the solution evolves from pale blue into a deep turquoise/blue slurry (Figure 1). The reaction is stirred (Note 27) at 20-22 °C for 20 h. Upon completion of the reaction (TLC monitoring) (Note 28), the mixture is transferred into a 1.0 L Erlenmeyer flask, diluted with dichloromethane (CH$_2$Cl$_2$) (250 mL) and quenched with an aqueous solution of 0.5 N HCl (100 mL) (Note 29). Afterwards, the mixture is transferred to a 1.0 L separatory funnel. Additional CH$_2$Cl$_2$ (50 mL) is used to assist transfer. The aqueous layer is separated and further extracted with CH$_2$Cl$_2$ (1 x 50 mL). The combined organic layers are washed with a saturated solution of sodium bicarbonate (NaHCO$_3$) (100 mL) (Note 30), and the aqueous layer is further extracted with CH$_2$Cl$_2$ (1 x 50 mL). The combined organic layers are then washed with brine (100 mL), dried over MgSO$_4$ (50 g) (Note 16), filtered through a glass filter funnel filled with cotton (Note 31), and concentrated using a rotary evaporator (30 °C water bath, 10 mmHg) to give a white solid. The crude product is purified by flash chromatography on silica gel to afford **2** (11.73 g, 82.4%) (Note 32) as a white foam (Note 33 and 34) (Figure 2).

Notes

1. *N,N'*-Carbonyldiimidazole (CDI) was purchased from Sigma-Aldrich (reagent grade) and used as received.
2. Dichloromethane was purchased from Sigma-Aldrich (certified 99.8%, containing 50-150 ppm amylene as stabilizer) and used as supplied. Caution should be taken with the CH_2Cl_2 quality in order to prevent the formation of undesired side compounds. CH_2Cl_2 stabilized with EtOH should be avoided, otherwise the formation of ethyl 1*H*-imidazole-1-carboxylate could also be detected on the first step.
3. Cooling is necessary in order to avoid the formation of symmetrical urea (Note 22).
4. The mixture is stirred at 500 rpm throughout the reaction.
5. L-Alanine methyl ester hydrochloride, 99% from Sigma-Aldrich, was used as received. This compound is very hygroscopic.
6. Diisopropylethylamine (DIPEA) (99.5%) was purchased from Sigma-Aldrich and used without further purification.
7. For small scale synthesis (<15 mmol) triethylamine (Et_3N) was used as base. For operational facility on large scale DIPEA was used, which, in contrast to Et_3N, generates a soluble salt in the presence of HCl•H_2N-Ala-OMe in CH_2Cl_2. This soluble salt is easier to add dropwise into the CDI pre-cooled suspension.
8. The flask containing the mixture of L-alanine methyl ester hydrochloride and DIPEA was rinsed twice with CH_2Cl_2 (2 x 5 mL). Then, the combined 10 mL were added to the dropping funnel.
9. A dropping funnel with a Teflon tap was used for the addition of the amino acid to the CDI suspension.
10. Calcium chloride (anhydrous, granular, ≤ 7.0 mm, ≥93.0%) from Sigma-Aldrich was used as received. The drying tube containing a sintered glass filter was used.
11. The dropping funnel containing the mixture of L-alanine methyl ester hydrochloride and DIPEA was rinsed twice with CH_2Cl_2 (2 x 5 mL).
12. The slow addition of L-alanine methyl ester solution is necessary in order to avoid the symmetric urea formation (Note 22).
13. A second reaction was allowed to proceed for 22 hours.
14. The formation of the desired product was observed by TLC (on Merck silica gel 60 F_{254} TLC aluminum plates) and visualized with UV light and ninhydrin staining solution (Note 35). R_f product: 0.20, eluent:

EtOAc 100%. Visualization was difficult with UV light when dilute samples were used.

15. Ethyl acetate (≥99.8% from Sigma-Aldrich) was used as received.

16. Magnesium sulfate (≥99.5% from Sigma-Aldrich) was used as received.

17. Additional EtOAc (2 x 20 mL) is used during filtration to assist transfer. An M grade glass filter was used for the filtration.

18. The crude reaction product (14.5 g) was adsorbed on silica (40 g of silica with 150 mL of CH_2Cl_2 followed by evaporation) and then loaded onto a column (diameter: 7 cm, height: 70 cm) packed with silica gel (300 g of silica, pore size 60Å, 230-400 mesh, 40-63 μm particle size, Fluka Analytical) slurry in EtOAc 100%. After 500 mL of initial elution, fraction collection (250 mL fractions) is begun, and elution is continued with 4.5 L of pure EtOAc. The desired α-activated amino ester is obtained in fractions 4-18, which are concentrated by rotary evaporation (30 °C, 10 mmHg) and then dried at 0.05 mmHg. Residual ethyl acetate was difficult to remove and could be observed by ^{1}H NMR after 1 week under house vacuum (<1%). Ethyl acetate did not hinder the formation and isolation of product during the next step.

19. Yields were adjusted for wt% of product as determined by QNMR (97.2% wt% and 98.2% wt% for Run 1 and 2, respectively).

20. The chemical yield of this step can be improved to 86% by using 1.5 equiv of CDI.

21. A second reaction at the same scale provided 10.52 g (73.2%, 98.2% wt%) of a white solid. Yields were adjusted for potency of product. Weight percent was determined by QNMR with benzyl benzoate as an internal standard. Characterization as follows: $[\alpha]_D^{26}$ +31.3 (c 0.7, $CHCl_3$); ^{1}H NMR (400 MHz, $CDCl_3$) δ: 1.54 (d, J = 7.2 Hz, 3 H), 3.81 (s, 3 H), 4.66 (quint, J = 7.2 Hz, 1 H), 7.01 (br d, J = 6.9 Hz, 1 H), 7.08 (s, 1 H), 7.41 (t, J = 1.4 Hz, 1 H), 8.18 (s, 1 H); ^{13}C NMR (100 MHz, $CDCl_3$) δ: 17.3, 49.5, 52.5, 116.4, 129.5, 136.1, 148.7, 172.8; IR (neat): 3138.9, 2969.5, 2878.8, 2809.5, 1740.3, 1711.7, 1553.7, 1484.4, 1454.5, 1376.9, 1288.4, 1256.4, 1213.2, 1150.6, 1104.2, 1072.7, 754.8 cm^{-1}; HRMS (ESI)$^+$ [M+H]$^+$ calcd for $C_8H_{12}N_3O_3$: 198.0879. Found: 198.0873. mp 90.0–92.5 °C; the melting point was lower than that reported by the submitters (mp 95.6–97.9 °C) due to residual solvent and impurities.

22. The symmetrical urea depicted below is generated from auto-condensation of the free α-aminoester and its activated form. Its formation is observed either when the reaction is run at higher temperatures (> 0 °C) or when the reagents are added at once. It might

be noted that on small-scale reactions, only small amounts (<7%) of this compound were observed in the crude material.[6] A peak in ^{1}H NMR was observed after column chromatography that is consistent with the 6H singlet expected at ~3.7 ppm. If this resonance is assigned correctly, then 1.7% and 1% of the symmetrical urea impurity was observed, respectively, in Runs 1 and 2 after column chromatography. The impurity was removed during the subsequent isolation in Step 2.

$$\text{MeO}_2\text{C}\underset{\text{(Me)}}{\overset{\text{Me}}{\diagup}}\overset{\text{O}}{\underset{\text{H}\quad\text{H}}{\overset{\|}{\text{N}}}\text{N}}\overset{\text{Me}}{\diagdown}\text{CO}_2\text{Me}$$

23. Charges were not adjusted for wt% of methyl (1H-imidazole-1-carbonyl)-L-alaninate.

24. The checkers performed this chemistry at ambient temperature (20-22 °C) in an unjacketed flask without temperature control. The submitters report that control of temperature is very important to ensure good reaction yields. Temperatures lower than 25 °C lead to the formation of dipeptides, with erosion of isolated yields.

25. The following reagents were purchased from commercial sources and used without further purification: Boc-Phe-OH, ≥99% from Aldrich; HOBt hydrate, wetted with not less than 14 wt.% water, 97% from Aldrich; CuBr$_2$, 99% from Aldrich.

26. The rubber septum is equipped with a needle (22G x 1½" 0.7 x 40 mm) in order to allow the removal of CO$_2$ that is formed during the reaction.

27. The mixture is stirred at 400 rpm throughout the reaction.

28. The formation of the dipeptide was monitored by TLC analysis on Merck silica gel 60 F$_{254}$ TLC aluminum plates and visualized with UV light and ninhydrin staining solution (Note 35). R$_f$ dipeptide: 0.3, eluent: pentane/EtOAc 7:3. Pentane (≥99% from Sigma-Aldrich) was used as received.

29. The mixture is swirled vigorously for 10 min while deep blue color faded to very pale light blue.

30. The addition of a saturated solution of NaHCO$_3$ to the organic layer was followed by the formation of an emulsion, which disappears after standing in the separatory funnel for approximately 15 minutes.

31. Additional CH$_2$Cl$_2$ (2 x 20 mL) is used during filtration to assist transfer.

32. Yields are based on two full scale runs and are adjusted for wt% of product as determined by QNMR (99.6% wt% and 100% wt% for Run 1 and 2, respectively).

33. The crude product (14.0 g) was adsorbed on silica (40 g of silica with 120 mL of CH_2Cl_2 followed by evaporation) and then was loaded onto a column (diameter: 7 cm, height: 70 cm) packed with a short pad of silica gel (200 g of silica pore size 60Å, 230-400 mesh, 40-63 μm particle size, Fluka Analytical) slurry in pentane:EtOAc 7:3. After 500 mL of initial elution, fraction collection (250 mL fractions) is begun, and elution is continued with 3.25 L of additional solvent. The desired dipeptide is obtained in fractions 3-13, which are concentrated by rotary evaporation (30 °C, 10 mm Hg) and then dried at 0.05 mmHg.

34. A second reaction at the same scale provided 11.86 g (83%, 99.6% wt%) of a white foam. Yields were adjusted for potency of product. Weight percent was determined by QNMR with benzyl benzoate as an internal standard. Characterization as follows: mp 108.0–110.5 °C; [α]$_D^{26}$ +0.40 (c 0.99, CHCl$_3$); ^{1}H NMR (400 MHz, CDCl$_3$) δ: 1.34 (d, J = 7.0 Hz, 3 H), 1.40 (s, 9 H), 3.02–3.12 (m, 2 H), 3.71 (s, 3 H), 4.26–4.44 (m, 1 H), 4.52 (quint, J = 7.1 Hz, 1 H), 5.08 (br. S, 1 H), 6.46–6.66 (m, 1 H), 7.18–7.32 (m, 5 H); ^{13}C NMR (100 MHz, CDCl$_3$) δ: 18.0, 28.1, 38.3, 47.9, 52.2, 55.4, 79.9, 126.7, 128.4, 129.3, 136.6, 155.3, 170.9, 172.8. IR (neat): 3323.4, 2984.1, 2946.6, 1751.7, 1692.8, 1655.5, 1522.4, 1445.9, 1385.8, 1366.2, 1250.5, 1159.4, 1049.5, 988.9, 859.6, 664.4 cm^{-1}; HRMS (ESI)$^+$ [M+H]$^+$ calcd for $C_{18}H_{27}N_2O_5$: 351.1920. Found: 351.1915.

35. The ninhydrin stain was prepared using 1.5 g of ninhydrin dissolved in 100 mL of n-butanol and 3.0 mL of AcOH.

Working with Hazardous Chemicals

The procedures in *Organic Syntheses* are intended for use only by persons with proper training in experimental organic chemistry. All hazardous materials should be handled using the standard procedures for work with chemicals described in references such as "Prudent Practices in the Laboratory" (The National Academies Press, Washington, D.C., 2011; the full text can be accessed free of charge at http://www.nap.edu/catalog.php?record_id=12654). All chemical waste should be disposed of in accordance with local regulations. For general guidelines for the management of chemical waste, see Chapter 8 of Prudent Practices.

In some articles in *Organic Syntheses*, chemical-specific hazards are highlighted in red "Caution Notes" within a procedure. It is important to recognize that the absence of a caution note does not imply that no significant hazards are associated with the chemicals involved in that procedure. Prior to performing a reaction, a thorough risk assessment should be carried out that includes a review of the potential hazards associated with each chemical and experimental operation on the scale that is planned for the procedure. Guidelines for carrying out a risk assessment and for analyzing the hazards associated with chemicals can be found in Chapter 4 of Prudent Practices.

The procedures described in *Organic Syntheses* are provided as published and are conducted at one's own risk. *Organic Syntheses, Inc.*, its Editors, and its Board of Directors do not warrant or guarantee the safety of individuals using these procedures and hereby disclaim any liability for any injuries or damages claimed to have resulted from or related in any way to the procedures herein.

Discussion

The amide functionality is of paramount relevance in organic and pharmaceutical chemistry, being present in natural and unnatural bioactive compounds, polymers and materials. Nowadays, around 25% of all drugs on the market hold at least one amide bond on their structure.[2] Although several different methods for amidation in general are described,[3] options are still more scarce in peptide chemistry.[4] The most widespread method for peptide bond formation relies on the use of a peptide coupling reagent[5] in order to transform a carboxylic acid function into a more reactive intermediate, being thus prone to react with a free amine. High yielding procedures are guaranteed by very effective peptide coupling reagents. However, this strategy suffers from major drawbacks, including racemization, low atom economy, high cost, and constraints with regard to the direction of the peptide synthesis (i.e., C→N direction).

Recently, we have developed a practical and efficient procedure for the formation of peptide bonds on the basis of a new mode of activation of amino acids.[6] The method relies on the use of activated α-aminoesters, (obtained from α-aminoesters and inexpensive *N,N'*-carbonyldiimidazole - CDI),[7] instead of classical carboxylic acid activation. These activated compounds are prepared under mild conditions, they are readily purified by simple filtration over silica gel and they are stable for months when stored at 4 °C (Scheme 1).

$$\text{HCl·H}_2\text{N} \overset{\text{R}^1}{\underset{}{\bigvee}} \text{CO}_2\text{R} \quad \xrightarrow[\substack{\text{CH}_2\text{Cl}_2 \\ 23\ °\text{C, 12-24 h}}]{\substack{\text{CDI (1.0 equiv)} \\ \text{Base (1.0 equiv)}}} \quad \overset{\text{O}\quad\text{R}^1}{N\diagdown N\diagup N\diagdown H \diagup CO_2R}$$

(1.0 equiv)   activated α-aminoesters

R = Me, Et, *t*-Bu, Allyl
R[1] = Proteogenic amino acid side chains

> Readily available and stable substrates
> Excellent amino acid residues tolerance
> Easy work-up
> High Yields
> More that 15 examples synthesized

Scheme 1. CDI-mediated synthesis of activated α-aminoesters[6]

From the optimized procedure, it was found that the peptide-bond formation is accomplished under very mild conditions, without the need for a base. Whereas the coupling reaction can proceed in the sole presence of a free amino acid residue and an activated α-aminoester in CH_2Cl_2 as solvent, it was found that catalytic amounts (10 mol%) of $CuBr_2$ and HOBt has a synergic effect on the condensation reaction, yielding the required dipeptides in improved isolated yields (Scheme 2).

Organic Syntheses

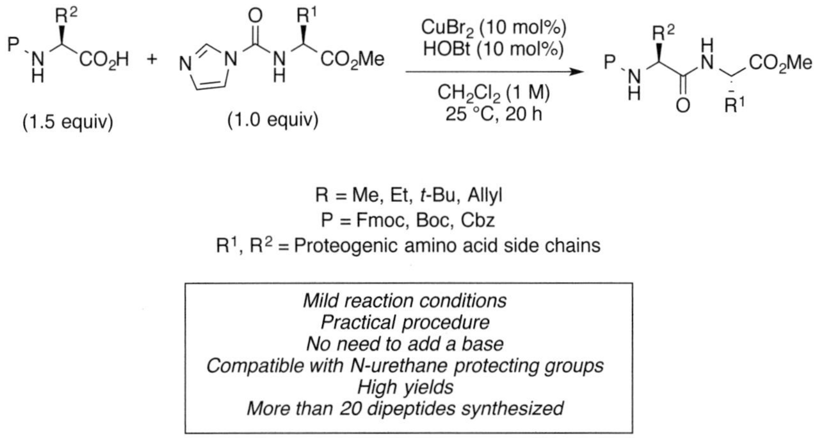

R = Me, Et, *t*-Bu, Allyl
P = Fmoc, Boc, Cbz
R^1, R^2 = Proteogenic amino acid side chains

Mild reaction conditions
Practical procedure
No need to add a base
Compatible with N-urethane protecting groups
High yields
More than 20 dipeptides synthesized

Scheme 2. Dipeptide syntheses via activated α-aminoesters

The above procedure is representative of the synthesis of diversely substituted dipeptides. Indeed, amino acids bearing the most common *N*-urethane protecting groups (e.g. Fmoc, Boc and Cbz) as well as amino acid residues with functionalized side-chains are compatible with the method. Moreover, the strategy features a convenient issue to the less common reverse N→C direction peptide synthesis (ribosomal peptide direction synthesis) as the preparation of an illustrative tetrapeptide model has been made.[6] Besides, this system presents new opportunities for the preparation of a key amide functional group in pharmaceutical chemistry.[8]

References

1. Institut Charles Gerhardt Montpellier (ICGM), UMR 5253 CNRS-UM-ENSCM, Ecole Nationale Supérieure de Chimie, 8 Rue de l'Ecole Normale, 34296 Montpellier Cedex 5, France. E-mail: jean-marc.campagne@enscm.fr and renata.marcia_de_figueiredo@enscm.fr. The Agence Nationale de la Recherche (ANR) under the Programme Jeunes Chercheuses Jeunes Chercheurs (JCJC) 2012, grant agreement ANR-12-JS07-0008-01 (NIPS project) is gratefully acknowledged for financial support.

2. a) Ghose, A. K; Viswanadhan, V. N.; Wendoloski J. J. *J. Combin. Chem.* **1999**, *1*, 55-68; b) Carey, J. S.; Laffan, D.; Thomson, C.; Williams, M. T. *Org. Biomol. Chem.* **2006**, *4*, 2337-2347.

3. Selected examples of amide formation throughout non-conventional carboxylic acid activation: a) Bode, J. W.; Fox, R. M.; Baucom, K. D. *Angew. Chem. Int. Ed.* **2006**, *45*, 1248–1252; b) Shen, B.; Makley, D. M.; Johnston, J. N. *Nature* **2010**, *465*, 1027–1032; c) Wang, T.; Danishefsky, S. J. *J. Am. Chem. Soc.* **2012**, *134*, 13244–13247; d) Kolakowski, R. V.; Shangguan, N.; Sauers, R. R.; Williams, L. J. *J. Am. Chem. Soc.* **2006**, *128*, 5695–5702; e) Wilson, R. M.; Stockdill, J. L.; Wu, X.; Li, X.; Vadola, P. A.; Park, P. K.; Wang, P.; Danishefsky, S. J. *Angew. Chem. Int. Ed.* **2012**, *51*, 2834–2848; f) Chen, W.; Shao, J.; Hu, M.; Yu, W.; Giulianotti, M. A.; Houghten, R. A.; Yu, Y. *Chem. Sci.* **2013**, *4*, 970–976; g) Schuemacher, A. C.; Hoffmann, R. W. *Synthesis* **2001**, 243–250; h) Sasaki, K.; Crich, D. *Org. Lett.* **2011**, *13*, 2256–2259; i) Crich, D.; Sasaki, K. *Org. Lett.* **2009**, *11*, 3514–3517; j) Nordstrøm, L. U.; Vogt, H.; Madsen, R. *J. Am. Chem. Soc.* **2008**, *130*, 17672–17673.

4. a) Jones, J. *The Chemical Synthesis of Peptides*, Oxford Science Publications, Oxford, **1991**; b) *Synthesis of Peptides and Peptidomimetics*, Goodman, M.; Felix, A.; Moroder, L.; Toniolo, C. *Houben-Weyl, Methods of Organic Chemistry*, *Vol E22a*, Thieme, Stuttgart, **2002**; c) Sewald, N.; Jakubke, H.-D. *Peptides: Chemistry and Biology*, Wiley-VCH, Weinheim, **2002**; d) Guzmán, F.; Barberis, S.; Illanes, A. *Electron. J. Biotechnol.* **2007**, *10*, 279–314.

5. Selected reviews: a) Albericio, F.; Chinchilla, R.; Dodsworth, D. J.; Nájera, C. *Org. Prep. Proced. Int.* **2001**, *33*, 203–303; b) Han, S.-Y.; Kim, Y.-A. *Tetrahedron* **2004**, *60*, 2447–2467; c) Montalbetti, C. A. G. N.; Falque, V. *Tetrahedron* **2005**, *61*, 10827–10852; d) Valeur, E.; Bradley, M. *Chem. Soc. Rev.* **2009**, *38*, 606–631; e) El-Faham, A.; Albericio, F. *Chem. Rev.* **2011**, *111*, 6557–6602; f) Joullié, M. M.; Lassen, K. M. *ARKIVOC* **2010**, 189; g) Amblard, M.; Fehrentz, J.-A.; Martinez, J.; Subra, G. *Mol. Biotechnol.* **2006**, *33*, 239–254; h) Lanigan, R. M.; Sheppard, T. D. *Eur. J. Org. Chem.* **2013**, 7453–7465.

6. Suppo, J.-S.; Subra, G.; Bergès, M.; de Figueiredo R. M.; Campagne, J.-M. *Angew. Chem. Int. Ed.* **2014**, *53*, 5389–5393.

7. a) Paul, R.; Anderson, G. W. *J. Am. Chem. Soc.* **1960**, *82*, 4596–4600; b) Staab, H. A. *Angew. Chem. Int. Ed.* **1962**, *1*, 351–367; c) Heller, S. T.; Sarpong, R. *Org. Lett.* **2010**, *12*, 4572–4575; d) Heller, S. T.; Fu, T.; Sarpong, R. *Org. Lett.* **2012**, *14*, 1970–1973.

8. Roughley, S. D.; Jordan, A. M. *J. Med. Chem.* **2011**, *54*, 3451–3479.

Appendix
Chemical Abstracts Nomenclature (Registry Number)

L-Alanine methyl ester hydrochloride; (2491-20-5)
Diisopropylethylamine; (7087-68-5)
N,N'-Carbonyldiimidazole; (530-62-1)
Boc-L-phenylalanine; (13734-34-4)
Copper(II) bromide; (7789-45-9)
1-Hydroxybenzotriazole hydrate; (123333-53-9)

Jean-Simon Suppo was born in Briançon, France, in 1989. He received his Master's degree in Chemistry from Aix-Marseille University in 2012 under the direction of Prof. Laurence Feray. Then, he moved to Montpellier and began his doctoral studies under the guidance of Prof. Jean-Marc Campagne and Dr. Renata Marcia de Figueiredo. His graduate research focuses on inverse peptide synthesis through amine activation.

Renata Marcia de Figueiredo was born in Boa Esperança-MG, Brazil. She received her Ph.D. degree from the University of Paris Sud (Orsay-France) in 2005. Then, she moved to Germany as a postdoctoral research fellow in the group of Prof. M. Christmann in RWTH-Aachen. In 2008, she was appointed CNRS researcher at the Ecole Nationale Supérieur de Chimie de Montpellier (ENSCM) where she has joined the group of Prof. J.-M. Campagne. Her research interests include the development and the application of catalytic asymmetric methodologies to the total synthesis of natural products and biologically active targets.

Jean-Marc Campagne was born in Pau, France, in 1967. After studies at the Ecole Nationale Supérieure de Chimie de Montpellier (ENSCM), he received his Ph.D. at the University of Montpellier in 1994. After post-doctoral training with Prof B. Trost (Stanford University, USA) and Prof. L. Ghosez (Université Catholique de Louvain, Belgium), he was appointed CNRS researcher at the Institut de Chimie des Substances Naturelles in Gif-sur-Yvette in 1998. Since 2005 he moved to the ENSCM where he was appointed as Professor. His current interests concern the development of catalytic asymmetric transformations and their application to the total synthesis of natural products.

John T. Colyer was born in Columbus, Indiana in 1977. In 2000 he earned his B. S. in chemistry from Indiana University-Purdue University Indianapolis working under the guidance of Professor William H. Moser. He completed his M.S. in 2004 at the University of Arizona under the guidance of Professor Michael P. Doyle, where he studied dirhodium (II) carboxamidate catalysis. In 2004 he joined Amgen in Thousand Oaks, CA and currently works in the Chemical Process Research and Development group.

Christopher J. Borths received a B.S. in Chemistry and Biology from the University of Kentucky in 1998. He began his graduate studies in Chemistry at the University of California, Berkeley, earning a M.S. in 2000, and in 2004 he obtained a Ph.D. from the California Institute of Technology for the development of novel organocatalytic methods with Prof. David MacMillan. He then joined the Chemical Process Research and Development group at Amgen where he is currently a Senior Scientist.

(S)-1,1-Diphenylprolinol Trimethylsilyl Ether

Robert K. Boeckman, Jr.*[1], Douglas J. Tusch, and Kyle F. Biegasiewicz

Department of Chemistry, University of Rochester, Rochester, N.Y. 15627

Checked by Eduardo V. Mercado-Marin and Richmond Sarpong

A.

PhMgBr / THF / RT, 1.5 h → NaOH, EtOH / reflux, 1 h → **1**

B.

1 → TMSOTf / Et₃N, DCM / –78 °C to RT / 3 h → **2**

Procedure

A. *(S)-1,1-Diphenylprolinol (1).* A 1-L 24/40 three-necked round-bottomed flask is equipped with an egg-shaped, Teflon-coated, magnetic stir bar (15 x 32 mm). The left neck is capped with a rubber septum, the center neck is fitted with a Friedrich's condenser with an inert gas inlet connected to a dry nitrogen source and a bubbler, and the right neck with a 250-mL pressure equalizing addition funnel also capped with a rubber septum (Note 1). The flask is charged by syringe with 250 mL (0.25 mol, 2.5 equiv) of a 1M solution of phenylmagnesium bromide in tetrahydrofuran (THF) (Note 2). A solution of N-Boc-L-proline methyl ester (22.9 g, 100 mmol, 1.00 equiv) (Note 3) in THF (200 mL) (Note 4), is charged to the dropping funnel via syringe and added dropwise to the solution of the Grignard reagent via the addition funnel over 45 min. The reaction is stirred at ambient temperature for 90 min, and then cooled to 0 °C in an ice

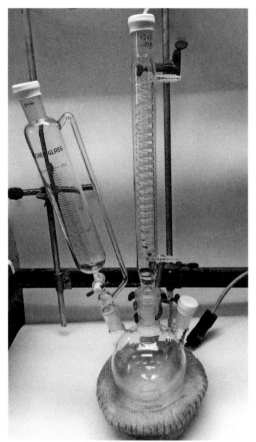

Figure 1. Glassware Assembly for Step A

water bath. The reaction is quenched over 5 min by the controlled addition of a saturated aqueous solution of ammonium chloride (150 mL) via addition funnel. The biphasic mixture is transferred to a 2-L separatory funnel and diluted with water (150 mL). The layers are separated, and the aqueous layer is extracted with diethyl ether (3 x 150 mL). The combined organic layers are dried over sodium sulfate (15 g) and gravity filtered through a large powder funnel equipped with a conical medium porosity filter paper into a 1-L single-necked (24/40) round-bottomed flask. The solvent is removed from the filtrate by rotary evaporation (30 mmHg) at room temperature, providing a clear colorless residue.

The flask containing the residue is equipped with an egg-shaped, Teflon-coated magnetic stir bar (15 x 32 mm). Ethanol (500 mL) (Note 5) is added to the flask, followed by sodium hydroxide (40.0 g, 1.00 mol, 10.0 equiv) (Note 6) and stirring is initiated. The flask is fitted with a Friedrich's condenser open to the atmosphere. The flask is heated to reflux using a 1-L electric heating mantle and allowed to stir at reflux for 1 h. Heating is discontinued, the Friedrich's condenser removed, and the reaction is concentrated by rotary evaporation using a 50 °C water bath (30 mmHg), providing a light yellow amorphous solid. The residue is dissolved in water (200 mL) and diethyl ether (200 mL), and the resulting biphasic mixture is transferred to a 1-L separatory funnel. The layers are separated and the aqueous layer is extracted with diethyl ether (2 x 200 mL). The combined organic layers are washed successively with water (200 mL) and a saturated aqueous solution of sodium chloride (200 mL). The organic layer is dried over sodium sulfate (15.0 g), and filtered through a large powder funnel containing a conical medium porosity filter paper into a 1-L round-bottomed flask. The solution is concentrated by rotary evaporation (50 mmHg) at room temperature to afford a light yellow solid. The solid is dissolved in 200 mL of boiling hexanes, 2.5 g decolorizing carbon is added, and the solution is filtered hot through a large powder funnel equipped with a conical medium porosity filter paper into a 500 mL Erlenmeyer flask, submerged in a water bath heated to 70 °C, that contains refluxing hexanes (30 mL). Once the filtration is complete, the flask is cooled to 0 °C and kept at 0 °C for 1 h, which results in crystallization (Note 7). The hexanes are decanted and the white crystals are washed with cold (0 °C) hexanes (3 x 20 mL), and the hexanes washes are successively decanted. The resulting moist crystalline solid is transferred to a 250-mL 24/40 single-necked round-bottomed flask that is then fitted with a vacuum adaptor and dried on a vacuum pump (0.15 mmHg) overnight (14 h) to provide 14.04–15.64 g (55–62%) of (S)-1,1-diphenylprolinol (1) as white crystals (Notes 8 and 9). A second crop of crystals can be obtained by concentrating the mother liquor and hexanes washes by rotary evaporation (30 mmHg) to dryness, dissolving the residue in 50 mL boiling hexanes, then cooling the solution to 0 °C and ageing the solution for 1 h at 0 °C. After decanting the hexanes, the resulting crystalline solids are washed with cold hexanes (3 x 10 mL), isolated by decantation, and dried under vacuum (0.15 mmHg) overnight (14 h) to give 3.09–4.44 g (12–18%) (Note 8) of white crystals of comparable purity to the first crop. The total yield is 17.44–19.16 g (69–76%) (Note 10) of (S)-1,1-diphenyl-prolinol (1) with a melting point of 73–74 °C.

B. (S)-1,1-Diphenylprolinol trimethylsilyl ether (2). A 1-L 24/40 single-necked round-bottomed flask is equipped with an egg-shaped, Teflon-coated magnetic stir bar (15 x 32 mm) and a 60 mL pressure equalizing addition funnel fitted with a nitrogen inlet. The flask is placed under an atmosphere of nitrogen and flame dried. After cooling to ambient temperature, the flask is charged with a solution of *(S)-1,1-diphenylprolinol* (**1**) (17.7 g, 70.0 mmol, 1.00 equiv) in dichloromethane (350 mL) (Note 11). The solution is cooled to –78 °C in a dry ice and acetone bath, followed by addition of triethylamine (12.7 mL, 9.20 g, 91.0 mmol, 1.30 equiv) (Note 12) in one portion by syringe. Trimethylsilyl trifluoromethanesulfonate (16.5 mL, 20.2 g, 91.0 mmol, 1.30 equiv) (Note 13) is added dropwise via the addition funnel over 30 min. The reaction mixture is stirred and allowed to warm to 0 °C over 2 h. The cooling bath is removed and the reaction mixture is allowed to warm to ambient temperature over 1 h. The reaction is quenched by addition of a solution of sat aq. sodium bicarbonate (100 mL) over 0.5 min. The mixture is diluted with water (100 mL) and transferred to a 1-L separatory funnel. The phases are separated, and the aqueous phase is extracted with dichloromethane (3 x 100 mL). The combined organic phases are dried over anhydrous sodium sulfate (15 g), then gravity filtered through a powder funnel equipped with a medium porosity conical filter paper. The filtrate is concentrated by rotary evaporation (30 mmHg) to afford the impure product as an orange oil. Purification by column chromatography with elution by 60% diethyl ether/hexanes yields 17.51–17.60 g (77–78%) of *(S)-1,1-diphenylprolinol trimethylsilyl ether* (**2**) as a light yellow oil (Note 14).

Notes

1. The submitters recommend that the apparatus be assembled under an atmosphere of nitrogen and flame dried.
2. The Grignard reagent was prepared using the following procedure: A 24/40 1-L three-necked round-bottomed flask equipped with an egg-shaped Teflon coated magnetic stir bar (15 x 32 mm), a Friedrich's condenser with an inert gas inlet in the middle neck, 250-mL pressure equalizing addition funnel capped with a rubber septum in the right neck. Magnesium turnings (6.7 g, 275 mmol, 2.75 equiv) are added to the flask through the remaining open neck. The flask is sealed by capping the open neck with a rubber septum and stirring is initiated. The apparatus is placed under an atmosphere of nitrogen and flame

dried. After allowing the apparatus to cool to ambient temperature, a crystal of iodine (50 mg, 0.2 mmol) is dissolved in 125 mL of anhydrous THF (Note 3) and added via syringe, resulting in a light brown transparent solution. Bromobenzene (26.3 mL, 39.3 g, 250 mmol, 2.5 equiv) is added to the addition funnel, and one quarter of the volume is added to the flask over a one minute period. The flask is heated by a 1-L electric heating mantle until the iodine color dissipates (~45 °C). Heating is stopped, and 125 mL of anhydrous THF is added to the addition funnel to dilute the bromobenzene. The bromobenzene solution is then added dropwise to the flask at a rate sufficient to maintain reflux in the flask (added over 20 min). The clear solution becomes cloudy and brown during the addition. When the addition is complete, heat is reapplied using a 1 L electric heating mantle and the flask is allowed to stir at reflux for 1 h. Heating is discontinued, and the brown cloudy reaction mixture containing unused magnesium is cooled to ambient temperature and used as obtained.

3. *N-(tert-Butoxycarbonyl)-L-proline methyl ester* was prepared by methyl esterification of *N-(tert-butoxycarbonyl)-L-proline*. A 2-L (24/40), single-necked round-bottomed flask, fitted with a 100 mL pressure equalizing addition funnel capped with rubber septum, and egg-shaped Teflon-coated magnetic stir bar (15 x 32 mm) is charged with a solution of *N*-(*tert*-butoxycarbonyl)-*L*-proline (110 g, 0.52 mol, 1 equiv) (prepared from *L*-proline, which was obtained from Spectrum Chemical and used as received, according to the procedures: Keller, O.; Keller, W. E.; van Look, G.; Wersin, G., *Org. Synth.* **1985**, *63*, 160.) in 1.33 L of dimethylformamide (≥99.8%, obtained from Sigma-Aldrich and used as received) and potassium carbonate (obtained from Spectrum Chemical and used as received) (78.4 g, 0.57 mol, 1.1 equiv). Stirring is initiated and the suspension is cooled to 0 °C in an ice water bath and methyl iodide (64.0 mL, 1.03 mol, 2.0 equiv) (99%, obtained from Sigma-Aldrich and used as received) is added dropwise via addition funnel over a period of 10 min. After the addition is complete, the reaction is allowed to warm to ambient temperature and stirring is continued for 15 h. The reaction mixture is filtered through a bed of 15 g of Celite® in a 100-mL coarse fritted funnel and the filter cake is washed with diethyl ether (750 mL). The filtrate is transferred to a 4-L separatory funnel, diluted with of water (750 mL) and extracted with diethyl ether (3 x 375 mL). The combined organic phases are washed successively with water (3 x 750 mL) and sat brine (750 mL), dried over magnesium sulfate

(25.0 g), and vacuum filtered. The filtrate is concentrated *in vacuo* using a 30 °C water bath (30 mmHg) providing the crude product ester (109 g, 92%) as a yellow oil which is used without further purification.

4. Tetrahydrofuran (inhibitor free, Optima®, 99.9%) was obtained from Fisher Scientific and degassed by argon-bubbling for 1 h before being passed through an activated alumina column using a GlassContour solvent system. The solvent is withdrawn from the receiver flask with a syringe.

5. Ethanol (anhydrous) was obtained from Spectrum Chemical and used as received.

6. Sodium Hydroxide (pellets, certified ACS) obtained from Spectrum Chemical and used as received.

7. In the event that crystallization did not occur, the solution was transferred to an etched 500 mL round-bottomed flask cooled to 0 °C and aged for 1 h at 0 °C, which resulted in crystallization.

8. Step A was checked three times, and the yields represent the range of the three trials.

9. Physical properties of *(S)-1,1-diphenylprolinol* (**1**): ^{1}H NMR (600 MHz, CDCl$_3$) δ: 1.52–1.82 (m, 5H), 2.89–2.99 (m, 1H), 2.99–3.10 (m, 1H), 4.26 (t, *J* = 7.7 Hz, 1H), 4.65 (s(br)), 1H), 7.11–7.23 (m, 2H), 7.24–7.36 (m, 4H), 7.52 (d, *J* = 8.2 Hz, 2H), 7.59 (d, *J* = 8.2 Hz, 2H). ^{13}C NMR (150 MHz, CDCl$_3$) δ: 25.6, 26.4, 46.9, 64.6, 77.2, 125.7, 126.0, 126.5, 126.6, 128.1, 128.3, 145.6, 148.3. IR (neat): 3354, 3057, 3024, 2945, 2870, 1597, 1491, 1448, 1397, 1171, 991, 746, 696, 659, 634 cm^{-1}. HRMS (ESI) calcd for C$_{17}$H$_{20}$NO [M+H$^+$] 254.1539; Found: 254.1535. Anal. Calcd for C$_{17}$H$_{19}$NO: C, 80.60; H, 7.56; N, 5.53. Found: C, 80.64; H, 7.63; N, 5.43.

10. Step A was checked three times with the following results. Run A: 1st Crop: 14.04 g; 2nd Crop: 4.44 g; Total: 18.48 g. Run B: 1st Crop: 15.64 g; 2nd Crop: 3.52 g; Total: 19.16 g. Run C: 1st Crop: 14.35 g; 2nd Crop: 3.09 g; Total: 17.44 g.

11. Dichloromethane (Methylene Chloride, HPLC grade) was obtained from Fisher Scientific and was distilled over calcium hydride prior to use.

12. Triethylamine (HPLC grade, 99%) was obtained from Fisher Scientific and used as received.

13. Trimethylsilyl trifluoromethanesulfonate (99%) was obtained from Oakwood Chemical and used as received.

14. TMS Prolinol **2** is purified on a column (9 x 55 cm) packed with 200 g of silica (obtained from EMD Millipore, 60Å pore size, 230 - 400 mesh) in

30% diethyl ether/hexanes. Fraction collection (500 mL fractions) begins immediately; the column is eluted with 1.5 liters 30% diethyl ether/hexanes, 2 liters 60% diethyl ether/hexanes, then flushed with 1.5 liters of diethyl ether. Fractions 2-7 are pooled and contain the desired product. The product has an R_f of 0.06 in ether but will streak up the plate significantly if concentrated. Physical properties of *(S)-1,1-diphenylprolinol trimethylsilyl ether* (**2**) are: ^{1}H NMR (600 MHz, CDCl$_3$) δ: –0.09 (s, 9H), 1.31 – 1.42 (m, 1H), 1.50 – 1.63 (m, 3H), 1.69 (s(br)), 1H), 2.75 – 2.82 (m, 1H), 2.82 – 2.88 (m, 1H), 4.03 (t, J = 7.0 Hz, 1H), 7.19 – 7.30 (m, 6H), 7.35 (d, J = 6.9 Hz, 2H), 7.46 (d, J = 6.9 Hz, 2H). ^{13}C NMR (150 MHz, CDCl$_3$) δ: 2.3, 25.2, 27.7, 47.3, 65.6, 83.3, 126.9, 127.0, 127.65, 127.72, 127.74, 128.6, 145.9, 147.0. IR (neat): 3059, 3025, 2953, 1492, 1446, 1402, 1313, 1100, 1069, 877, 833, 749, 698 cm^{-1}. HRMS (ESI) calcd for C$_{20}$H$_{28}$NOSi [M+H${}^+$] 326.1935; Found: 326.1927. Anal. Calcd for C$_{20}$H$_{27}$NOSi: C, 73.79, H, 8.36, N, 4.30, found: C, 73.82, H, 8.33, N, 4.26.

Working with Hazardous Chemicals

The procedures in *Organic Syntheses* are intended for use only by persons with proper training in experimental organic chemistry. All hazardous materials should be handled using the standard procedures for work with chemicals described in references such as "Prudent Practices in the Laboratory" (The National Academies Press, Washington, D.C., 2011; the full text can be accessed free of charge at http://www.nap.edu/catalog.php?record_id=12654). All chemical waste should be disposed of in accordance with local regulations. For general guidelines for the management of chemical waste, see Chapter 8 of Prudent Practices.

In some articles in *Organic Syntheses*, chemical-specific hazards are highlighted in red "Caution Notes" within a procedure. It is important to recognize that the absence of a caution note does not imply that no significant hazards are associated with the chemicals involved in that procedure. Prior to performing a reaction, a thorough risk assessment should be carried out that includes a review of the potential hazards associated with each chemical and experimental operation on the scale that is planned for the procedure. Guidelines for carrying out a risk assessment and for analyzing the hazards associated with chemicals can be found in Chapter 4 of Prudent Practices.

The procedures described in *Organic Syntheses* are provided as published and are conducted at one's own risk. *Organic Syntheses, Inc.*, its Editors, and its Board of Directors do not warrant or guarantee the safety of individuals using these procedures and hereby disclaim any liability for any injuries or damages claimed to have resulted from or related in any way to the procedures herein.

Discussion

The proline based catalyst (*S*)-1,1-diphenylprolinol trimethylsilyl ether (**2**) has been used extensively in organic chemistry to successfully affect a variety of reactions, including α, β, and even γ-activation of aldehydes by enamine and iminium-ion chemistry.[2] The robustness of this catalyst system is demonstrated by its application to multicomponent cascades to form highly substituted piperidines,[3] cyclohexenes,[4] and hydropyrans.[5]

(*S*)-1,1-Diphenylprolinol trimethylsilyl ether (**2**) is readily prepared by the silylation of (*S*)-1,1-diphenylprolinol (**1**), however, commercial sources of diphenylprolinol can be prohibitively expensive, and current methods for the synthesis of diphenylprolinol and its derivatives are low yielding, practically challenging on scale, or include hazardous reagents such as phosgene.[6] Because of the prevalence of the diphenylprolinol catalysts and the lack of a practical synthesis, we desired an efficient, scalable synthesis of diphenylprolinol (**1**) and its derivatives.

The method reported provides access to diphenylprolinol (**1**) in two steps from the Boc-protected methyl ester of proline, and the trimethylsilyl ether of diphenylprolinol (**2**) via a single transformation from diphenyl prolinol (**1**).

References

1. Department of Chemistry, University of Rochester, Rochester, NY 14627, email rkb@rkbmac.chem.rochester.edu.

2. For a review of diarylprolinol silyl ether chemistry, see: Jensen, K. L.; Dickmeiss, G.; Jiang, H.; Albrecht, Ł.; Jørgensen, K. A. *Acc. Chem. Res.* **2011**, *45* (2), 248–264.
3. Urushima, T.; Sakamoto, D.; Ishikawa, H.; Hayashi, Y. *Org. Lett.* **2010**, *12* (20), 4588–4591.
4. Palomo, C.; Mielgo, A. *Angew. Chem. Int. Ed.* **2006**, *45* (47), 7876–7880.
5. McGarraugh, P. G.; Johnston, R. C.; Martínez-Muñoz, A.; Cheong, P. H.-Y.; Brenner-Moyer, S. E. *Chem. Eur. J.* **2012**, *18* (34), 10742–10752.
6. Xavier, L. C.; Mohan, J. J.; Mathre, D. J.; Thompson, A. S.; Carroll, J. D.; Corley, E. G.; Desmond, R. *Org. Synth.* **1997**, *74*, 50–71.

Appendix
Chemical Abstracts Nomenclature (Registry Number)

Phenylmagnesium bromide solution; (100-58-3)
N-Boc-L-proline methyl ester; (59936-29-7)
(S)-1,1-Diphenylprolinol: (*S*)-(–)-2-(Diphenylhydroxymethyl)pyrrolidine;
(112068-01-6)
Triethylamine; (121-44-8)
Trimethylsilyl trifluoromethanesulfonate; (27607-77-8)
(S)-1,1-Diphenylprolinol trimethylsilyl ether: (S)-(–)-α,α-Diphenyl-2-
pyrrolidinemethanol trimethylsilyl ether; (848821-58-9)
Magnesium turnings; (7439-95-4)
Bromobenzene; (108-86-1)
N-(*tert*-Butoxycarbonyl)proline; (15761-39-4)
Methyl iodide: Iodomethane; (74-88-4)

O rganic
S yntheses

Robert K. Boeckman, Jr. received his Bachelors of Science degree in Chemistry in 1966 from Carnegie Institute of Technology (now Carnegie Mellon University). He moved on to Brandeis University where he received the Ph.D. degree under the supervision of James B. Hendrickson and Ernest Grunwald in 1971. He joined the research group of Gilbert Stork in 1970 as an NIH postdoctoral fellow. He began his academic career at Wayne State University in 1972, where he rose to the rank of Professor in 1979. In 1980, he joined the faculty of the University of Rochester where he is currently Marshall D. Gates Jr. Professor of Chemistry. His research interests lie primarily in the area of synthetic organic chemistry, both the development of new synthetic methodology and the total synthesis of complex substances of biological interest.

Douglas J. Tusch received his Bachelors of Science degree in Biochemistry from the Rochester Institute of Technology in 2010 where he researched alkaloid natural product synthesis under the guidance of Professor Jeremy Cody. He then joined the group of Professor Robert Boeckman at the University of Rochester and earned his M.S. in chemistry in 2012. He is currently pursuing a Ph.D. in Professor Boeckman's group studying the areas of total synthesis and organocatalysis.

Kyle F. Biegasiewicz received his Bachelors of Science in Chemistry from Niagara University in 2010, completing undergraduate research with Professor Ronny Priefer. He is currently a chemistry graduate student at the University of Rochester working under the tutelage of Professor Robert K. Boeckman, Jr. His graduate research areas include the total synthesis of natural products and organocatalysis.

Eduardo V. Mercado-Marin received his Bachelors of Science in Chemistry from the University of California, Santa Barbara in 2011, completing undergraduate research under the guidance of Professor Thomas R. R. Pettus. He is currently a chemistry graduate student at the University of California, Berkeley working with Professor Richmond Sarpong. His graduate research is focused on a unified approach to the total syntheses of prenylated indole alkaloid natural products.

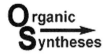

Organocatalyzed Direct Asymmetric α–Hydroxymethylation of Aldehydes

Robert K. Boeckman, Jr.*[1], Douglas J. Tusch, and Kyle F. Biegasiewicz

Department of Chemistry, University of Rochester, Rochester, N.Y. 15627

Checked by Eduardo V. Mercado-Marin and Richmond Sarpong

Procedure

(R)-2-(Hydroxymethyl)-3-methylbutanoic acid (1). A 250-mL 24/40 single-necked round-bottomed flask is equipped with a cylindrical Teflon-coated magnetic stir bar (7 x 25 mm). The flask is charged with *(S)*-1,1-diphenylprolinol trimethylsilyl ether (1.63 g, 5 mmol, 0.1 equiv) (Note 1), toluene (100 mL) (Note 2), pH 7 buffer (5 g) (Note 3), and 37% aqueous formaldehyde solution (11.2 mL, 12.2 g, 0.150 mol, 3.0 equiv) (Note 4). Vigorous stirring is initiated (Note 5), and isovaleraldehyde (5.49 mL, 4.31 g, 0.050 mol, 1.0 equiv) (Note 6) is added in one portion to the flask. The flask is capped with a yellow hard plastic Caplug® (Note 7), which is sealed with parafilm. The flask is allowed to stir at ambient temperature for 36 h (Note 8). The mixture is transferred to a 500 mL separatory funnel, the flask is rinsed with a minimum of toluene (~ 15 mL) and the rinsate added to the separatory funnel. The phases are separated and the aqueous phase is extracted with 30 mL toluene. The combined organic phases are added to a 24/40 1-L single-necked round-bottomed flask and concentrated by rotary evaporation (30 mmHg) in a 30 °C water bath to a clear colorless oil.

Org. Synth. **2015**, *92*, 320-327
DOI: 10.15227/orgsyn.92.0320

Published on the Web 11/10/2015
© 2015 Organic Syntheses, Inc.

The 1-L flask containing the residue is equipped with a cylindrical Teflon-coated magnetic stir bar (7 x 25 mm) and a 250-mL pressure equalizing addition funnel caped with a septum and a nitrogen inlet (Figure 1). *tert*-Butanol (250 mL) (Note 9) is added to the flask, followed by 2-methyl-2-butene (53.0 mL, 35.0 g, 0.50 mol, 10 equiv) (Note 10). The flask is then submerged in an ice water bath and cooled to 0 °C. A solution of sodium chlorite (18.1 g, 0.20 mol, 4 equiv) (Note 11) and sodium monobasic hydrogen phosphate monohydrate (27.6 g, 0.20 mol, 4 equiv) (Note 12) in 125 mL water is prepared by stirring until all solids dissolve. This yellow solution is then added via addition funnel to the cooled flask over 6 min, resulting in a slight exotherm. The reaction mixture turns dark yellow. The reaction mixture is allowed to warm to ambient temperature and stirred over a 6 h period. During this time, the color of the reaction mixture fades to a light yellow. The reaction mixture is then evaporated via rotary evaporator (30 mmHg) in a 50 °C water bath to remove *tert*-butanol. The resulting aqueous phase (approx. 140 mL) is diluted with 200 mL of ethyl acetate, 75 mL of sat aq sodium chloride solution, and 75 mL of 2.7 M aq hydrochloric acid, and transferred to a 1-L separatory funnel and mixed. After phase separation, the phases are separated and the aq phase is extracted with ethyl acetate (3 x 200 mL). The combined organic phases are dried over anhydrous sodium sulfate (15 g), gravity filtered through a powder funnel equipped with conical shaped medium porosity filter paper and concentrated via rotary evaporation (30 mmHg) in a 50 °C water bath to a yellow oil. Column chromatography provided the 6.06–6.32 g (92–96%, 93.0–93.5% ee) of (R)-2-(hydroxymethyl)-3-methylbutanoic acid (**1**) (Notes 13 and 14) as a white crystalline solid.

Figure 1. Reaction Set up

Organic
Syntheses

Notes

1. *(S)*-1,1-Diphenylprolinol trimethylsilyl ether was prepared using the procedure outlined in the preceding article.[2]
2. Toluene (99.9%, Certified ACS) was obtained from Fisher Scientific and used as received.
3. A stock bottle of pH 7 buffer was prepared by mixing potassium dibasic hydrogen phosphate, which was obtained from Spectrum Chemical and used as received, (64.9 g, 0.477 mol, 1 equiv) and potassium monobasic hydrogen phosphate, which was obtained from Spectrum Chemical and used as received, (91.1 g, 0.523 mol, 1.10 equiv) in a grinder and processing until a free flowing uniform solid formed. The solid is added directly to the reaction vessel.
4. Formaldehyde (36.5–38% aqueous formaldehyde) was obtained from EMD and used as received.
5. Vigorous stirring is required for efficient mixing of the phases. Stirring is sufficiently vigorous if the reaction appears homogenous, such that no distinct phases or phase boundary can be observed.
6. Isovaleraldehyde (≥97%) was obtained from Sigma-Aldrich and was distilled and stored under nitrogen prior to use. The presence of even trace amounts of isovaleric acid was found to catalyze elimination during the hydroxymethylation.
7. The Caplug® (part number WW12) is comprised of low density polyethylene. When rubber septa were used on this apparatus, a decrease in yield was noted, presumably due to septum leaching.
8. Reaction monitored by ¹H NMR. An aliquot (~0.20 mL) is taken from the toluene layer, diluted in 0.8 mL CDCl₃, then examined by NMR. The reaction is judged complete by disappearance of the aldehyde triplet of isovaleraldehyde at 9.75 ppm.
9. *tert*-Butyl alcohol (>99%, ACS Reagent) was obtained from EMD and was typically melted by submerging in warm water (bath temp ≈ 35 °C) for 15 min prior to use.
10. 2-Methyl-2-butene (>90.0%, remainder mainly 2-methyl-1-butene) was obtained from TCI and used as received.
11. Sodium chlorite (80%) was obtained from Alfa Aesar and used as received.
12. Sodium phosphate monobasic monohydrate (Baker analyzed ACS Reagent) was obtained from Spectrum Chemical and used as received.

13. Acid **1** is purified on a column (3.5 x 40 cm) packed with 75 g of silica (obtained from EMD Millipore, 60Å pore size, 230 - 400 mesh) packed in dichloromethane. Fraction collection (25 mL fractions) begins immediately; the column is eluted with 200 mL 1% methanol/dichloromethane, 200 mL 2% methanol/dichloromethane, 300 mL 3% methanol/dichloromethane, 200 mL 5% methanol/dichloromethane, 200 mL 7.5% methanol/dichloromethane, then flushed with 200 mL of 10% methanol/dichloromethane. Fractions 21-52 are pooled and contain the desired product. The product has an R$_f$ of 0.28 in 10% methanol/dichloromethane. Physical properties of *(R)-2-(hydroxymethyl)-3-methylbutanoic acid* (**1**) are: ^{1}H NMR (600 MHz, CDCl$_3$) δ: 0.99 (t, *J* = 6.9 Hz, 6H), 2.02 (h, *J* = 6.9 Hz, 1H), 2.42 (ddd, *J* = 8.7, 7.2, 4.0 Hz, 1H), 3.79 (dd, *J* = 11.2, 4.0 Hz, 1H), 3.87 (dd, *J* = 11.2, 8.7 Hz, 1H), 6.76 (bs, 1H). ^{13}C NMR (150 MHz, CDCl$_3$) δ: 20.3, 20.7, 27.8, 54.4, 61.6, 180.1. [α]$_D^{20}$ –5.50 (c 5.50, CHCl$_3$). IR (neat) 3371, 2964, 2879, 1707, 1468, 1392, 1270, 1198, 1065, 1014 cm^{-1}. mp 71–73 °C; HRMS (ESI) calcd for C$_6$H$_{11}$O$_3$ [M – H$^+$] 131.0714, Found 131.0713. Anal. calcd for C$_6$H$_{12}$O$_3$: C, 54.53, H, 9.15. Found: C, 54.68, H, 9.05.

14. Enantiomeric excess determined chiral CG of the methyl ester of the acid. Esterification procedure as follows: A small sample of the acid (50 mg, 0.38 mmol, 1 equiv) is added to a 5-mL single-necked round-bottomed flask equipped with cylindrical Teflon-coated magnetic stir bar (3.5 x 12 mm) capped with a rubber septum. DMF (1 mL) is added, followed by K$_2$CO$_3$ (105 mg, 0.76 mmol, 2 equiv). Stirring is initiated, and methyl iodide (26 μL, 0.42 mmol, 1.1 equiv) is added. The flask is let stir at ambient temperature 14 h. The reaction diluted with water (1 mL), transferred to a 25 mL separatory funnel, and extracted with diethyl ether (1 mL). The ether extract can be directly injected on the GC. A racemate for comparison was synthesized by substituting pyrrolidine for *(S)-1,1-diphenylprolinol trimethylsilyl ether* in the procedure discussed above. GC conditions: Column: Agilent CycloSil-B, 30 m, 0.25 mm, 0.25 μm, 112-6632 Agilent part # 19091J. GC: Agilent Technologies 7820A. Injector temp: 250 °C. Detector temp: 275 °C. Flow: 2 mL/min. Initial temp: 40 °C. Final Temp: 200 °C. Initial time: 1 min. Rate: 4 °C/min. T$_{minor}$: 21.186 min. T$_{major}$: 21.462 min.

Working with Hazardous Chemicals

The procedures in *Organic Syntheses* are intended for use only by persons with proper training in experimental organic chemistry. All hazardous materials should be handled using the standard procedures for work with chemicals described in references such as "Prudent Practices in the Laboratory" (The National Academies Press, Washington, D.C., 2011; the full text can be accessed free of charge at http://www.nap.edu/catalog.php?record_id=12654). All chemical waste should be disposed of in accordance with local regulations. For general guidelines for the management of chemical waste, see Chapter 8 of Prudent Practices.

In some articles in *Organic Syntheses*, chemical-specific hazards are highlighted in red "Caution Notes" within a procedure. It is important to recognize that the absence of a caution note does not imply that no significant hazards are associated with the chemicals involved in that procedure. Prior to performing a reaction, a thorough risk assessment should be carried out that includes a review of the potential hazards associated with each chemical and experimental operation on the scale that is planned for the procedure. Guidelines for carrying out a risk assessment and for analyzing the hazards associated with chemicals can be found in Chapter 4 of Prudent Practices.

The procedures described in *Organic Syntheses* are provided as published and are conducted at one's own risk. *Organic Syntheses, Inc.*, its Editors, and its Board of Directors do not warrant or guarantee the safety of individuals using these procedures and hereby disclaim any liability for any injuries or damages claimed to have resulted from or related in any way to the procedures herein.

Discussion

α-Substituted β-hydroxy aldehydes and carboxylic acids are an important class of chiral building blocks that have proven useful in the synthesis of many biologically relevant molecules.[3] Current methods for the synthesis of such intermediates typically rely on the use of chiral auxiliaries as pioneered by Evans.[4] These methods can provide good stereoselectivity but requires multiple manipulations and recycling of stoichiometric auxiliary, therefore, limiting scalability.

Direct asymmetric α-hydroxymethylation of aldehydes would provide one-step access to chiral α-substituded β-hydroxy aldehydes. Current methods to affect a direct asymmetric α-hydroxymethylation of an aldehyde rely on gaseous formaldehyde.[5] Other attempts at direct organocatalyzed hydroxymethylation have led to formation of elimination products owing to the unrecognized need to control the pH during the process and avoid the presence of acid.[6] Initial studies by our group indicated (S)-1,1-diphenylprolinol trimethylsilyl ether can catalyze the direct asymmetric α-hydroxymethylation of aldehydes.[7] Herein we report the direct asymmetric α-hydroxymethylation of isovaleraldehyde catalyzed by (S)-1,1-diphenylprolinol trimethylsilyl ether. Our method is mild, and tolerates a variety of functional groups including alkynes, aromatics, silylated alcohols and esters. The method uses easily handled aqueous formaldehyde, and employs a catalytic amount of a chiral catalyst, affording in one step a synthetically versatile chiral intermediate for further elaboration.

References

1. Department of Chemistry, University of Rochester, Rochester, NY 14627, email rkb@rkbmac.chem.rochester.edu.
2. Boeckman, Jr., R. K.; Tusch, D. J.; Biegasiewicz, K. F. *Org. Synth.* **2015**, *92*, 309-319.
3. For select examples, see: a) Evans, D. A.; Dow, R. L.; Shih, T. L.; Takacs, J. M.; Zahler, R. *J. Am. Chem. Soc.* **1990**, *112*, 5290–5313; b) Crimmins, M. T.; Carroll, C. A.; King, B. W. *Org. Lett.* **2000**, *2*, 597–599; c) Evans, D. A.; Connell, B. T. *J. Am. Chem. Soc.* **2003**, *125*, 10899–10905; d) Lister, T.; Perkins, M. V. *Angew. Chem., Int. Ed.* **2006**, *45*, 2560–2564.
4. a) Evans, D. A.; Ennis, M. D.; Mathre, D. J. *J. Am. Chem. Soc.* **1982**, *104*, 1737–1739; b) Evans, D. A.; Urpi, F.; Somers, T. C.; Clark, J. S.; Bilodeau, M. T. *J. Am. Chem. Soc.* **1990**, *112*, 8215–8216.
5. Casas, J.; Sundén, H.; Córdova, A. *Tetrahedron Lett.* **2004**, *45*, 6117–6119.
6. a) Erkkilä, A.; Pihko, P. M. *Eur. J. Org. Chem.* **2007**, 4205-4216.; b) Benohoud, M.; Erkkilä, A; Pihko, P. M. *Org. Synth.* **2010**, *87*, 201-208.
7. Boeckman, R. K.; Miller, J. R. *Org. Lett.* **2009**, *11* (20), 4544-4547.

Appendix
Chemical Abstracts Nomenclature (Registry Number)

(S)-1,1-Diphenylprolinol trimethylsilyl ether: (S)-(–)-α, α-Diphenyl-2-pyrrolidinemethanol trimethylsilyl ether; (848821-58-9)
pH 7 buffer: potassium phosphate dibasic and potassium phosphate monobasic; (7558-11-4) and (7778-77-0)
37% Aqueous formaldehyde solution; (50-00-0)
Isovaleraldehyde: 3-Methylbutyraldeyde; (590-86-3)
2-Methyl-2-butene: 2-methylbut-2-ene; (513-35-9)
Sodium chlorite; (7758-19-2)
Sodium phosphate, monobasic, monohydrate; (10049-21-5)
(R)-2-(Hydroxymethyl)-3-methylbutanoic acid; (72604-80-9)

Robert K. Boeckman, Jr. received his Bachelors of Science degree in Chemistry in 1966 from Carnegie Institute of Technology (now Carnegie Mellon University). He moved on to Brandeis University where he received the Ph.D. degree under the supervision of James B. Hendrickson and Ernest Grunwald in 1971. He joined the research group of Gilbert Stork in 1970 as an NIH postdoctoral fellow. He began his academic career at Wayne State University in 1972, where he rose to the rank of Professor in 1979. In 1980, he joined the faculty of the University of Rochester where he is currently Marshall D. Gates Jr. Professor of Chemistry. His research interests lie primarily in the area of synthetic organic chemistry, both the development of new synthetic methodology and the total synthesis of complex substances of biological interest.

Organic
Syntheses

Douglas J. Tusch received his Bachelors of Science degree in Biochemistry from the Rochester Institute of Technology in 2010 where he researched alkaloid natural product synthesis under the guidance of Professor Jeremy Cody. He then joined the group of Professor Robert Boeckman at the University of Rochester and earned his M.S. in chemistry in 2012. He is currently pursuing a Ph.D. in Professor Boeckman's group studying the areas of total synthesis and organocatalysis.

Kyle F. Biegasiewicz received his Bachelors of Science in Chemistry from Niagara University in 2010, completing undergraduate research with Professor Ronny Priefer. He is currently a chemistry graduate student at the University of Rochester working under the tutelage of Professor Robert K. Boeckman, Jr. His graduate research areas include the total synthesis of natural products and organocatalysis.

Eduardo V. Mercado-Marin received his Bachelors of Science in Chemistry from the University of California, Santa Barbara in 2011, completing undergraduate research under the guidance of Professor Thomas R. R. Pettus. He is currently a chemistry graduate student at the University of California, Berkeley working with Professor Richmond Sarpong. His graduate research is focused on a unified approach to the total syntheses of prenylated indole alkaloid natural products.

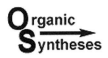

4-Methyl-1-(2-(phenylsulfonyl)ethyl)-2,6,7-trioxabicyclo[2.2.2]octane

Michael J. Di Maso, Michael A. St. Peter, and Jared T. Shaw*[1]

Department of Chemistry, University of California, Davis, One Shields Avenue, Davis, California, 95616, United States of America

Checked by Sunna Jung and Keisuke Suzuki

A.

PhSO$_2$Na, AcOH
———————————→
H$_2$O, reflux

1

B.

PhO$_2$S⌒⌒OH

EDC, DMAP
CH$_2$Cl$_2$, 0 °C to rt

2

C.

PhO$_2$S

BF$_3$·OEt$_2$
———————————→
CH$_2$Cl$_2$, 0 °C

3

Procedure

A. *3-(Phenylsulfonyl)propanoic acid (**1**).* To a 1-L three-necked round-bottomed flask equipped with a 4.5 cm, oval Teflon-coated magnetic stirring bar is added maleic anhydride (10.9 g, 111 mmol, 1.0 equiv) in deionized water (370 mL). The mixture is cooled to 0 °C in an ice-water bath and, in order, benzenesulfinic acid sodium salt (20.0 g, 122 mmol, 1.1 equiv) and glacial acetic acid (8 mL, 140 mmol, 1.2 equiv) are added (Notes 1 and 2).

Org. Synth. **2015**, *92*, 328-341
DOI: 10.15227/orgsyn.092.0328

Published on the Web 11/10/2015
© 2015 Organic Syntheses, Inc.

The flask is equipped with a reflux condenser, and the reaction is heated to reflux in an oil bath (bath temp = 110 °C) open to the air for 16 h (Note 3) (Figure 1). The hot reaction mixture is vacuum filtered through a

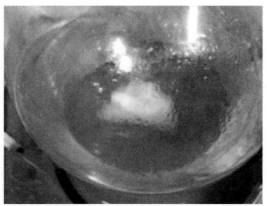

Figure 1. Reaction mixture formed in Step A

Büchner funnel into a 1-L Erlenmeyer flask to remove small oil droplets, if present (Note 4). The filtrate is cooled to 0 °C with an ice water bath and then acidified to pH ~1 with concentrated HCl (approx. 25 mL). The precipitate is filtered on a Büchner funnel and washed with cold hexanes (150 mL). The product is transferred to a 200-mL beaker and dried in a vacuum oven (60 °C, 0.5 mmHg) overnight to give 16.7 g (70%) of 3-(phenylsulfonyl)-propanoic acid **1** as a white solid (Notes 5, 6 and 7) (Figure 2).

Figure 2. Product formed in Step A

B. *(3-Methyloxetan-3-yl)methyl 3-(phenylsulfonyl)propanoate* *(2)*. A 2-L, three-necked round-bottomed flask is equipped with a 4.5 cm, oval Teflon-coated magnetic stirring bar. Two necks are fitted with rubber septa and the middle neck is fitted with a nitrogen stopcock inlet. The flask is flame dried under vacuum and backfilled with nitrogen. The septum is removed to add 3-(phenylsulfonyl)propanoic acid *(1)* (13.5 g, 64.0 mmol, 1 equiv). The septum is replaced, and 315 mL of methylene chloride is added by syringe (Note 8). The septum is removed to add dimethylaminopyridine (2.31 g, 18.9 mmol, 0.3 equiv) in one portion. The septum is reinserted and (3-methyloxetan-3-yl)methanol (9.9 mL, 0.10 mol, 1.6 equiv) (Note 9) is added to the reaction mixture by syringe over one min. The reaction mixture is cooled to 0 °C in an ice water bath. After cooling, the septum is removed to allow for the addition of 1-ethyl-3-(3-dimethylaminopropyl)carbodiimide hydrochloride (14.5 g, 75.6 mmol, 1.2 equiv) (Note 10) in one portion (Figure 3). The septum is quickly

Figure 3. Color of reaction after addition of EDC

replaced, and the reaction is allowed to warm slowly to room temperature by allowing the ice water bath to warm to room temperature overnight (Note 11). When the reaction is determined to be complete by TLC (~16 h) (Note 12), the reaction mixture is concentrated on a rotary evaporator (20 mmHg, 40 °C). The crude mixture is partitioned between diethyl ether (200 mL) and water (200 mL), and the phases are separated in a 500-mL separatory funnel. The aqueous phase is extracted with diethyl ether (2 x 200 mL). The combined organic layers are washed with 1M HCl (2 x 250 mL) and brine (300 mL) in a 1-L separatory funnel. The organic layer is dried over sodium sulfate (~10 g), gravity filtered into a tared 300-mL round-bottomed flask (Note 13), and concentrated on a rotary evaporator (40 °C, 20 mmHg) to give a viscous oil that crystallizes after several times of co-evaporation with hexanes (250 mL) (Note 14). The solid is dried overnight at 0.01 mmHg to give 15.8 g (84%) of (3-methyloxetan-3-yl)methyl 3-(phenylsulfonyl)propanoate (Notes 15, 16 and 17).

   C.   *4-Methyl-1-(2-(phenylsulfonyl)ethyl)-2,6,7-trioxabicyclo[2.2.2]octane (3).* (3-Methyloxetan-3-yl)methyl 3-(phenylsulfonyl)propanoate *(2)* (13.5 g, 46.3 mmol, 1 equiv) and 60 mL of methylene chloride (Note 8) are added to a flame-dried three-necked 300-mL round-bottomed flask equipped with a 4.5 cm Teflon-coated, oval magnetic stirring bar. Two necks are fitted with rubber septa and the middle neck is fitted with a nitrogen stopcock inlet. The flask is evacuated and backfilled with nitrogen. The solution is cooled to 0 °C with an ice water bath. Once cooled, boron trifluoride diethyl etherate (1.5 mL, 12 mmol, 0.25 equiv) (Note 18) is added by syringe over a period of 15 sec and the reaction is stirred at 0 °C (Note 19). When the reaction is complete (Note 20) (~2 h), triethylamine (1.6 mL, 12 mmol, 0.25 equiv) (Note 21) is added by syringe over a period of fifteen sec (Note 22). The reaction is diluted with diethyl ether (200 mL) and water (200 mL). The mixture is transferred to a 500-mL separatory funnel and the layers are separated. The aqueous layer is extracted with diethyl ether (2 x 200 mL). The combined organics are dried over sodium sulfate (~15 g), gravity filtered into a 1-L, round-bottomed flask, and concentrated on a rotary evaporator to give 15.84 g of crude 4-methyl-1-(2-(phenylsulfonyl)ethyl)-2,6,7-trioxabicyclo[2.2.2]octane. After adding hexanes (200 mL) and ethyl acetate (200 mL), the suspension is heated until the solid dissolves completely to form a clear solution (oil bath temperature 85 °C). The solution is allowed to cool to room temperature and then cooled to –20 °C overnight. The resulting crystals are collected by suction filtration on a Büchner funnel (Note 23), washed with 200 mL of cold hexanes, transferred to a 200 mL beaker and dried overnight at 0.01 mmHg to provide 8.8 g (65%) of 4-methyl-1-(2-(phenylsulfonyl)ethyl)-2,6,7-trioxabicyclo[2.2.2]-octane as a white powder (Notes 24, 25, and 26).

Notes

1. Maleic anhydride briquettes (99%) were purchased from Sigma Aldrich and were pulverized in a mortar and pestle before use. Benzenesulfinic acid, sodium salt (97%) was purchased from Sigma Aldrich (checkers) and was used without further purification. Glacial acetic acid was purchased from Tokyo Chemical Industry Co., Ltd. (checkers) and used without further purification.
2. The procedure was revised during the checking process to include cooling of the aqueous solution during the addition of reagents. In the absence of cooling the checkers observed formation of a dark brown suspension, which could be avoided by cooling the reaction solution.
3. The submitters report no change in yield when reaction times varied from 12 to 24 h.
4. The mixture is filtered while still nearly boiling (80-100 °C). Advantec qualitative filter paper, Grade 1 (checkers) or Whatman qualitative filter paper, Grade 1 (submitters) was used on the Büchner funnel. Oil droplets, which are believed to be polymeric byproducts, can be formed in the procedure. The filtration is performed to remove these oil droplets (Figure 4).

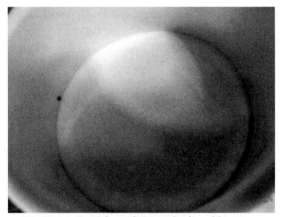

Figure 4. Residue obtained after filtration

5. The purity of 3-(phenylsulfonyl)propanoic acid (**1**) was checked by quantitative NMR using 1,3,5-trimethoxybenzene (purified by sublimation: 60 °C, 0.40 mmHg) as an internal standard. The purity of the white powder was found to be 99+% pure by weight.

6. The checkers received yields of **1** that ranged from 13.7–16.7 g (58–70%).

7. 3-(Phenylsulfonyl)propanoic acid (**1**) is bench stable. Physical properties: mp = 121.4–122.8 °C; ^{1}H NMR (600 MHz, CD$_3$OD) δ: 2.66 (t, J = 7.4 Hz, 2H), 3.50 (t, J = 7.4 Hz, 2H), 7.66 (t, J = 7.5 Hz, 2H), 7.75 (t, J = 7.5 Hz, 1H), 7.95 (d, J = 7.5 Hz, 2H); ^{13}C NMR (150 MHz, CD$_3$OD) δ: 28.5, 52.4, 129.3, 130.6, 135.3, 140.0, 173.4; IR(ATR) 2935 (broad), 1700, 1447, 1435, 1413; HRMS (ESI-TOF) m/z calcd for C$_9$H$_9$O$_4$S (M-H)$^-$ 213.0227, found 213.0222; Anal. Calcd. for C$_9$H$_{10}$O$_4$S: C, 50.46; H, 4.70; S, 14.97. Found: C, 50.17; H, 4.67; S, 14.71. Proton NMR was also taken in dimethylsulfoxide to test for an exchangeable proton: ^{1}H NMR (600 MHz, DMSO-d_6) δ: 2.52 (t, J = 7.4 Hz, 2H), 3.52 (t, J = 7.4 Hz, 2H), 7.67 (d, J = 7.6 Hz, 2H), 7.77 (t, J = 7.6 Hz, 1H), 7.90 (d, J = 7.6 Hz, 2H), 12.57 (s, 1H).

8. The checkers purchased methylene chloride from Kanto Chemical Co., Inc., (dehydrated, ≥99.5%), and the submitters purchased methylene chloride from Fischer Scientific and purified the solvent by passage through a bed of activated alumina.

9. 4-(Dimethylamino)pyridine (99%) was purchased from Tokyo Chemical Industry Co., Ltd. (checkers) or from Acros Organics (submitters) and was used without further purification. 3-Hydroxymethyloxetane (98%) was purchased from Tokyo Chemical Industry Co., Ltd. (checkers) or from AK Scientific (submitters) and was used without further purification.

10. 1-(3-Dimethylaminoproply)-3-ethylcarbodiimide hydrochloride (98%) was purchased from Tokyo Chemical Industry Co., Ltd. (checkers) or from TCI America (submitters). In both cases the reagent was used without further purification.

11. The reaction mixture turned into a dark brown solution after addition of 1-(3-dimethylaminoproply)-3-ethylcarbodiimide hydrochloride (EDC) (Figure 3).

12. The progress of the reaction was followed by TLC analysis on silica gel with 5:95 methanol:methylene chloride and visualized with a UV-lamp. 3-(Phenylsulfonyl)propanoic acid (**1**), R_f = 0.2; (3-Methyloxetan-3-yl)methyl 3-(phenylsulfonyl)propanoate (**2**), R_f = 0.6.

13. Portions of the organic layer are filtered into the 300-mL, round-bottomed flask and concentrated repetitively to transfer the product into the flask. Alternatively, the entire organic layer can be filtered into a 1-L, round-bottomed flask for evaporation, after which the oil can be transferred to a smaller flask using methylene chloride.

14. The co-evaporations were performed three to five times using 250 mL of hexanes each time. The submitters report that the viscous oil crystallizes on standing.

15. The purity of (3-methyloxetan-3-yl)methyl 3-(phenylsulfonyl)-propanoate (2) was determined by quantitative NMR using 1,3,5-trimethoxybenzene (purified by sublimation: 60 °C, 0.40 mmHg) as an internal standard. The purity of 2 was found to be 96% pure by weight.

16. (3-Methyloxetan-3-yl)methyl 3-(phenylsulfonyl)propanoate (2) is bench stable. Physical properties: mp = 48.8-50.8 °C; ^{1}H NMR (600 MHz, CDCl$_3$) δ: 1.31 (s, 3H), 2.81 (t, J = 7.6 Hz, 2H), 3.45 (t, J = 7.6 Hz, 2H), 4.15 (s, 2H), 4.38 (d, J = 6.0 Hz, 2H), 4.47 (d, J = 6.0 Hz, 2H), 7.60 (t, J = 7.6 Hz, 2H), 7.69 (t, J = 7.6 Hz, 1H), 7.93 (d, J = 7.6 Hz, 2H); ^{13}C NMR (150 MHz, CDCl$_3$) δ: 21.1, 27.6, 38.9, 51.3, 69.6, 79.4, 128.1, 129.5, 134.1, 138.4, 170.1; IR (ATR) 2951, 2936, 2874, 1726, 1299 cm^{-1}; HRMS (ESI-TOF) m/z calcd for C$_{14}$H$_{19}$O$_5$S (M+H)$^+$ 299.0948, found 299.0954; Anal. Calcd. for C$_{14}$H$_{18}$O$_5$S: C, 56.36; H, 6.08; S, 10.75. Found: C, 56.16; H, 6.00; S, 10.47.

17. The yields of the reaction ranged from 79-84%.

18. Boron trifluoride diethyl etherate was purchased from Sigma Aldrich (checkers and submitters) and was distilled following a modified procedure of Armarego and Chai prior to use.[2] To a 50-mL, round-bottomed flask were added calcium hydride (~2 g) and boron trifluoride diethyl etherate. The flask was fitted with a short path vacuum distillation head equipped with two receiving flasks and placed under an argon atmosphere. The flask is heated with an oil bath. Several drops are collected at 63 °C (50-51 mmHg) before changing the receiving flask and collecting the main drops (63 °C, 50-51 mmHg) as a colorless liquid. When the distillation is complete, the flask containing boron trifluoride diethyl etherate is filled with argon and then sealed. The submitters report that using undistilled reagent leads to lower yields and difficulty in purifying the final product.

19. The checkers observed that the reaction turns light yellow after addition of boron trifluoride diethyl etherate.

20. The progress of the reaction was followed by TLC analysis on silica gel with 50:50 ethyl acetate:hexanes and visualized with potassium permanganate stain. (3-Methyloxetan-3-yl)methyl 3-(phenylsulfonyl)-propanoate (**2**), R$_f$ = 0.3. 4-Methyl-1-(2-(phenylsulfonyl)ethyl)-2,6,7-trioxabicyclo[2.2.2]octane (**3**), R$_f$ = 0.5.

21. Triethylamine (99.5%) was purchased from Tokyo Chemical Industry Co., Ltd. (checkers) or from Fisher Scientific (submitters) and was used without further purification.

22. The checkers observed that the reaction turns light orange upon addition of triethylamine.

23. The checkers report that concentrating the filtrate and recrystallizing the resulting solid (from 100 mL of hexane and 10 mL of ethyl acetate), additional product (**3**) of lower purity can be obtained.

24. Purity of **3** was determined by quantitative NMR using 1,3,5-trimethoxybenzene (99+%) as an internal standard. 4-Methyl-1-(2-(phenylsulfonyl)ethyl)-2,6,7-trioxabicyclo[2.2.2]octane (**3**) was found to be 98 % pure by weight.

25. Physical properties: mp = 161.3–162.6 °C; ^{1}H NMR (600 MHz, CDCl$_3$) δ: 0.77 (s, 3H), 2.01–2.04 (m, 2H), 3.25–3.27 (m, 2H), 3.83 (s, 6H), 7.55 (t, J = 7.6 Hz, 2H), 7.64 (t, J = 7.6 Hz, 1H), 7.88 (d, J = 7.6 Hz, 2H); ^{13}C NMR (151 MHz, CDCl$_3$) δ: 14.4, 30.4 (x2), 51.3, 72.7, 107.7, 128.2, 129.3, 133.7, 138.7; IR(ATR) 2958, 2936, 2882, 1273, 1229, 1144 cm^{-1}; HRMS (ESI-TOF) m/z calcd for C$_{14}$H$_{19}$O$_5$S (M+H)$^+$ 299.0948, found 299.0955; Anal. Calcd. for C$_{14}$H$_{18}$O$_5$S: C, 56.36; H, 6.08; S, 10.75. Found: C, 56.39; H, 6.04; S, 10.44. Two carbons overlap in CDCl$_3$. A spectrum in deuterated acetonitrile provided better resolution: ^{13}C NMR (150 MHz, CD$_3$CN) δ: 14.2, 31.0, 31.4, 51.7, 73.3, 108.4, 128.9, 130.5, 134.9, 139.9.

26. The yields of the bench stable 4-methyl-1-(2-(phenylsulfonyl)ethyl)-2,6,7-trioxabicyclo[2.2.2]octane (**3**) ranged from 56-65%.

Working with Hazardous Chemicals

The procedures in *Organic Syntheses* are intended for use only by persons with proper training in experimental organic chemistry. All hazardous materials should be handled using the standard procedures for work with chemicals described in references such as "Prudent Practices in the Laboratory" (The National Academies Press, Washington, D.C., 2011;

the full text can be accessed free of charge at
http://www.nap.edu/catalog.php?record_id=12654). All chemical waste
should be disposed of in accordance with local regulations. For general
guidelines for the management of chemical waste, see Chapter 8 of Prudent
Practices.

In some articles in *Organic Syntheses*, chemical-specific hazards are
highlighted in red "Caution Notes" within a procedure. It is important to
recognize that the absence of a caution note does not imply that no
significant hazards are associated with the chemicals involved in that
procedure. Prior to performing a reaction, a thorough risk assessment
should be carried out that includes a review of the potential hazards
associated with each chemical and experimental operation on the scale that
is planned for the procedure. Guidelines for carrying out a risk assessment
and for analyzing the hazards associated with chemicals can be found in
Chapter 4 of Prudent Practices.

The procedures described in *Organic Syntheses* are provided as
published and are conducted at one's own risk. *Organic Syntheses, Inc.*, its
Editors, and its Board of Directors do not warrant or guarantee the safety of
individuals using these procedures and hereby disclaim any liability for any
injuries or damages claimed to have resulted from or related in any way to
the procedures herein.

Discussion

In 1986, Leon Ghosez published the synthesis of his orthoester sulfone
reagent (Scheme 1).[3] This reagent was synthesized in a two pot sequence to
yield the trimethyl orthoester from cheap, readily available starting
materials. The orthoester-sulfone provides two key reactive sites. The
sulfone imparts nucleophilic character to the terminal carbon while the
orthoester carbon can be deprotected *in situ* to reveal an electrophilic center.
This reagent thus functions as a useful d^3-synthon or homoenolate
equivalent as well as a masked electrophilic carbonyl in the acid oxidation
state. The reagent has found special use with enantiomerically pure
epoxides and aziridines to form enantiomerically pure lactones and lactams,
respectively.[4,5]

Scheme 1. Uses of the Ghosez Reagent in organic synthesis

While this reagent has been useful in many natural product syntheses, it suffers from well documented reproducibility issues in the published synthetic route (Scheme 2A).[5] While an alternative synthetic route to the trimethyl orthoester is available from the Parham laboratory and was recently used in the total synthesis of alstonerine,[6] the preparation warns that the product decomposes on attempting to purify the product via vacuum distillation (Scheme 2B).[7] While this seems to solve the reproducibility problem, it requires significantly more expensive reagents to arrive at the desired trimethyl orthoester. The low yields increase the overall cost and render this synthetic sequence economically unviable for continued large-scale preparation. In addition to the difficulties of synthesizing this trimethyl-orthoester, the reagent suffers from often needing to be used in 2,[8] 3,[9] or even 4 equivalents[10] upon optimization of the desired reaction.

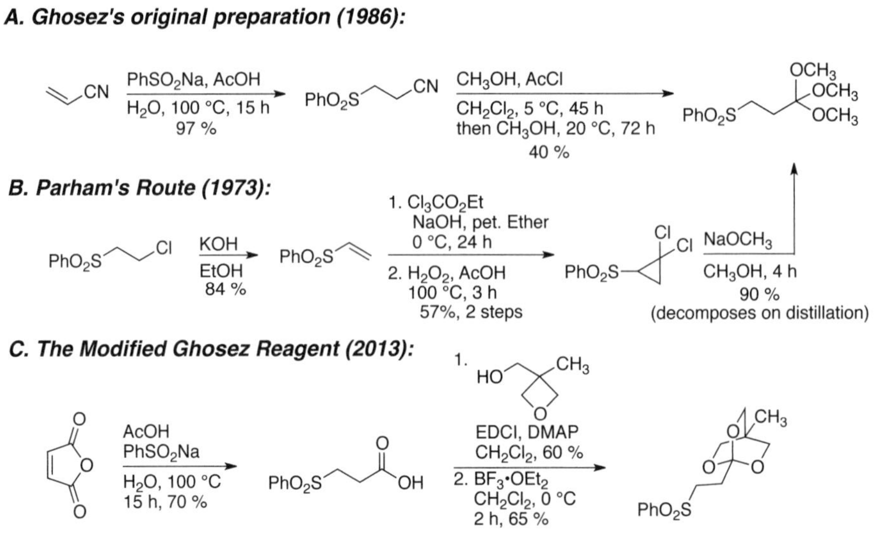

A. Ghosez's original preparation (1986):
CN
PhSO2Na, AcOH
H2O, 100 °C, 15 h
97 %
PhO2S
CN
CH3OH, AcCl
CH2Cl2, 5 °C, 45 h
then CH3OH, 20 °C, 72 h
40 %
OCH3
OCH3
OCH3
PhO2S
B. Parham's Route (1973):
PhO2S
Cl
KOH
EtOH
84 %
PhO2S
1. Cl3CO2Et
NaOH, pet. Ether
0 °C, 24 h
2. H2O2, AcOH
100 °C, 3 h
57%, 2 steps
Cl
Cl
PhO2S
NaOCH3
CH3OH, 4 h
90 %
(decomposes on distillation)
C. The Modified Ghosez Reagent (2013):
O
O
O
AcOH
PhSO2Na
H2O, 100 °C
15 h, 70 %
O
PhO2S
OH
1.
HO
CH3
O
EDCI, DMAP
CH2Cl2, 60 %
2. BF3•OEt2
CH2Cl2, 0 °C
2 h, 65 %
CH3
O
O
O
PhO2S

Scheme 2. Current Routes to γ-Ortho-ester Sulfones

In our studies to synthesize 6,6′-binaphthopyranone natural products, we encountered difficulties in synthesizing Ghosez's reagent and developed a reliable route to an analogous reagent[11] Replacing the trimethyl orthoester with a 2,6,7-trioxabicyclo[2.2.2]octane (OBO) orthoester, originally developed by Corey,[12] allowed for a facile synthesis from mostly commodity chemicals on multi-gram scale to provide a reagent that is bench stable for months without degradation.

References

1. Department of Chemistry, University of California, Davis, One Shields Avenue, Davis, California, 95616, jtshaw@ucdavis.edu. We thank the NIH/NIAID (R01AI08093) for support. M.J.D. thanks the University of California, Davis for Bradford Borge and Dow/Corson Fellowships and the Department of Education for a GAANN fellowship.
2. Armarego, W. L. F.; Chai, C. L. L. *Purification of Laboratory Chemicals (Sixth Edition)*, Chai, W. L. F. A. L. L., Ed. Butterworth-Heinemann: Oxford, 2009.

3. De, L. S.; Nemery, I.; Roekens, B.; Carretero, J. C.; Kimmel, T.; Ghosez, L. *Tetrahedron Lett.* **1986**, *27*, 5099-102.
4. Carretero, J. C.; De, L. S.; Ghosez, L. *Tetrahedron Lett.* **1987**, *28*, 2135-8.
5. Craig, D.; Lu, P.; Mathie, T.; Tholen, N. T. H. *Tetrahedron* **2010**, *66*, 6376-6382.
6. Craig, D.; Goldberg, F. W.; Pett, R. W.; Tholen, N. T. H.; White, A. J. P. *Chem. Commun.* **2013**, *49*, 9275-9277.
7. Parham, W. E.; McKown, W. D.; Nelson, V.; Kajigaeshi, S.; Ishikawa, N. *J. Org. Chem.* **1973**, *38*, 1361-5.
8. Surivet, J.-P.; Vatele, J.-M. *Tetrahedron Lett.* **1996**, *37*, 4373-4376.
9. Mulzer, J.; Ohler, E. *Angew. Chem., Int. Ed.* **2001**, *40*, 3842-3846.
10. Nicolaou, K. C.; Patron, A. P.; Ajito, K.; Richter, P. K.; Khatuya, H.; Bertinato, P.; Miller, R. A.; Tomaszewski, M. J. *Chem. - Eur. J.* **1996**, *2*, 847-868.
11. Grove, C. I.; Di Maso, M. J.; Jaipuri, F. A.; Kim, M. B.; Shaw, J. T. *Org. Lett.* **2012**, *14*, 4338-4341.
12. Corey, E. J.; Raju, N. *Tetrahedron Lett.* **1983**, *24*, 5571-5574.

Appendix
Chemical Abstracts Nomenclature (Registry Number)

Maleic Anhydride: 2,5-Furandione; (108-31-6)
Benzenesulfinic acid sodium salt: Benzenesulfinic acid, sodium salt; (873-55-2)
Glacial acetic acid: acetic acid; (64-19-7)
3-(Phenylsulfonyl)propanoic acid; (10154-71-9)
Dimethylaminopyridine: 4-Pyridinamine, *N,N*-dimethyl-; (1122-58-3)
(3-Methyloxetan-3-yl)methanol: 3-Methyl-3-oxetanemethanol; (3143-02-0)
1-Ethyl-3-(3-dimethylaminopropyl)carbodiimide: 1,3-Propanediamine, *N'*-(ethylcarbonimidoyl)-*N,N*-dimethyl-, monohydrochloride; (25952-53-8)
(3-Methyloxetan-3-yl)methyl 3-(phenylsulfonyl)propanoate; (1394135-60-4)
Boron trifluoride diethyl etherate: Boron, trifluoro[1,1'-oxybis[ethane]]-, (T-4)-; (109-63-7)
Triethylamine: Ethanamine, *N,N*-diethyl-; (121-44-8)
4-Methyl-1-(2-(phenylsulfonyl)ethyl)-2,6,7-trioxabicyclo[2.2.2]octane; (1394135-62-6)

Organic
Syntheses

Jared Shaw received his Ph. D. from Keith Woerpel at UC Irvine in 1999 and then moved to Harvard as an NIH postdoctoral fellow with David Evans. Dr. Shaw became an institute fellow at the Institute for Chemistry and Cell Biology (ICCB) at Harvard Medical School where he helped found the Center for Chemical Methodology and Library Development (CMLD) in 2003, which later became part of the Broad Institute of Harvard and MIT. In 2007, Jared joined the faculty of the University of California, Davis as an assistant professor and was promoted to associate professor in 2012. He works on the development of new methods for the synthesis of natural products and other molecules that modulate biological phenomena.

Michael J. Di Maso received his B.A./M.S. in chemistry from Northwestern University, where he worked under the supervision of Professor Karl A. Scheidt. In 2011, he began graduate research under the supervision of Jared Shaw at the University of California, Davis. His graduate research focuses on synthetic methodology and alkaloid natural product total synthesis.

Michael St. Peter is an undergraduate researcher at the University of California, Davis majoring in Pharmaceutical Chemistry and Food Science. He began working in Prof. Shaw's lab in 2014 on synthetic methodology and natural product total synthesis.

Sunna Jung was born in 1988 in Seoul, Korea. She received her B. Sc. degree in dept. of chemistry from Pohang University of Science and Technology (POSTECH) in 2010. Then she moved to Tokyo Institute of Technology and obtained M. Sc. Degree in 2012 under the supervision of Prof. Keisuke Suzuki. Currently, she is continuing her Ph. D. research in the same group working on cyclophane chemistry.

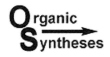

1,3-Dimethylimidazoyl-2-ylidene borane

Sean Gardner, Takuji Kawamoto and Dennis P. Curran[*1]
Department of Chemistry, University of Pittsburgh, Pittsburgh, PA 15208

Checked by Fumihiro Wakita and Keisuke Suzuki

A.

$1 \xrightarrow[\substack{CH_2Cl_2 \\ rt-40\ ^\circ C}]{CH_3I} 2$

B.

$2 \xrightarrow[\substack{toluene \\ reflux}]{NaBH_4} 3$

Procedure

A. *1,3-Dimethyl-1H-imidazol-3-ium iodide* (**2**). A three-necked, 300 mL round-bottomed flask is equipped with a 3.0 cm × 0.6 cm octagonal Teflon-coated magnetic stir bar. The center neck is fitted with a reflux condenser, which is sealed with a rubber septum. One of the side necks is fitted with a rubber septum, and the other is fitted with a three-way stopcock. A nitrogen inlet hose is connected to the three-way stopcock and a nitrogen outlet needle is put into the septum. The flask is dried with a heat gun under vacuum and backfilled with nitrogen. With nitrogen gas flowing, *N*-methyl imidazole **1** (4.11 g, 50 mmol) in dichloromethane (10 mL) is added *via* syringe (Note 1). Iodomethane (3.74 mL, 60 mmol) is taken up in a syringe, and the needle is inserted into the flask *via* the septum. Initially, about 1 mL of iodomethane is added, upon which the solution warms and boils gently.

The remaining iodomethane is added periodically in portions over about 10 min to maintain a gentle boil. After the addition is complete, the mixture is stirred for 1 h, during which time the solution cools to room temperature (Note 2). The three-way stopcock is replaced with a rubber septum. Without removing the stir bar, the needle and the condenser are removed and the solvent is concentrated by rotary evaporation (40 °C, 40 mmHg) to give a pale yellow solid of salt **2** (11.3 g) (Note 3), which is used in the next step without further purification (Note 4).

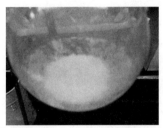

Figure 1. 1,3-Dimethyl-1H-imidazol-3-ium iodide **2**

B. *1,3-Dimethylimidazoyl-2-ylidene borane* (**3**). The same three-necked flask is fitted with a reflux condenser on the center neck. A three-way stopcock with a hose, leading to the back of the fume hood, is placed on top of condenser to vent away the hydrogen gas that is generated. One of the side necks is sealed with a rubber septum. The other side neck is fitted with a three-way stopcock connected to a nitrogen inlet hose. After removing the septum, sodium borohydride (2.27 g, 60 mmol) is added by powder funnel. Toluene (50 mL) is added (Notes 5 and 6) to the solution through the funnel, rinsing the solid residues into the flask, and stirring is initiated. The side neck is closed with the septum.

Under a flow of nitrogen gas, the flask is immersed in a 125 °C oil bath. As the mixture nears the reflux point, the imidazolium salt begins to melt. A second ionic liquid phase forms and begins to bubble (Note 7). As the solution reaches reflux, an ionic liquid phase is visible as a white viscous material on the bottom and the top of the toluene, with droplets periodically floating up and sinking down through the toluene, creating exchange between the sunken and floating layers. Over about 8 h, the white liquid phase becomes more viscous and gradually adheres to the inside of the flask. Eventually the top phase and the exchanging droplets disappear. After 20 h at reflux, the mixture is cooled to give a largely clear toluene phase with a white, glue-like material adhering to the inside of the flask.

Figure 2. Reaction Set-up for Step B

The toluene phase contains most of the product, and it is carefully decanted from the insoluble viscous material while still hot (Note 8). Additional toluene (25 mL) is added to the reaction flask to cover the residue, then the condenser is put back in place and the mixture is heated to reflux to further extract remaining product from the mixture. After 15 min of stirring, the hot toluene is decanted again. The hot extraction process is repeated once more with fresh toluene (25 mL). The toluene extracts are combined, cooled, and concentrated to dryness by rotary evaporation (40 °C,

40 mmHg) to give the crude product (4.38–4.69 g) as a white solid (Note 9). The solid is recrystallized from water (about 14 mL/g) (Note 10) and the resulting colorless, needle-shaped crystals of **3** are isolated by vacuum filtration on a Büchner funnel and washed with aliquots of ice-cold water (3 x 5 mL). The crystals are dried under high vacuum (0.3 mmHg) for 12 h at 50 °C to yield 2.58–2.78 g (47–51%) of compound **3** (Note 11).

Figure 3. After refluxing 20 h.

Figure 4. After decantation.

Figure 5. Crude product.

Figure 6. Recrystallization.

Notes

1. Methyl iodide (stabilized with copper chips, 99.5%) and 1-methylimidazole (99%) were purchased from Tokyo Chemical Industry Co., Ltd. and used as received. Dichloromethane (dehydrated Super) was purchased from Kanto Chemical Co., Inc., and was purified under argon by using Organic Solvent Pure Unit (Wako Pure Chemical Industries, Ltd.) The submitters used as received methyl iodide (stabilized, 99%) purchased from Acros Organics and 1-methylimidazole (99%) purchased from Sigma-Aldrich. Dichloromethane (99.9%) was purchased from Fisher Scientific and

passed through a column of activated alumina under nitrogen prior to use.

2. A small aliquot is typically removed at this point by capillary pipette and several small drops are added to CDCl$_3$ (0.5 mL) in an NMR tube. A ^{1}H NMR spectrum is recorded to ensure that N-methyl imidazole is absent and the salt formation is complete.

3. Due to the extreme hygroscopic nature of the intermediate, the checkers did not determine the melting point of the solid, although the submitters report the mp for crude **2** to be 66–69 °C. Spectral data for crude **2**: ^{1}H NMR (600 MHz, CD$_3$OD) δ: 3.98 (s, 6 H), 7.628 (s, 2 H), 8.98 (s, 1 H). ^{13}C NMR (150 MHz, CD$_3$OD) δ: 37.01, 37.03, 124.87, 124.91, 138.65.

4. The imidazolium iodide (**2**) is hygroscopic and becomes sticky or even liquefies during transfer under a humid atmosphere. Thus, the flask for the first reaction is used for the second reaction without material transfer.

5. Sodium borohydride (95%) was purchased from Tokyo Chemical Industry Co., Ltd. and used as received. Toluene (dehydrated–Super Plus) was purchased from Kanto Chemical Co., Inc. and used as received. The submitters purchased sodium borohydride (98%+) from Acros Organics and toluene (ACS reagent grade) from JT Baker. Both were used as received.

6. A spatula was used to scrape the clumpy solid down beneath the surface of the toluene, if needed.

7. When the reaction is performed on this scale using a 300 mL flask, the gas bubbles dissipate on the surface of the toluene during the early stage of the reaction with headspace to spare. With smaller flasks the bubbles tend to foam more, and the foam may rise into the condenser.

8. The clamp is removed from the clamp holder and the clamp is used as a handle to carefully pour the hot toluene through a glass funnel into another 300 mL round-bottomed flask. The product is not fully soluble in toluene at ambient temperature and will begin to precipitate if the solution cools prior to the first decantation. Toluene is a potential recrystallization solvent for this NHC-borane, but water (Note 10) gives higher recovery and better quality product in this method of synthesis, where the byproducts are likely salts.

9. The crude product may be sufficiently pure for some uses. Good quality ^{1}H and ^{11}B NMR spectral data are provided, though the melting point (128–134 °C) is somewhat depressed. If recrystallization is bypassed,

traces of sodium iodide and other salts can be removed from the crude product by suspension in a small amount of water at room temperature, followed by brief stirring, filtration and vacuum drying.

10. For the recrystallization of the crude material, the checkers used a three-necked, 300 mL round-bottomed flask, in which the crude material and a 3.0 cm × 0.6 cm octagonal Teflon-coated magnetic stir bar were placed, along with about 75% of the calculated volume of water (g of crude material x 14 mL/g x 0.75). The center neck is fitted with a reflux condenser. The side necks are fitted with rubber septa. The flask is open to the atmosphere; the mixture is brought to an oil bath and heated to 100 °C. Small portions of water are repeatedly added through the condenser, while allowing the mixture to come back to a boil between each addition. The additions are repeated until the solid is completely dissolved. The flask with the clear liquid is allowed to cool slowly. The submitters report that the product can also be purified by sublimation; 90–100 °C (heating bath temperature), 2 mmHg.

11. Characterization data of **3**: mp 140–141 °C; ^{1}H NMR (600 MHz, CDCl$_3$) d: 1.01 (m, 3 H, $J_{11_{B-H}}$ = 86.2 Hz, $J_{10_{B-H}}$= 29.3 Hz), 3.73 (s, 6 H), 6.80 (s, 2 H) ppm. ^{13}C NMR (150 MHz, CDCl$_3$) δ: 36.1, 120.0, 172.2 (q, J_{B-C} = 51.9 Hz). ^{11}B NMR (128 MHz, CDCl$_3$) δ: –38.5 (q, J_{H-B} = 86.3 Hz). IR (NaCl, DR) 3167, 3132, 2999, 2944, 2275, 1575, 1479, 1448, 1235, 1216, 1121, 860, 754, 666, 624 cm^{-1}. HRMS (ESI) m/z [C$_5$H$_{10}$BN$_2$]$^+$ calcd 109.0933, found 109.0935. Anal calcd for C$_5$H$_{11}$BN$_2$: C, 54.61; H, 10.08; N 25.47, found C, 54.42; H, 9.73; N, 25.57.

Working with Hazardous Chemicals

The procedures in *Organic Syntheses* are intended for use only by persons with proper training in experimental organic chemistry. All hazardous materials should be handled using the standard procedures for work with chemicals described in references such as "Prudent Practices in the Laboratory" (The National Academies Press, Washington, D.C., 2011; the full text can be accessed free of charge at http://www.nap.edu/catalog.php?record_id=12654). All chemical waste should be disposed of in accordance with local regulations. For general guidelines for the management of chemical waste, see Chapter 8 of Prudent Practices.

In some articles in *Organic Syntheses*, chemical-specific hazards are highlighted in red "Caution Notes" within a procedure. It is important to recognize that the absence of a caution note does not imply that no significant hazards are associated with the chemicals involved in that procedure. Prior to performing a reaction, a thorough risk assessment should be carried out that includes a review of the potential hazards associated with each chemical and experimental operation on the scale that is planned for the procedure. Guidelines for carrying out a risk assessment and for analyzing the hazards associated with chemicals can be found in Chapter 4 of Prudent Practices.

The procedures described in *Organic Syntheses* are provided as published and are conducted at one's own risk. *Organic Syntheses, Inc.*, its Editors, and its Board of Directors do not warrant or guarantee the safety of individuals using these procedures and hereby disclaim any liability for any injuries or damages claimed to have resulted from or related in any way to the procedures herein.

Discussion

Interest in the chemistry of N-heterocyclic carbene boranes has increased in recent years following the realization that such compounds are typically stable solids that are convenient to handle in ambient lab conditions and convenient to purify by crystallization or chromagraphy.[2]

1,3-Dimethylimidazoyl-2-ylidene borane **3** was first reported in 2010[3] but has quickly become one of the most commonly used NHC-boranes. Several of its reactions are summarized in Figure 1. It is a precursor to many interesting substituted NHC-boranes, including metal-complexed boranes[3b] and boriranes (boracyclopropanes).[4] It undergoes BH insertion reactions with reactive carbenes derived from diazo compounds.[5] It has diverse reduction chemistry as a radical hydrogen atom donor[6] and a neutral hydride source.[7] The compound hydroborates benzyne,[8] but is inert towards standard alkenes and alkynes. However, alkene and alkyne hydroboration can be catalyzed by borenium ions (or their reactive equivalents).[9]

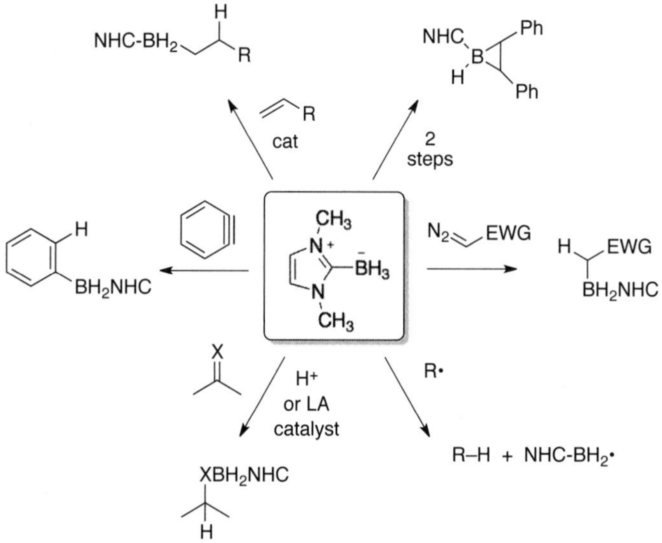

Scheme 1. Selected reactions of NHC-borane 3

Previous routes to 1,3-dimethylimidazoyl-2-ylidene borane **3** are summarized in Figure 2. In essence they all involve complexation of the N-heterocyclic carbene **5** with a borane source. We first made **3** by the deprotonation of **2** with sodium hexamethyldisilylazide to make the NHC **5** in situ, followed by the addition of $BH_3 \bullet THF$.[3a] Later we used trimethylamine borane, which is easier to handle then $BH_3 \bullet THF$.[10] However, heating is needed with the amine-borane because of its stability (presumably triethylamine must be released from the amine-borane to allow complexation with the NHC).

Braunschweig made **3** from the NHC-carboxylate **4** and $BH_3 \bullet SMe_2$.[3b] These methods all require exclusion of air and moisture either because of the base, the borane, or both. The products from these preparations were purified by flash chromatography or precipitation with hexane.

The procedure described above is fashioned after reactions of some amine salts and with alkali metal borohydrides to make amine-boranes.[11] Compared to the borane sources, the borohydride is less expensive and easier to handle. The iodide counterion is used because the imidazolium iodide is easy to make and because it melts below the boiling point of toluene. The iodide is made by a variant of Zoller's procedure[12] where slow addition of methyl iodide starting at ambient temperature replaces rapid addition with cooling. The cooling method can be problematic because the methylation is not fast at 0 °C. If all the methyl iodide is added but the reaction does not progress in the cold, then a very exothermic reaction can occur suddenly as the solution is warmed to room temperature.

Scheme 2. Prior preparations of 3

The solvent toluene is an important component in Step B even though neither of the precursors is very soluble, even at reflux. Extensive frothing and foaming occurred when the iodide salt was melted without solvent in the presence of sodium borohydride. When the reaction was conducted by heating in dioxane, the NHC-borane was typically formed in much lower yields (10–20% isolated) alongside large amounts of water-soluble byproducts that appear to be amines or polyamines (many resonances between 2–3 ppm in the ^{1}H NMR spectrum). Lesser amounts of these byproducts form in this procedure, and they remain trapped in the glue-like phase that adheres to the flask.

Compared to the neat (solvent-free) reaction, toluene helps to prevent foaming. During the early stage of the reaction in toluene, an ionic liquid phase forms as the imidazolium iodide melts. This phase is simultaneously present above, below, and suspended in the toluene. The droplets suspended within the toluene are variously floating up or sinking down. We call this the "lava lamp" stage of the reaction, and it lasts several hours.

The initial ionic liquid phase presumably contains the melted imidazolium iodide and sodium borohydride. In principle, this phase is more dense than toluene. However, we speculate that as the reaction occurs and dihydrogen and the NHC-borane are formed, the local density of the ionic liquid phase decreases in places. Droplets break off and float to the surface. On the way to or at the surface, the hydrogen gas is released to the atmosphere and the NHC-borane is released to the toluene. The density of the ionic liquid droplet then increases, so it sinks back down.

Regardless of whether this procedure or another is followed to make **3**, we recommend the use of water as a recrystallization solvent because it gives sparkling white crystals in high purity and with good recovery. That **3** can actually be recrystallized from hot water is a clear testament to its water stability. Indeed, the structure of **3** seems to be a kind of sweet-spot for water recrystallization. Addition of more or larger alkyl or aryl substituents quickly decreases the solubility of NHC-boranes in hot water. Going in the other direction, a triazolium analog of **3** has increased water solubility.[13]

NHC complexes of borane are not pyrophoric, and we routinely weigh and transfer **3** and other NHC-boranes in air. The autoxidation of **3** has not been studied, but it could well occur under suitable conditions in solution.[14] However, solid samples of **3** have been stored at ambient temperatures without exclusion of air for long periods (> 1 year) without evidence of significant decomposition.

This and other carbene boranes are potential bases and reducing agents, and should be stored and handled as such. Acids with pKa's less than about 1–2 react instantly with typical NHC-boranes to form dihydrogen.[15] Likewise, as mild reductants[7c] NHC-boranes are expected to react vigorously with strong oxidants. To verify this, we added one drop of 70% nitric acid to a few milligrams of **3**. In various trials it either ignited instantly with a flash of flame and an audible pop, or sizzled and smoked vigorously for a few seconds. Accordingly, **3** and presumably other NHC-boranes are potentially hypergolic (spontaneously ignite on contact) with strong oxidants.[16]

Like their borohydride and amine-borane cousins, NHC-boranes should be stored separately from acids or oxidants. And they should not be added to any waste with acids or oxidants. Even if no initial reaction is noticed, it is likely that pressure will build in sealed containers over time due to formation of gases (minimally dihydrogen and perhaps other gases depending on the acid or oxidant).

A short study of the scope and limitations of this procedure has recently appeared.[17] Briefly, 1,3-dialkylimidazolium salts form NHC-boranes provided that the melting point of the salt is below the boiling point of toluene. The ionic liquid phase does not form if the salt does not melt.

References

1. We thank the National Science Foundation for funding of this work. Contact information: email, curran@pitt.edu; address, Department of Chemistry, University of Pittsburgh, Pittsburgh, PA 15260 USA.
2. Curran, D. P.; Solovyev, A.; Makhlouf Brahmi, M.; Fensterbank, L.; Malacria, M.; Lacôte, E. *Angew. Chem. Int. Ed.* **2011**, *50*, 10294-10317.
3. (a) Walton, J. C.; Makhlouf Brahmi, M.; Fensterbank, L.; Lacôte, E.; Malacria, M.; Chu, Q.; Ueng, S.-H.; Solovyev, A.; Curran, D. P. *J. Am. Chem. Soc.* **2010**, *132*, 2350-2358; (b) Bissinger, P.; Braunschweig, H.; Kupfer, T.; Radacki, K. *Organometallics* **2010**, *29*, 3987-3990.
4. (a) Braunschweig, H.; Claes, C.; Damme, A.; Dei; Dewhurst, R. D.; Horl, C.; Kramer, T. *Chem. Commun.* **2015**, *51*, 1627-1630; (b) Bissinger, P.; Braunschweig, H.; Kraft, K.; Kupfer, T. *Angew. Chem. Int. Ed.* **2011**, *50*, 4704-4707.
5. (a) Li, X.; Curran, D. P. *J. Am. Chem. Soc.* **2013**, *135*, 12076-12081; (b) Chen, D.; Zhang, X.; Qi, W.-Y.; Xu, B.; Xu, M.-H. *J. Am. Chem. Soc.* **2015**, *137*, 5268-5271
6. (a) Ueng, S.-H.; Fensterbank, L.; Lacôte, E.; Malacria, M.; Curran, D. P. *Org. Lett.* **2010**, *12*, 3002-3005; (b) Pan, X.; Lacôte, E.; Lalevée, J.; Curran, D. P. *J. Am. Chem. Soc.* **2012**, *134*, 5669-5675; (c) Tehfe, M.-A.; Monot, J.; Makhlouf Brahmi, M.; Bonin-Dubarle, H.; Curran, D. P.; Malacria, M.; Fensterbank, L.; Lacôte, E.; Lalevée, J.; Fouassier, J.-P. *Polym. Chem.* **2011**, *2*, 625-631.
7. (a) Lamm, V.; Pan, X.; Taniguchi, T.; Curran, D. P. *Beilstein J. Org. Chem.* **2013**, *9*, 675-680; (b) Taniguchi, T.; Curran, D. P. *Org. Lett.* **2012**, *14*, 4540-4543; (c) Horn, M.; Mayr, H.; Lacôte, E.; Merling, E.; Deaner, J.; Wells, S.; McFadden, T.; Curran, D. P. *Org. Lett.* **2012**, *14*, 82-85.
8. Taniguchi, T.; Curran, D. P. *Angew. Chem. Int. Ed.* **2014**, *53*, 13150-13154.
9. (a) Prokofjevs, A.; Boussonnière, A.; Li, L.; Bonin, H.; Lacôte, E.; Curran, D. P.; Vedejs, E. *J. Am. Chem. Soc.* **2012**, *134*, 12281-12288; (b) Boussonnière, A.; Pan, X.; Geib, S. J.; Curran, D. P. *Organometallics* **2013**, *32*, 7445-7450; (c) Pan, X.; Boussonnière, A.; Curran, D. P. *J. Am. Chem. Soc.* **2013**, *135*, 14433-14437.
10. Makhouf Brahmi, M.; Monot, J.; Desage-El Murr, M.; Curran, D. P.; Fensterbank, L.; Lacôte, E.; Malacria, M. *J. Org. Chem.* **2010**, *75*, 6983-6985.

rganic
yntheses

11. (a) Beachley, O. T.; Washburn, B. *Inorg. Chem.* **1975**, *14*, 120-123; (b) Schaeffer, G. W.; Anderson, E. R. *J. Am. Chem. Soc.* **1949**, *71*, 2143-2145.
12. Zoller, U. *Tetrahedron* **1988**, *44*, 7413-7426.
13. Tehfe, M.-A.; Monot, J.; Malacria, M.; Fensterbank, L.; Fouassier, J.-P.; Curran, D. P.; Lacôte, E.; Lalevée, J. *ACS Macro Lett.* **2012**, *1*, 92-95.
14. (a) Lacôte, E.; Curran, D. P.; Lalevée, J. *Chimia* **2012**, *66*, 382-385; (b) Tehfe, M.-A.; Makhlouf Brahmi, M.; Fouassier, J.-P.; Curran, D. P.; Malacria, M.; Fensterbank, L.; Lacôte, E.; Lalevée, J. *Macromolecules* **2010**, *43*, 2261-2267.
15. Solovyev, A.; Chu, Q.; Geib, S. J.; Fensterbank, L.; Malacria, M.; Lacôte, E.; Curran, D. P. *J. Am. Chem. Soc.* **2010**, *132*, 15072-15080.
16. Li, S.; Gao, H.; Shreeve, J. M. *Angew. Chem. Int. Ed.* **2014**, *53*, 2969-2972.
17. Gardner, S.; Kawamoto, T.; Curran, D. P. *J. Org. Chem.* **2015**, *80*, 9794-9797.

Appendix
Chemical Abstracts Nomenclature (Registry Number)

1-Methylimidazole: *H*-Imidazole, 1-methyl-; (616-47-7)
Methyl iodide: Methane, iodo-; (74-88-4)
1,3-Dimethyl-1*H*-imidazol-3-ium iodide: 1*H*-Imidazolium, 1,3-dimethyl-, iodide: (4333-63-4)
Sodium borohydride: Borate(1-), tetrahydro-, sodium (1:1); (16940-66-2)
Toluene: Benzene, methyl-; (108-88-3)
1,3-Dimethylimidazol-2-ylidene borane: boron, (1,3-dihyro-1,3-dimethyl-2*H*-imidazol-2-ylidene)trihydro-; (1211417-77-4)

Organic
Syntheses

Sean R. Gardner received his B.S. in chemistry from Drexel University in 2008, where he worked under the supervision of Dr. Elizabeth T. Papish. He is currently pursuing his graduate studies under the supervision of Professor Dennis P. Curran. His research focuses on the synthesis of N-heterocyclic carbene borane complexes and development of novel reaction methodologies.

Takuji Kawamoto obtained his B.S. degree (2009), M.S. degree (2011), and Ph.D. degree (2014) from Osaka Prefecture University (Japan) under supervision of Professor Ilhyong Ryu. Then he studied as a postdoctoral fellow with Professor Dennis P. Curran at University of Pittsburgh (USA). His postdoctoral research mainly focused on carbon-boron bond forming reactions via radical intermediates. He is currently working at Yamaguchi University (Japan) as an assistant professor.

Dennis P. Curran was born in Easton, Pennsylvania in 1953. He obtained a BS in 1975 from Boston College, then a Ph.D. in 1979 from the University of Rochester (with A. S. Kende). After postdoctoral work at the University of Wisconsin (with B. M. Trost), he joined the University of Pittsburgh in 1981. He is currently a Distinguished Service Professor and the Bayer Professor of Chemistry. He is the editor of Volume 83 of Organic Syntheses.

Organic
Syntheses

Fumihiro Wakita was born in 1990 in Amami Ōshima, Japan. He received his M.Sc. degree from Waseda University in 2015. In the same year, he joined the research group of Prof. Keisuke Suzuki at Tokyo Institute of Technology. His research focuses on the synthesis of natural products.

Organic
Syntheses

Copper-Catalyzed Electrophilic Amination of Heteroaromatic and Aromatic C–H Bonds via TMPZnCl•LiCl Mediated Metalation

Stacey L. McDonald, Charles E. Hendrick, Katie J. Bitting, Qiu Wang[*1]

Department of Chemistry, Duke University Durham, NC 27708-0346 (USA)

Checked by Joshua D. Sieber and Chris H. Senanayake

A.

HN—NBoc (1.2 equiv)

$\xrightarrow{\text{BPO (1.0 equiv)}}{\text{Na}_2\text{HPO}_4 \text{ (1.5 equiv), DMF, rt}}$

BzO–N—NBoc **1**

B.

(structure with Me Me groups) NH (1.0 equiv)

1) *n*-BuLi (1.0 equiv), THF, 0 °C to rt
2) ZnCl₂ (1.0 equiv), THF, 0 °C to rt

(structure) NZnCl•LiCl **2** (TMPZnCl•LiCl)

C.

(pyridine with F, 1.0 equiv)

1) TMPZnCl•LiCl **2** (1.0 equiv), THF, rt
2) Cu(OAc)₂ (5 mol%), THF, 50 °C
BzO–N—NBoc **1** (1.2 equiv)

(product pyridine with F and NBoc) **3**

Procedure

Caution! Reactions and subsequent operations involving peracids and peroxy compounds should be run behind a safety shield. For relatively fast reactions, the rate of addition of the peroxy compound should be slow enough so that it reacts rapidly and no significant unreacted excess is allowed to build up. The reaction mixture should be stirred efficiently while the peroxy compound is being added, and cooling should generally be provided since many reactions of peroxy compounds are exothermic. New or unfamiliar reactions, particularly those run at elevated temperatures, should be run first on a small scale. Reaction products should never be recovered from the final reaction mixture by distillation until all residual active oxygen compounds (including unreacted peroxy compounds) have been destroyed. Decomposition of active oxygen compounds may be accomplished by the procedure described in Korach, M.; Nielsen, D. R.; Rideout, W. H. Org. Synth. 1962, 42, 50 (Org. Synth. 1973, Coll. Vol. 5, 414). [Note added January 2011].

A. *tert-Butyl-4-(benzoyloxy)piperazine-1-carboxylate (1).²* A 500-mL three-necked 24/40 round-bottomed flask is equipped with a 5 cm ellipsoid-shaped, Teflon-coated magnetic stirbar. Two necks are capped with rubber septa, and to the third neck is attached a nitrogen inline adapter. The flask is evacuated under high vacuum (~1 mmHg) and dried with a heat gun for 5 min. The flask is then allowed to cool to room temperature under vacuum. The flask is then back-filled with nitrogen and a thermocouple is charged through one of the necks. Under a slight positive pressure of nitrogen, and

without stirring, one septum is removed, and solid benzoyl peroxide (24.2 g, 99.9 mmol, 1.0 equiv) (Note 1) and solid sodium phosphate dibasic (21.3 g, 150 mmol, 1.5 equiv) (Note 2) are added *via* a plastic powder funnel. Dimethylformamide (250 mL) (Note 3) is added (Note 4). The resulting heterogeneous white slurry is allowed to stir at room temperature and solid *N*-Boc-piperazine (22.3 g, 120 mmol, 1.2 equiv) (Note 5) is added in one portion *via* a plastic powder funnel. The septum is replaced and the reaction is allowed to stir at room temperature under nitrogen. Slight heat (Note 6) was evolved, and the reaction mixture gradually turned from a heterogeneous white slurry to a bright yellow slurry (Figure 1).

Figure 1. Yellow Slurry formed in Step A

After 1 h (Note 7), the reaction is poured directly into a 2-L Erlenmeyer flask containing deionized water (400 mL) and a 5 cm ellipsoid-shaped, Teflon-coated magnetic stirbar. The reaction flask is rinsed with additional deionized water (100 mL). To this solution is added ethyl acetate (350 mL) (Note 8). The resulting mixture is allowed to stir for 10 min. The organic and aqueous layers are decanted into a 2-L separatory funnel (Note 9), and the residue rinsed with an additional 100 mL of EtOAc. The organic layer is collected and washed with a saturated aqueous sodium bicarbonate solution (2 × 200 mL) (Note 10). The aqueous layers are combined and extracted with ethyl acetate (3 × 200 mL) (Notes 8 and 11). The combined organic portions are washed with deionized water (4 × 250 mL) (Note 12), followed by brine (200 mL). The organic layer is dried over anhydrous magnesium sulfate (50 g) (Note 13), and vacuum filtered through a pad of celite (25 g) in a 12 cm funnel with #1 filter paper, which was subsequently rinsed with ethyl acetate (2 x 100 mL) (Note 8). The filtrate is concentrated by rotary evaporation (22 °C, 70 – 30 mmHg) to give an off-white to yellow solid (Note 14). The crude residue is purified by flash chromatography on silica gel (Note 15) to afford *tert*-butyl-4-(benzoyloxy)piperazine-1-carboxylate as a crystalline white solid (22.2–22.8 g, 73 – 75%) (Note 16).

B. *2,2,6,6-Tetramethylpiperidyl–ZnCl•LiCl (TMPZnCl•LiCl) (2).*[3] An oven-dried 250-mL three-necked, 24/40 round-bottomed flask is equipped with a 4 cm ellipsoid-shaped, Teflon-coated magnetic stir bar. One neck is capped with a rubber septum, the center neck is equipped with a 50-mL addition funnel capped with a rubber septum, and the third neck is equipped with a nitrogen inline adapter. The apparatus is subjected to three cycles of evacuation and refilling with nitrogen, and a thermocouple is then added through one of the necks. Tetrahydrofuran (40 mL) (Note 17) and 2,2,6,6-tetramethylpiperidine (9.96 g, 11.9 mL, 70.5 mmol, 1.0 equiv) (Note 18) are added via a plastic syringe to the round-bottomed flask, and the flask is immersed in an ice-water bath. A solution of *n*-butyllithium (2.55 M in hexanes, 27.7 mL, 70.6 mmol, 1.0 equiv) (Note 19) is transferred to the addition funnel via cannula then added dropwise over ca. 20 min to the reaction mixture while maintaining an internal reaction temperature of < 10 °C (Note 20). The resulting solution turns golden in color immediately upon the addition of *n*-butyllithium. The resultant mixture is removed from the ice-water bath after 1 h and allowed to warm to room temperature over 30 min. The mixture is then allowed to stir an additional 1.5 h at room temperature.

A second oven-dried, 250-mL, 24/40 three-necked round-bottomed flask equipped with a 2 cm ellipsoid-shaped stir bar is charged with zinc chloride (ZnCl₂) (9.55 g, 70.0 mmol, 1.0 equiv) (Note 21). Two necks are capped with rubber septa. The third neck is equipped with a nitrogen inline adapter. The flask is then evacuated using high vacuum (~1 mmHg) and dried using a heat gun for 5 min. The flask was then allowed to cool to ambient temperature under vacuum. This drying procedure was repeated two more times. After the final cooling to room temperature, the flask is refilled with nitrogen, and a thermocouple is added through one of the necks. Tetrahydrofuran (60 mL) (Note 17) is then added *via* a plastic syringe. The resulting white suspension is cooled in an ice-water bath. The LiTMP solution is then added dropwise via cannula transfer over ca. 10 min to the ZnCl₂ suspension (Figure 2) while maintaining an internal batch temperature of 10–15 °C.

Figure 2. Addition of LiTMP solution to ZnCl₂

After complete addition of the LiTMP, the batch is stirred for 30 min, and the gold colored reaction mixture is warmed to room temperature over 30 min and allowed to stir continuously at room temperature under a slight positive pressure of nitrogen. After 14 h, the agitation is stopped to allow

Figure 3. Addition of 2,2,6,6-
tetramethylpiperidyl–ZnCl•LiCl

solids to settle. The resulting brown-orange solution is titrated (Note 22) (0.40–0.41 M).

C. *tert-Butyl-4-(3-fluoropyridin-2-yl)piperazine-1-carboxylate (3)*. An oven-dried, 1-L three-necked, 24/40 round-bottomed flask is equipped with a 5 cm ellipsoid-shaped, Teflon-coated magnetic stir bar. One neck is capped with a rubber septum. The second neck is equipped with a 125-mL addition funnel that is capped with a rubber septum. The third neck is equipped with a reflux condenser with a nitrogen inline adapter. The apparatus is subjected to three cycles of evacuation and refilling with nitrogen. A thermocouple is then added to the flask through one of the necks. Tetrahydrofuran (125 mL) is added to the round-bottomed flask by cannula and 3-fluoropyridine (4.3 mL, 4.9 g, 50.0 mmol, 1.0 equiv) (Note 23) is added *via* a plastic syringe. 2,2,6,6-Tetramethylpiperidyl–ZnCl•LiCl (**2**) (0.41 M, 122 mL, 112.0 g, 50 mmol, 1.0 equiv) is transferred by cannula to the addition funnel and then slowly added dropwise over ca. 15 min to the reaction at room temperature (Note 24) (Figure 3). The resulting orange solution is stirred at room temperature for 4 h.

An oven-dried, 250-mL, 24/40 pear-shaped flask equipped with a 2 cm ellipsoid-shaped Teflon-coated magnetic stir bar is charged *tert*-butyl-4-(benzoyloxy)piperazine-1-carboxylate (**1**) (18.4 g, 60.0 mmol, 1.2 equiv) and copper(II) acetate (454 mg, 2.50 mmol, 0.05 equiv) (Note 25). The flask is capped with a rubber septum and subjected to three cycles of evacuation and refilling with nitrogen. Tetrahydrofuran (125 mL) (Note 17) is transferred by cannula to the flask. The resulting mixture is stirred at room temperature for 10 min and is then cannula transferred over ca. 10 min to the aryl zincate mixture at room temperature (Note 26). The reaction flask is inserted into an oil bath at 50 °C. The reaction is allowed to stir at 50 °C for

18.5 h and is then removed from heat and cooled to room temperature over 20 min (Notes 26 and 27). The orange-brown solution is directly vacuum filtered (Note 28) through neutral alumina (166 g) (Note 29) in a medium porosity fritted glass funnel, washed with ethyl acetate (500 mL) (Note 8), and concentrated by rotary evaporation (Note 30) to give a thick brown oil. The crude residue is purified by flash chromatography on silica gel (Note 31) to afford *tert*-butyl-4-(3-fluoropyridin-2-yl)piperazine-1-carboxylate (**3**) (7.05–7.77 g, 50–53%) as a yellow oil that crystallizes very slowly at 4 °C (Notes 32 and 33).

Notes

1. Benzoyl peroxide (Luperox® A98, reagent grade, 98%) was purchased from Sigma Aldrich and used as received. The reaction was not stirred until solvent was added due to shock sensitivity of benzoyl peroxide.
2. Sodium phosphate dibasic (BioReagent, ≥99%) was purchased from Sigma Aldrich and used as received.
3. Dimethylformamide (anhydrous, 99.8%) was purchased from Sigma Aldrich and used as received. The submitters used dimethylformamide (ACS grade, ≥99.8%) purchased from Sigma Aldrich and used it directly.
4. The internal reaction temperature decreases from 20 °C to 15 °C upon the addition of dimethylformamide.
5. N-Boc-piperazine (≥98.0%) was purchased from Sigma Aldrich and used as received. The submitters purchased this material from ArkPharm, Inc. (98%) and used it directly.
6. The internal reaction temperature reached a maximum of 37 °C in ~10 min and then began to slowly cool back to room temperature.
7. Completeness of reaction is judged by the disappearance of benzoyl peroxide by thin-layer chromatography (TLC), performed on glass backed pre-coated silica gel plates (DC-Kieselgel 60 F_{254}) with a UV254 indicator, using 20% ethyl acetate–hexanes as the eluent (R_f of benzoyl peroxide = 0.43; R_f of the product = 0.19). The product is visualized with a 254 nm UV lamp and $KMnO_4$ stain. The submitter's utilized TLC plates on aluminum backed pre-coated

silica gel plates (250 μm, Agela Technologies) with a UV254 indicator.

8. Ethyl acetate (ACS reagent, ≥99.7%) was purchased from Sigma Aldrich and used as received. The submitters used ethyl acetate (Chromasolv®, for HPLC, ≥99.7%) purchased from Sigma Aldrich and used it directly.

9. A portion of solid material would not dissolve in the organic or aqueous phases. This insoluble material was not transferred to the separatory funnel.

10. Sodium bicarbonate (ACS reagent grade, 99.7–100.3%) was purchased from Sigma Aldrich and used as received.

11. During the ethyl acetate extraction, a solid began to crystallize in the aqueous layer. After shaking the separatory funnel, the aqueous portion containing this solid was removed from the funnel quickly to avoid this solid material settling to the bottom, which results in clogging of the separatory funnel.

12. The first water wash gave fast layer separation. The remaining three water washes gave slow phase separations (~30 min needed to obtain good phase cuts).

13. Magnesium sulfate (ACS reagent, >99.5%, anhydrous) was purchased from Sigma Aldrich and used as received. The submitters dried the material using sodium sulfate (131 g, ACS reagent, >99.0%, anhydrous, granular, Sigma Aldrich).

14. The organic layer (and the aqueous layer) is tested for the presence of peroxides using a peroxide test strip before concentration (or disposal of the aqueous layer). No evidence for peroxides was observed.

15. The crude solid is dissolved in 20% ethyl acetate/hexanes with enough methylene chloride added to dissolve all the solids. This material is then dry-loaded onto 56 g (125 mL) of silica gel. The volatile organics are then removed by rotary evaporation (35 mmHg, bath temperature = 22 °C). Chromatography is then performed using a 6 cm diameter flash chromatography column packed with 208 g (400 mL) of silica gel (Grade 60, 230–400 mesh, Fisher Scientific). The column is eluted using 20% ethyl acetate/hexanes (1700 mL) followed by 35% ethyl acetate/hexanes (700 mL) by collecting 100 mL fraction sizes using 125 mL Erlenmeyer flasks. The product is typically found in fractions 6–21.

TLCs were eluted with 35% ethyl acetate-hexanes (product R_f = 0.35).

16. The physical properties of *tert*-butyl-4-(benzoyloxy)piperazine-1-carboxylate (**1**) are: mp 103–105 °C; R_f = 0.19 (Note 7); ^{1}H NMR (400 MHz, CDCl$_3$) δ: 1.48 (s, 9H), 2.92 (br s, 2H), 3.32 (br s, 2H), 3.44 (br s, 2H), 4.04 (br s, 2H), 7.45 (t, J = 7.9 Hz, 2H), 7.58 (tt, J = 7.4, 1.4 Hz, 1H), 8.03–7.98 (m, 2H); ^{13}C NMR (100 MHz, CDCl$_3$) δ: 28.3, 41.9 (br s), 55.7, 80.1, 128.4, 129.0, 129.3, 133.1, 154.3, 164.4; IR (thin flim): cm^{-1} 2977, 2905, 2864, 2848, 1737, 1691, 1599, 1583, 1447, 1409, 1363, 1274, 1248, 1228, 1169; HRMS (ESI) [M + H] calcd for C$_{16}$H$_{22}$N$_2$O$_4$: 307.1658; found 307.1651; QHMNR (400 MHz, CDCl$_3$, dimethyl fumarate standard (Sigma-Aldrich, 96.9%)): 97–99%.

17. Anhydrous tetrahydrofuran was purchased from Sigma Aldrich and used as received. The submitter's used tetrahydrofuran (Chromasolv®, for HPLC, 99.9%) purchased from Sigma Aldrich and dried the solvent using an Innovative Technologies solvent purification system before use.

18. 2,2,6,6-Tetramethylpiperidine (≥99%) was purchased from Sigma Aldrich and used as received. The submitters used 2,2,6,6-tetramethylpiperidine (98%) purchased from Matrix Scientific, and distilled the material over calcium hydride under nitrogen atmosphere (152 °C) prior to use.

19. *n*-Butyllithium solution (*n*-BuLi) (2.5 M in hexanes) was purchased from Sigma Aldrich. The *n*-BuLi was titrated at 2.55 M prior to use using the following protocol. A 2-necked, 14/20 round-bottomed flask with a 2-cm ellipsoid shaped Teflon coated magnetic stir-bar, a nitrogen inlet adapter and a septum was evacuated using high vacuum (~1 mmHg) and dried using a heat gun for ~3 min. The flask was allowed to cool to ambient temperature under vacuum. The flask was refilled with nitrogen, the septum was removed, and 223.8 mg (1.016 mmol) of 2,6-di-*tert*-butyl-4-methylphenol (BHT, 99.0%, Sigma-Aldrich) and 3 mg of 1,10-phenanthroline (>99%, Sigma-Aldrich) were charged to the flask under a positive pressure of nitrogen. The septum was replaced, and anhydrous THF (3.0 mL, Note 17) was charged. The resultant solution was agitated at ambient temperature, and the *n*-BuLi solution was charged dropwise by syringe until the color of the mixture changed from

yellow to black indicating the endpoint (0.398 mL). The submitters titrated *n*-BuLi using an alternate literature procedure.[4]

20. The addition takes ca. 20 min.

21. Zinc chloride (ZnCl$_2$) (anhydrous, >97%, ACS, Redi-Dri^{TM}) was purchased from Sigma Aldrich. The submitters purchased ZnCl$_2$ (anhydrous, 97%, ACS) from Strem Chemicals, Inc.

22. 2,2,6,6-Tetramethylpiperidyl–ZnCl•LiCl was titrated according to the following literature procedure.[3] A 2-necked, 14/20 round-bottomed flask with a 2 cm ellipsoid shaped Teflon coated magnetic stir-bar, a nitrogen inlet adapter and a septum was evacuated using high vacuum (~1 mmHg) and dried using a heat gun ~3 min. The flask was then allowed to cool to ambient temperature under vacuum. The flask was refilled with nitrogen, the septum was removed, and 217.6 mg (1.782 mmol) of benzoic acid (Sigma-Aldrich, >99.5%) and 3 mg of 4-(phenylazo)diphenylamine (Sigma-Aldrich, 97%) were charged to the flask under a positive pressure of nitrogen. The septum was replaced, and anhydrous THF (3.0 mL, Note 17) was charged. The resultant orange solution was stirred at 0 °C, and the TMPZnCl•LiCl solution was charged dropwise by syringe until the color of the mixture changed from orange to a persistent red color indicating the endpoint (4.35 mL).

23. 3-Fluoropyridine (99%) was purchased from Oakwood Chemicals and fractionally distilled (107 °C) before use. The submitters used material from Matrix Scientific and also purified the material by distillation prior to use.

24. The internal reaction temperature remained at 21 °C throughout the addition.

25. Copper(II) acetate (anhydrous, 97%) was purchased from Strem Chemicals, Inc. and used as received.

26. Completeness of reaction is judged by the disappearance of *tert*-butyl-4-(benzoyloxy)piperazine-1-carboxylate by thin-layer chromatography (TLC), on glass backed pre-coated silica gel plates (DC-Kieselgel 60 F$_{254}$) with a UV254 indicator, using 20% ethyl acetate–hexanes as the eluent (R$_f$ of *tert*-butyl-4-(benzoyloxy)piperazine-1-carboxylate = 0.19; R$_f$ of the product = 0.35). The product is visualized with a 254 nm UV lamp and KMnO$_4$ stain. The submitter's utilized TLC plates on aluminum backed pre-

coated silica gel plates (250 µm, Agela Technologies) with a UV254 indicator.

27. Conversion of the limiting reagent (3-fluoropyridine) was determined by ^{19}F NMR spectroscopy by dissolving an aliquot of the reaction mixture in CDCl$_3$ followed by analysis. Despite the fact that the *tert*-butyl-4-(benzoyloxy)piperazine-1-carboxylate is consumed, 3-fluoropyridine is present as determined by ^{19}F NMR spectroscopy.

28. The reaction mixture is poured slowly.

29. Aluminum oxide (activated, neutral, Brockman grade I, 58 Å) was purchased from Alfa Aesar and used as received.

30. Crude mixture was dried on high vacuum (1 mmHg) for ~20 min after rotary evaporation (bath temperature = 22 °C, 35 mmHg) to remove excess ethyl acetate and 2,2,6,6-tetramethylpiperidine.

31. The crude oil is dissolved in methylene chloride and then dry-loaded onto 54 g (~125 mL) of silica gel. The volatile organics are then removed by rotary evaporation (35 mmHg, bath temperature = 22 °C). Chromatography is then performed using a 6 cm diameter flash chromatography column packed with 223 g (~450 mL) of silica gel (Grade 60, 230–400 mesh, Fisher Scientific). The column is eluted using 15% ethyl acetate/hexanes (1200 mL) followed by 22% ethyl acetate/hexanes (825 mL) followed by 27% ethyl acetate/hexanes (525 mL) by collecting 75 mL fraction sizes using 125 mL Erlenmeyer flasks. The product is typically found in fractions 11-26. TLC's were eluted with 20% ethyl acetate-hexanes (product R$_f$ = 0.35). The submitter's used the following chromatography conditions: Column diameter: 8 cm, silica: 800 mL (Silicycle, SiliaFlash® P60, 230–400 mesh, 60 Å), eluent: 3500 mL (20% ethyl acetate–hexanes), fraction size: 25 mL, 18 x 150 mm test tubes. The crude product was dry-loaded on 100 mL of silica and 800 mL eluent flushed through column before collecting fractions. The product is typically found in fractions 17–60.

32. The product solidified very slowly after isolation. However, by adding a small amount of product seeds to subsequent batches of the product, the material solidified rapidly at room temperature.

33. The physical properties of *tert*-butyl-4-(3-fluoropyridin-2-yl)piperazine-1-carboxylate are: mp 54–55 °C; R$_f$ = 0.35 (TLC, Note 26); ^{1}H NMR (500 MHz, CDCl$_3$) δ: 8.00 (dt, J = 4.8, 1.5 Hz, 1H), 7.24 (ddd, J = 13.0, 8.0, 1.5 Hz, 1H), 6.77 (ddd, J = 8.0, 4.8, 3.0 Hz,

1H), 3.58–3.53 (m, 4H), 3.47–3.41 (m, 4H), 1.48 (s, 9H); ^{13}C NMR (100 MHz, CDCl$_3$) δ: 154.8, 150.1 (d, J = 255.1 Hz), 149.8 (d, J = 6.4 Hz), 142.8 (d, J = 5.5 Hz), 123.2 (d, J = 18.8 Hz), 116.1 (d, J = 2.6 Hz), 79.8, 47.5 (d, J = 5.2 Hz), 43.6 (br s), 28.4; the submitters reported the following ^{13}C NMR values: (125 MHz, CDCl$_3$, 60 °C) δ: 154.6, 149.8 (J_{C-F} = 255.0 Hz), 149.6 (J_{C-F} = 6.2 Hz), 142.6 (J_{C-F} = 5.2 Hz), 122.9 (J_{C-F} = 18.9 Hz), 115.7, 79.5, 47.3 (J_{C-F} = 5.0 Hz), 43.6, 28.3; IR (thin film): 2976, 2929, 2859, 1692, 1605, 1469, 1453, 1417, 1365, 1266, 1238, 1216, 1165 cm^{-1}; HRMS (ESI) [M + H] calcd for C$_{14}$H$_{20}$FN$_3$O$_2$: 282.1618; found 282.1624; QHMNR (500 MHz, CDCl$_3$, dimethyl fumarate standard (Sigma-Aldrich, 96.9%)): 96–98%.

Working with Hazardous Chemicals

The procedures in *Organic Syntheses* are intended for use only by persons with proper training in experimental organic chemistry. All hazardous materials should be handled using the standard procedures for work with chemicals described in references such as "Prudent Practices in the Laboratory" (The National Academies Press, Washington, D.C., 2011; the full text can be accessed free of charge at http://www.nap.edu/catalog.php?record_id=12654). All chemical waste should be disposed of in accordance with local regulations. For general guidelines for the management of chemical waste, see Chapter 8 of Prudent Practices.

In some articles in *Organic Syntheses*, chemical-specific hazards are highlighted in red "Caution Notes" within a procedure. It is important to recognize that the absence of a caution note does not imply that no significant hazards are associated with the chemicals involved in that procedure. Prior to performing a reaction, a thorough risk assessment should be carried out that includes a review of the potential hazards associated with each chemical and experimental operation on the scale that is planned for the procedure. Guidelines for carrying out a risk assessment and for analyzing the hazards associated with chemicals can be found in Chapter 4 of Prudent Practices.

The procedures described in *Organic Syntheses* are provided as published and are conducted at one's own risk. *Organic Syntheses, Inc.*, its Editors, and its Board of Directors do not warrant or guarantee the safety of individuals using these procedures and hereby disclaim any liability for any injuries or damages claimed to have resulted from or related in any way to the procedures herein.

Discussion

Heteroaromatic and aryl amines are functionally and biologically important molecules.[5] The importance of nitrogen-containing compounds continues to drive the development of new C–N bond-forming transformations.[6] Among different amination strategies (Scheme 1),[7] C–H amination offers a direct method to introduce amino groups into molecules without stepwise functional group manipulations. It would be greatly desirable to develop a catalytic amination system effective for sp² C–H bonds with broader applications in complex molecule synthesis. For example, a range of catalyst systems have been developed, including Pd, Ru, and Cu.[8]

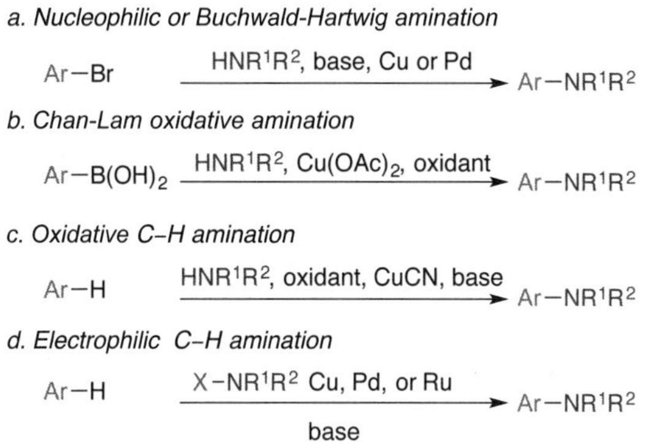

a. Nucleophilic or Buchwald-Hartwig amination

$$Ar-Br \xrightarrow{HNR^1R^2, \text{ base, Cu or Pd}} Ar-NR^1R^2$$

b. Chan-Lam oxidative amination

$$Ar-B(OH)_2 \xrightarrow{HNR^1R^2, \text{ Cu(OAc)}_2, \text{ oxidant}} Ar-NR^1R^2$$

c. Oxidative C–H amination

$$Ar-H \xrightarrow{HNR^1R^2, \text{ oxidant, CuCN, base}} Ar-NR^1R^2$$

d. Electrophilic C–H amination

$$Ar-H \xrightarrow[\text{base}]{X-NR^1R^2 \text{ Cu, Pd, or Ru}} Ar-NR^1R^2$$

Scheme 1. Amination Strategies to Access Ar–NR¹R²

We have recently developed a facile electrophilic C–H amination that can be achieved *via* organozinc intermediates derived from C–H bonds,[9] including a broad range of heteroaromatic and aryl substrates (Table 1). Such organozinc reagents can be generated *in situ* using the strong and non-nucleophilic base TMPZnCl•LiCl.[3] Additionally, readily available *O*-benzoylhydroxylamines have been demonstrated as an effective electrophilic nitrogen source in this C–H amination reaction, similar to those electrophilic aminations of organometallic reagents reported previously.[2,8,10]

Table 1. Direct amination of heteroaromatic and aryl amines using TMPZnCl·LiCl base[a,b]

$$\text{Ar−H} \quad \xrightarrow[\substack{\text{2) Cu(OAc)}_2 \text{ (5 mol\%), THF, rt} \\ \text{BzO−NR}^1\text{R}^2 \text{ (1.2 equiv)}}]{\text{1) TMPZnCl·LiCl (1.0 equiv), THF, rt, 1h}} \quad \text{Ar−NR}^1\text{R}^2$$

(1.0 equiv)

65%[c]   60%   81%[c]

76%   90%   65%[d]

98%   90%[e]   86%

94%   86%   89%

52%   61%   71%[f]

[a]Isolated yields. [b]All reactions run at a 0.20 mmol scale and 10 mol% Cu(OAc)$_2$ used. [c]Amination run at 50 °C. [d]Reaction run in CH$_2$Cl$_2$ due to low solubility of starting material in THF. [e]Deprotonation run at 65 °C. [f]Deprotonation run for 1.5 h.

In summary, this H−Zn exchange/amination strategy offers a rapid and powerful way to access a variety of highly functionalized complex aromatic amines. It is especially attractive with the use of a low cost copper catalyst and readily available reagents. Additionally, the mild reactivity of organozinc reagents allows for good compatibility with different functional groups, such as esters, amides, and halides.

References

1. Department of Chemistry, Duke University, Durham, NC 27708, qiu.wang@duke.edu. The authors thank Duke University for financial support to the research. Fellowships from the Burroughs Wellcome Endowment (S.L.M.) and Duke University Pharmacological Sciences Training Program (C.E.H.) are gratefully acknowledged. The authors thank Dr. George R. Dubay for performing HRMS experiments.
2. (a) Berman, A. M.; Johnson, J. S. *J. Am. Chem. Soc.* **2004**, *126*, 5680; (b) Berman, A. M.; Johnson, J. S. *J. Org. Chem.* **2005**, *70*, 364; (c) Berman, A. M.; Johnson, J. S. *Synthesis* **2005**, *11*, 1799; (d) Berman, A. M.; Johnson, J. S. *J. Org. Chem.* **2006**, *71*, 219.
3. (a) Mosrin, M.; Knochel, P. *Org. Lett.* **2009**, *11*, 1837; (b) Mosrin, M.; Bresser, T.; Knochel, P. *Org. Lett.* **2009**, *11*, 3406.
4. Burchat, A. F.; Chong, J. M.; Nielsen, N. *J. Organomet. Chem.* **1997**, *542*, 281.
5. (a) Bansal, Y.; Silakari, O. *Bioorgan. Med. Chem.* **2012**, *20*, 6208; (b) Eckhardt, M.; Langkopf, E.; Mark, M.; Tadayyon, M.; Thomas, L.; Nar, H.; Pfrengle, W.; Guth, B.; Lotz, R.; Sieger, P.; Fuchs, H.; Himmelsbach, F. *J. Med. Chem.* **2007**, *50*, 6450; (c) Gallardo-Godoy, A.; Gever, J.; Fife, K. L.; Silber, B. M.; Prusiner, S. B.; Renslo, A. R. *J. Med. Chem.* **2011**, *54*, 1010; (d) Jin, C. Y.; Burgess, J. P.; Rehder, K. S.; Brine, G. A. *Synthesis* **2007**, 219; (e) Massari, S.; Daelemans, D.; Barreca, M. L.; Knezevich, A.; Sabatini, S.; Cecchetti, V.; Marcello, A.; Pannecouque, C.; Tabarrini, O. *J. Med. Chem.* **2010**, *53*, 641; (f) Orjales, A.; Mosquera, R.; Labeaga, L.; Rodes, R. *J. Med. Chem.* **1997**, *40*, 586; (g) Zhang, T. Z.; Yan, Z. H.; Sromek, A.; Knapp, B. I.; Scrimale, T.; Bidlack, J. M.; Neumeyer, J. L. *J. Med. Chem.* **2011**, *54*, 1903; (h) Zimmermann, G.; Papke, B.; Ismail, S.; Vartak, N.; Chandra, A.; Hoffmann, M.; Hahn, S. A.; Triola, G.; Wittinghofer, A.; Bastiaens, P. I.; Waldmann, H. *Nature* **2013**, *497*, 638.
6. Hili, R.; Yudin, A. K. *Nat. Chem. Biol.* **2006**, 284.

7. (a) Guram, A. S.; Rennels, R. A.; Buchwald, S. L. *Angew. Chem. Int. Ed.* **1995**, 1348; (b) Louie, J.; Hartwig, J. F. *Tetrahedron Lett.* **1995**, *36*, 3609; (c) Chan, D. M. T.; Monaco, K. L.; Wang, R. P.; Winters, M. P. *Tetrahedron Lett.* **1998**, *39*, 2933; (d) Lam, P. Y. S.; Clark, C. G.; Saubern, S.; Adams, J.; Winters, M. P.; Chan, D. M. T.; Combs, A. *Tetrahedron Lett.* **1998**, *39*, 2941; (e) Monguchi, D.; Fujiwara, T.; Furukawa, H.; Mori, A. *Org. Lett.* **2009**, *11*, 1607; (f) Wang, Q.; Schreiber, S. L. *Org. Lett.* **2009**, *11*, 5178; (g) Greck, C.; Genet, J. P. *Synlett* **1997**, 741; (h) Dembech, P.; Seconi, G.; Ricci, A. *Chem. Eur. J.* **2000**, *6*, 1281; (i) Seebach, D. *Angew. Chem. Int. Ed.* **1979**, *18*, 239; (j) Erdik, E.; Ay, M. *Chem. Rev.* **1989**, *89*, 1947; (k) Barker, T. J.; Jarvo, E. R. *Synthesis* **2011**, 3954.

8. (a) Matsuda, N.; Hirano, K.; Satoh, T.; Miura, M. *Angew. Chem. Int. Ed.* **2012**, *51*, 3642; (b) Miki, Y.; Hirano, K.; Satoh, T.; Miura, M. *Org. Lett.* **2013**, *15*, 172; (c) Rucker, R. P.; Whittaker, A. M.; Dang, H.; Lalic, G. *Angew. Chem. Int. Ed.* **2012**, *51*, 3953; (d) Xiao, Q.; Tian, L. M.; Tan, R. C.; Xia, Y.; Qiu, D.; Zhang, Y.; Wang, J. B. *Org. Lett.* **2012**, *14*, 4230; (e) Yoo, E. J.; Ma, S.; Mei, T.-S.; Chan, K. S. L.; Yu, J.-Q. *J. Am. Chem. Soc.* **2011**, *133*, 7652; (f) Shang, M.; Zeng, S.-H.; Sun, S.-Z.; Dai, H.-X.; Yu, J.-Q. *Org Lett.* **2013**, *15*, 5286.

9. (a) McDonald, S. L.; Hendrick, C. E.; Wang, Q. *Angew. Chem. Int. Ed.* **2014**, *53*, 4667. (b) McDonald, S. L.; Wang, Q. *Chem. Commun.* **2014**, *50*, 2535. (c) McDonald, S. L.; Wang, Q. *Angew. Chem. Int. Ed.* **2014**, *53*, 1867.

10. (a) Rucker, R. P.; Whittaker, M.; Dang, H.; Lalic, G. *J. Am. Chem. Soc.* **2012**, *134*, 6571; (b) Yan, X. Y.; Chen, C.; Zhou, Y. Q.; Xi, C. J. *Org. Lett.* **2012**, *14*, 4750; (c) Hirano, K.; Satoh, T.; Miura, M. *Org. Lett.* **2011**, *13*, 2395.

Appendix
Chemical Abstracts Nomenclature (Registry Number)
Benzoyl peroxide; (94-36-0)
Sodium phosphate dibasic; (7558-79-4)
N-Boc-piperazine: *tert*-Butyl piperazine-1-carboxylate; (57260-71-6)
2,2,6,6-Tetramethylpiperidine; (768-66-1)
n-Butyllithium; (109-72-8)
Zinc chloride; (7646-85-7)
3-Fluoropyridine; (372-47-4)
Copper(II) acetate: Cupric acetate (142-71-2)

Qiu Wang received her Ph.D. in organic chemistry from Emory University under the direction of Professor Albert Padwa (2005). She undertook postdoctoral training with Professor Andrew Myers at Harvard University (2005–2007) and Professor Stuart Schreiber at the Broad Institute of Harvard and MIT (2007–2011). She started her independent career in 2011 as an assistant professor of chemistry at Duke University. Her research interests focus on the synthesis and studies of biologically important nitrogen-containing molecules.

Stacey L. McDonald is originally from Fort Mill, SC. In 2009, she received her B.S. in chemistry from Wofford College in Spartanburg, SC. She received her Ph.D. in 2015 from Duke University under the direction of Professor Qiu Wang. Her graduate studies focused on the copper-catalyzed electrophilic amination of sp^2 and sp^3 C–H bonds.

Charles E. Hendrick hails from Kernersville, NC. He received his B.S. in chemistry in 2011 from Wake Forest University under the supervision of Professor Lindsay Comstock-Ferguson. He is currently a graduate student in the Wang group at Duke University where his research focuses on aryl amination strategies employing nitrogen-heteroatom bonds.

Organic
⟶
Syntheses

Katie J. Bitting originates from Elizabethtown, PA. She received her B.S. in biochemistry and molecular biology from Sweet Briar College in 2013. She is currently a graduate student at Duke University where she works on zincate-mediated functionalization of arenes in the Wang group.

Dr. Joshua D. Sieber obtained a B.S. in chemistry from Penn State University with honors in chemistry in 2003. He then received a Ph.D. from Boston College in 2008 where he worked under the direction of Professor James P. Morken and was an ACS Division of Organic Chemistry Fellow. Subsequently, he was an American Cancer Society postdoctoral fellow in the laboratory of Professor Barry M. Trost. After this period, he joined Chemical Development at Boehringer Ingelheim Pharmaceuticals in 2011 where he is currently a Senior Scientist. His research interests include the development and application of catalytic asymmetric reactions.

Direct C7 Functionalization of Tryptophan. Synthesis of Methyl (S)-2-((tert-Butoxycarbonyl)amino)-3-(7-(4,4,5,5-tetramethyl-1,3,2-dioxaborolan-2-yl)-1H-indol-3-yl)propanoate.

Kazuma Amaike, Richard P. Loach, and Mohammad Movassaghi[*1,2]

Department of Chemistry, Massachusetts Institute of Technology, Cambridge, MA 02139, United States

Checked by Danilo Pereira de Sant'Ana and Richmond Sarpong

![Reaction scheme: N-Boc-L-tryptophan methyl ester (1) bearing CO2Me and NHBoc groups, plus HBPin (2), treated with [Ir(cod)OMe]2, dtbpy, THF, 60 °C; then Pd(OAc)2, AcOH, 30 °C to give the C7-borylated product 3 with BPin group.]

Procedure

Methyl (S)-2-((tert-Butoxycarbonyl)amino)-3-(7-(4,4,5,5-tetramethyl-1,3,2-dioxa-borolan-2-yl)-1H-indol-3-yl)propanoate (**3**). A flame-dried, 500-mL two-necked round-bottomed flask, equipped with a 3.5 cm football-shaped magnetic stir bar and thermometer, is charged with N-Boc-L-tryptophan methyl ester (**1**, 6.31 g, 19.8 mmol, 1.0 equiv), (1,5-cyclooctadiene)(methoxy)iridium(I) dimer (328 mg, 0.500 mmol, 0.025 equiv), and 4,4'-di-*tert*-butyl-2,2'-bipyridine (266 mg, 0.991 mmol, 0.05 equiv) (Note 1). The flask is sealed with a rubber septum secured by copper wire and placed under a nitrogen atmosphere after three successive vacuum–argon cycles conducted slowly using a needle inlet through the septum (Figure 1). Fresh anhydrous tetrahydrofuran (180 mL) (Note 2) is introduced into the flask via a syringe to afford a dark brown solution.

Org. Synth. **2015**, *92*, 373-385
DOI: 10.15227/orgsyn.092.0373

Published on the Web 12/15/2015
©2015 Organic Syntheses, Inc.

Using a syringe, 4,4,5,5-tetramethyl-1,3,2-dioxaborolane (**2**, 14.4 mL, 99.1 mmol, 5.00 equiv) (Note 1) is added in a single portion, whereupon the solution rapidly changes color from brown to dark red. This reaction solution is stirred and maintained at 60 °C. After 13 h, TLC analysis indicates complete consumption of starting material **1** (Notes 3 and 4).

Figure 1. Reaction Apparatus (photo provided by checkers)

The reaction solution is cooled to 23 °C, and is concentrated under reduced pressure (20 mmHg, 30 °C) to afford a dark brown residue. Acetic acid (20.0 mL) (Note 1) is slowly added to this residue to give a brown solution, followed by addition of palladium(II) acetate (223 mg, 0.991 mmol, 0.05 equiv) (Note 1) in a single-portion. The mixture is stirred under a nitrogen atmosphere at 30 °C for 12 h (Note 5), at which time TLC analysis indicates complete consumption of the 2,7-diboronated intermediate. The reaction mixture is then cooled to 23 °C, filtered through Celite using a glass-sintered funnel (9 cm diameter, 4 cm height), and the filter cake is rinsed with ethyl acetate (3 × 150 mL). The filtrate is washed with saturated aqueous sodium bicarbonate (500 mL), the layers are separated, and the aqueous layer is extracted with ethyl acetate (2 × 300 mL). The organic layers are combined, dried over anhydrous sodium sulfate (15 g), filtered, and concentrated under reduced pressure (20 mmHg, 30 °C). The resulting brown residue is purified by flash column chromatography on silica gel (eluent: 5% acetone, 15% dichloromethane, 80% hexanes) (Note 6) to provide a light yellow solid. The solid is recrystallized (Note 7) to afford *N*-

Boc-7-boro-L-tryptophan methyl ester **3** as a white powdery solid 4.09–4.29 g (46.5–48.8%) (Notes 8 and 9) (Figure 2).

Figure 2. Reaction Product **3** (photo provided by authors)

Notes

1. *N*-Boc-L-Tryptophan methyl ester (99%) was used as purchased from Chem-Impex International, Inc. 4,4,5,5-Tetramethyl-1,3,2-dioxaborolane (97%), (1,5-cyclooctadiene)(methoxy)iridium(I) dimer and 4,4'-di-*tert*-butyl-2,2'-bipyridine (98%) were used as purchased from Sigma Aldrich Chemical Company, Inc. Palladium(II) acetate (98%) was used as purchased from Strem Chemicals, Inc. Acetic acid, glacial (ACS grade) was used as purchased from Fisher Scientific.

2. THF was purchased from Fisher Scientific and purified by the method of Grubbs et al.[3] under positive argon pressure.

3. Thin layer chromatography was performed using pre-coated (0.25 mm) silica gel 60 F-254 plates purchased from SiliCycle (eluent: 5% acetone, 15% dichloromethane, 80% hexanes): Compound **1** R_f = 0.06 (CAM, UV), Compound **3** R_f = 0.17 (CAM, UV), Compound **4** R_f = 0.26 (CAM, UV).

4. The intermediate *N*-Boc-2,7-diborotryptophan methyl ester (**4**, see Scheme 2) could be isolated in ca. 88% yield as a white solid, by flash column chromatography over silica gel (eluent: 5% acetone, 15% dichloromethane, 80% hexanes) (Note 6) of the crude mixture after the diboronation step (see Discussion section, Scheme 2). This product contained trace impurities (<5%) but was not subjected to further

chromatographic purification due to its sensitivity toward C2 protodeboronation. For reference, data for N-Boc-2,7-diborotryptophan methyl ester (**4**) is as follows: ^{1}H NMR (500 MHz, CDCl$_3$, 20 °C) δ: 1.33 (s, 9H), 1.39 (s, 18H), 1.40 (s, 6H), 3.33 (dd, J = 14.0, 10.0 Hz, 1H), 3.46 (dd, J = 14.0, 4.5 Hz, 1H), 3.70 (s, 3H), 4.36–4.28 (m, 1H), 6.00 (d, J = 6.8 Hz, 1H), 7.11 (t, J = 7.2 Hz, 1H), 7.70 (d, J = 6.8 Hz, 1H), 7.77 (d, J = 8.0 Hz, 1H), 9.21 (br-s, 1H); ^{13}C NMR (125 MHz, CDCl$_3$, 20 °C) δ: 24.9, 25.2, 27.4, 28.2, 28.5, 52.3, 55.5, 79.4, 84.0, 84.6, 119.4, 123.1, 123.2, 127.0, 131.9, 143.1, 155.9, 173.7; FTIR (neat) cm^{-1}: 3451 (br-s), 2979 (s), 1751 (s), 1718 (s), 1596 (m), 1558 (s), 1436 (m), 1368 (m), 1293 (s), 1167 (m), 1135 (s), 1050 (m), 851 (m); HRMS (ESI, TOF) (m/z) calc'd for C$_{29}$H$_{44}$B$_2$N$_2$O$_8$Na [M+Na]$^+$: 593.3202, found: 593.3204; mp 105–106 °C;

5. Longer exposure to these protodeboronation conditions led to isolation of trace amounts (<5%) of N-Boc-L-tryptophan methyl ester (**1**), resulting from proto-deboronation of product **3**.

6. Flash column chromatography (9.0 cm diameter, 17 cm height) was performed using silica gel (60-Å pore size, 40–63 μm, standard grade, Zeochem). The residue was loaded using dichloromethane (15 mL). After 500 mL of initial elution, fraction collection (50 mL fractions) is begun, and elution is continued with 2.7 L of eluent (5% acetone, 15% dichloromethane, 80% hexanes). The compound **3** is obtained in fractions 21-54.

7. The chromatographed product was poured into a 125 mL Erlenmeyer flask and 30 mL of hexanes/chloroform (3:1) was added. The mixture was heated to its boiling point (70 °C), and 5 mL portions of hexanes/chloroform (3:1) were added until the total volume was 55 mL (the solid was not completely dissolved). The mixture was cooled to 23 °C, capped and left to stand for 13 h, then placed in a fridge at 4 °C for 48 h. The recrystallized solid was then filtered with a glass-sintered funnel, washing 3 times with cooled hexanes.

8. The analytical data for tryptophan derivative **3** is as follows: ^{1}H NMR NMR (500 MHz, CDCl$_3$) δ: 1.39 (s, 12H), 1.43 (s, 9H), 3.30 (d, J = 5.4 Hz, 1H), 3.67 (s, 3H), 4.63 (dt, J = 8.5, 5.3 Hz, 1H) 5.06 (d, J = 8.2 Hz, 1H), 7.06 (s, 1H), 7.13 (t, J = 7.5 Hz, 1H), 7.64 (d, J = 7.0 Hz, 1H), 7.67 (d, J = 7.9 Hz, 1H), 9.13 (br-s, 1H, N$_1$H); ^{13}C NMR (125 MHz, CDCl$_3$) δ: 25.1, 28.0, 28.4, 52.3, 54.4, 79.8, 83.9, 109.6, 119.2, 122.4, 122.9, 126.7, 129.6, 141.4, 155.3, 172.8. FTIR (neat) cm^{-1}: 3384 (br-s), 2979 (m), 1740 (s), 1701 (s), 1592 (m), 1503 (s), 1438 (m), 1371 (m), 1334 (m), 1165 (m), 1130 (s),

1052 (m), 970 (s), 920 (s), 849 (m), 760 (s), 684(s); HRMS (ESI, TOF) (m/z) calc'd for $C_{23}H_{33}BN_2O_6$ [M+H]$^+$: 445.2510, found: 445.2508; mp 179–181 °C; Anal. calc'd for $C_{23}H_{33}BN_2O_6$: C, 62.17; H, 7.49; N, 6.30, found: C, 61.80; H, 7.23; N, 6.12. TLC (5% acetone, 15% dichloromethane, 80% hexanes), $R_f = 0.17$ (CAM, UV).

9. Mosher ester analysis provided an enantiomeric excess of >98% for the alcohol obtained from reduction of a sample of ester **1** that had been made by protodeboronation of 7-borotryptophan **3** (see Note 5). This is in full agreement with the expectation that this procedure does not erode the enantiopurity of 7-borotryptophan **3** with respect to tryptophan **1**.

Handling and Disposal of Hazardous Chemicals

The procedures in this article are intended for use only by persons with prior training in experimental organic chemistry. All hazardous materials should be handled using the standard procedures for work with chemicals described in references such as "Prudent Practices in the Laboratory" (The National Academies Press, Washington, D.C., 2011 www.nap.edu). All chemical waste should be disposed of in accordance with local regulations. For general guidelines for the management of chemical waste, see Chapter 8 of Prudent Practices.

In some articles in Organic Syntheses, chemical-specific hazards are highlighted in red "Caution Notes" within a procedure. It is important to recognize that the absence of a caution note does not imply that no significant hazards are associated with the chemicals involved in that procedure. Prior to performing a reaction, a thorough risk assessment should be carried out that includes a review of the potential hazards associated with each chemical and experimental operation on the scale that is planned for the procedure. Guidelines for carrying out a risk assessment and for analyzing the hazards associated with chemicals can be found in Chapter 4 of Prudent Practices.

These procedures must be conducted at one's own risk. *Organic Syntheses, Inc.*, its Editors, and its Board of Directors do not warrant or guarantee the safety of individuals using these procedures and hereby disclaim any liability for any injuries or damages claimed to have resulted from or related in any way to the procedures herein.

Organic Syntheses

Discussion

Indole derivatives are prevalent in many natural products and pharmaceutical compounds, mostly in the form of often-complex tryptophan and tryptamine-derived motifs.[4] The demand for such diversely substituted indole structures has led to the development of a wide range of methods for indole functionalization.[5] With regards to tryptophan and tryptamine derivatives, selective functionalization at indole C7 has proven especially difficult, with few methods available that are direct and readily scalable.[6] We sought to explore a direct C7 functionalization method for 3-substituted indoles by utilizing arene C–H boronation as a means to this end.[7] Of particular relevance to us were reports into iridium-catalyzed indole boronations,[8] which Smith had initially shown in 2006 would proceed selectively at C7 with C2-substituted indoles.[8a] These studies inspired us to investigate a more streamlined process for direct C7 boronation of tryptophan and tryptamine substrates in a single operation. By taking advantage of the more nucleophilic/basic C2 position of C3-substituted indoles, our two-step single-flask procedure provides expedient access to the corresponding C7-boronated compounds on multi-gram scale through direct C7 activation of non-functionalized tryptamines and tryptophans.[9] The premise behind this diboronation/protodeboronation sequence was our recognition of the high propensity of five-membered heterocycles to undergo rapid C2 protodemetalation.[10]

Examination of conditions for the diboronation of various N-protected tryptamines demonstrated that exposure to excess pinacolborane (5 equiv), catalytic amounts of [Ir(cod)OMe]₂, and 4,4′-di-*tert*-butylbipyridine in tetrahydrofuran at 60 °C, was sufficient to ensure full boronation at C2 and C7.[9] Consistent with our C2 protodeboronation hypothesis, these 2,7-diboronated indoles can be dissolved in dichloromethane followed by addition of trifluoroacetic acid to cleanly afford the desired 7-boronated indole derivatives.

As we were interested in converting these two steps into a single-flask operation, we first explored simple acidification of the reaction medium once the diboronation was complete. However, mere addition of an

Table 1. Rapid synthesis of C7-boronated 3-substituted indole derivatives

[Ir(cod)OMe]$_2$ (2.5 mol%)
4,4'-di-*tert*-butyl-2,2'-bipyridine
(5.0 mol%), HBpin (5.0 equiv)
THF, 60 °C;

CH$_2$Cl$_2$, TFA, 0 → 23 °C

Entry	Substrate	Product	Yield[a]
1[b]	NMeSO$_2$Mes	NMeSO$_2$Mes	60%
2	NHCO$_2$Me	NHCO$_2$Me	66%
3	Me	Me	54% 58%[c]
4	CO$_2$Me NHCO$_2$Me	CO$_2$Me NHCO$_2$Me	84% 82%[d] 72%[c]
5	OH	OH	63%

[a] Isolated yield after purification. [b] Boronation conducted at 80 °C.
[c] 2nd step: Pd(OAc)$_2$ (5 mol%), AcOH, 30 °C. [d] Gram-scale reaction.

equivolume (with respect to tetrahydrofuran) of trifluoroacetic acid at 0 °C
led to global protodeboronation and recovery of starting material.
Gratifyingly, dilution of the reaction mixture with dichloromethane

followed by the addition of trifluoroacetic acid at 0 °C, led to the desired C2-protodeboronated tryptamine in 60% yield for the two-step process (entry 1, Table 1).[9]

We then focused our attention on expanding the substrate scope (Table 1) to other 3-substituted indoles. In all but one case (entry 1), a temperature of 60 °C and reaction time of 4–7 h was found to be ideal for the diboronation reactions (entries 2–5, Table 1).[9] These results highlighted the general compatibility of our protocol with alcohol, carbamate, ester, and sulfonamide functional groups. The excellent yield obtained for a C7-boronated tryptophan derivative (entry 4, Table 1) and its ready conversion to the corresponding 7-halo, 7-hydroxy and 7-aryl tryptophan derivatives (Scheme 1) further highlight the versatility of this chemistry.[9]

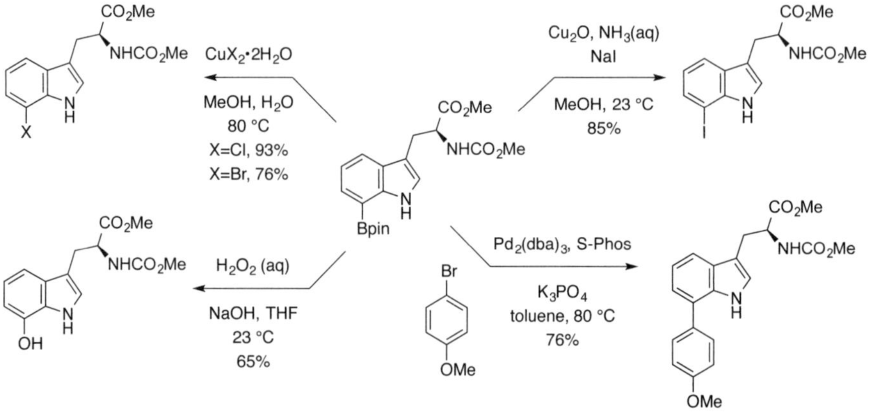

Scheme 1. Representative derivatization of a C7-boronated tryptophan

The 2,7-diborotryptophan **4** (Scheme 2), resulting from C2/C7 diboronation of N-Boc tryptophan methyl ester (**1**) using our standard iridium-catalyzed conditions,[9] could be isolated chromatographically (ca. 88%, Note 4) from the crude reaction mixture after removal of volatiles.

rganic
Syntheses

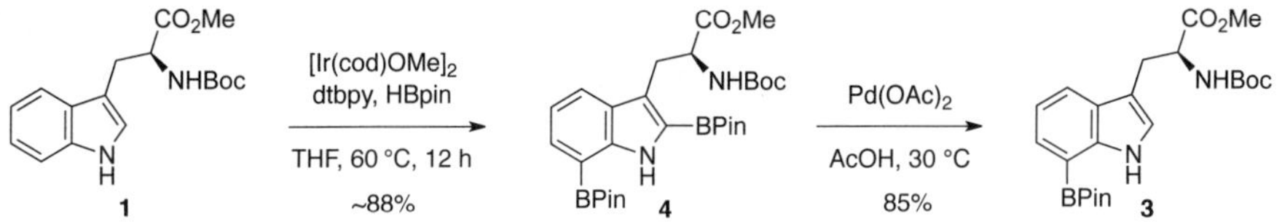

CO₂Me structure with NHBoc (compound **1**) → [Ir(cod)OMe]₂, dtbpy, HBpin, THF, 60 °C, 12 h, ~88% → BPin structure **4** → Pd(OAc)₂, AcOH, 30 °C, 85% → BPin structure **3**

Scheme 2. Two step sequence for C7 boronation of N-Boc tryptophan 1

We have developed milder conditions than those described in Table 1 for the C2 protodeboronation of the intermediate 2,7-diborotryptophan **4**. After exploring a range of additives in acetic acid, it was found that C2-selective protodeboronation of 2,7-diborotryptophan **4** could be engendered by inclusion of catalytic quantities of palladium(II) acetate. The optimal temperature for this step was found to be 30 °C, delivering the desired N-Boc 7-borotryptophan methyl ester **3** in 85% yield (Scheme 2). Further still, the dried crude mixture from the diboration step could also be subjected to the same palladium-catalyzed protodeboronation conditions, once more rendering this a two-step single-flask operation as described above.[9]

References

Address: Massachusetts Institute of Technology, Department of Chemistry, 77 Massachusetts Avenue, 18-290, Cambridge, MA 02139. E-mail: movassag@mit.edu
We acknowledge financial support by NIH-NIGMS (GM089732 and GM074825) and the NSF under CCI Center for selective C–H functionalization (CHE-1205646). R.P.L. thanks the Fonds de Recherche du Québec – Nature et Technologies for a postdoctoral fellowship. K.A. acknowledges support from the Institute of Transformative Bio-Molecules, Nagoya University, and the NSF program for Science Across Virtual Institutes for a summer fellowship.

Pangborn, A. B.; Giardello, M. A.; Grubbs, R. H.; Rosen, R. K.; Timmers, F. J. *Organometallics* **1996**, *15*, 1518.
For reviews, see: (a) Sundberg, R. J. Indoles; Academic Press: London, 1996. (b) Sumpter, W. G.; Miller, F. M. (eds) In *Heterocyclic Compounds with Indole and Carbazole Systems, Volume 8: Natural Products Containing the Indole*

Nucleus; John Wiley & Sons, Inc.: Hoboken, NJ, 2008. (c) Gribble, G. W. In *Top. Heterocycl. Chem.*, *"Heterocyclic Scaffolds II: Reactions and Applications of Indoles"* Vol. *26*; Springer-Verlag: Berlin, Heidelberg, 2010. (d) Vicente, R. *Org. Biomol. Chem.* **2011**, *9*, 6469. (e) Shiri, M. *Chem. Rev.* **2012**, *112*, 3508. (f) Kaushik, N. K.; Kaushik, N.; Attri, P.; Kumar, N.; Kim, C. H.; Verma, A. K.; Choi, E. H. *Molecules* **2013**, *18*, 6620. (g) Ishikura, M.; Abe, T.; Choshi, T.; Hibino, S. *Nat. Prod. Rep.* **2013**, *30*, 694.

5. For reviews, see: (a) Bandini, M.; Eichholzer, A. *Angew. Chem. Int. Ed.* **2009**, *48*, 9608. (b) Xie, Y.; Zhao, Y.; Qian, B.; Yang, L.; Xia, C.; Huang, H. *Angew. Chem. Int. Ed.* **2011**, *50*, 5682. (c) Davies, H. M. L.; Du Bois, J.; Yu, J.-Q. *Chem. Soc. Rev.* **2011**, *40*, 1855. (d) Broggini, G.; Beccalli, E. M.; Fasana, A.; Gazzola, S. *Beilstein J. Org. Chem.* **2012**, *8*, 1730.

6. (a) Allen, M. C.; Brundish, D. E.; Wade, R. *J. Chem. Soc., Perkin Trans. 1* **1980**, 1928. (b) Konda-Yamada, Y.; Okada, C.; Yoshida, K.; Umeda, Y.; Arima, S.; Sato, N.; Kai, T.; Takayanagi, H.; Harigaya, Y. *Tetrahedron* **2002**, *58*, 7851. (c) Kaiser, M.; Siciliano, C.; Assfalg-Machleidt, I.; Groll, M.; Milbradt, A. G.; Moroder, L. *Org. Lett.* **2003**, *5*, 3435. (d) Foo, K.; Newhouse, T.; Mori, I.; Takayama, H.; Baran, P. S. *Angew. Chem. Int. Ed.* **2011**, *50*, 2716. (e) Payne, J. T.; Andorfer, M. C.; Lewis, J. C. *Angew. Chem. Int. Ed.* **2013**, *52*, 5271. (f) Berthelot, A.; Piguel, S.; Le Dour, G.; Vidal, J. *J. Org. Chem.* **2003**, *68*, 9835. (g) Teng, X.; Degterev, A.; Jagtap, P.; Xing, X.; Choi, S.; Denu, R.; Yuan, J.; Cuny, G. D. *Bioorg. Med. Chem. Lett.* **2005**, *15*, 5039.

7. For reviews and representative reports, see: (a) Cho, J. Y.; Iverson, C. N.; Smith, M. R., III *J. Am. Chem. Soc.* **2000**, *122*, 12868. (b) Chen, H.; Schlecht, S.; Semple, T. C.; Hartwig, J. F. *Science* **2000**, *287*, 1995. (c) Cho, J. Y.; Tse, M. K.; Holmes, D.; Maleczka, R. E., Jr.; Smith, M. R., III *Science* **2002**, *295*, 305. (d) Ishiyama, T.; Takagi, J.; Hartwig, J. F.; Miyaura, N. *Angew. Chem. Int. Ed.* **2002**, *41*, 3056. (e) Boller, T. M.; Murphy, J. M.; Hapke, M.; Ishiyama, T.; Miyaura, N.; Hartwig, J. F. *J. Am. Chem. Soc.* **2005**, *127*, 14263. (f) Harrison, P.; Morris, J.; Marder, T. B.; Steel, P. G. *Org. Lett.* **2009**, *11*, 3586. (g) Fischer, D. F.; Sarpong, R. *J. Am. Chem. Soc.* **2010**, *132*, 5926. (h) Mkhalid, I. A. I.; Barnard, J. H.; Marder, T. B.; Murphy, J. M.; Hartwig, J. F. *Chem. Rev.* **2010**, *110*, 890. (i) Hartwig, J. F. *Chem. Soc. Rev.* **2011,** *40*, 1992. (j) Tajuddin, H.; Harrisson, P.; Bitterlich, B.; Collings, J. C.; Sim, N.; Batsanov, A. S.; Cheung, M. S.; Kawamorita, S.; Maxwell, A. C.; Shukla, L.; Morris, J.; Lin, Z.; Marder, T. B.; Steel, P. G. *Chem. Sci.* **2012**, *3*, 3505. (k) Hartwig, J. F. *Acc. Chem. Res.* **2012**, *45*, 864. (l) Preshlock, S. M.; Ghaffari, B.; Maligres, P. E.; Krska, S. W.; Maleczka, R. E., Jr.; Smith, M. R., III. *J. Am. Chem. Soc.* **2013**, *135*, 7572. (m) Robbins, D. W.; Hartwig, J. F. *Angew. Chem. Int. Ed.* **2013**, *52*, 933. (n) Larsen,

M. A.; Hartwig, J. F. *J. Am. Chem. Soc.* **2014**, *136*, 4287. (o) Sadler, S. A.; Tajuddin, H.; Mkhalid, I. A. I.; Batsanov, A. S.; Albesa-Jove, D.; Cheung, M. S.; Maxwell, A. C.; Shukla, L.; Roberts, B.; Blakemore, D. C.; Lin, Z.; Marder, T. B.; Steel, P. G. *Org. Biomol. Chem.* **2014**, *12*, 7318. (a) Paul, S.; Chotana, G. A.; Holmes, D.; Reichle, R. C.; Maleczka, R. E., Jr.; Smith, M. R., III. *J. Am. Chem. Soc.* **2006**, *128*, 15552. (b) Kallepalli, V. A.; Shi, F.; Paul, S.; Onyeozili, E. N.; Maleczka Jr., R. E.; Smith, M. R., III *J. Org. Chem.* **2009**, *74*, 9199. (c) Meyer, F. M.; Liras, S.; Guzman-Perez, A.; Perreault, C.; Bian, J.; James, K. *Org. Lett.* **2010**, *12*, 3870. (d) Robbins, D. W.; Boebel, T. A.; Hartwig, J. F. *J. Am. Chem. Soc.* **2010**, *132*, 4068. (e) Cho, S. H.; Hartwig, J. F. *J. Am. Chem. Soc.* **2013**, *135*, 8157. (f) Cho, S. H.; Hartwig, J. F. *Chem. Sci.* **2014**, *5*, 694 (g) Homer, J. A.; Sperry, Jonathan *Tetrahedron Lett.* **2014**, *55*, 5798. (h) Pitts, A. K.; O'Hara, F.; Snell, R. H.; Gaunt, M. J. *Angew. Chem. Int. Ed.* **2015**, *54*, 5451. (i) Feng, Y.; Holte, D.; Zoller, J.; Umemiya, S.; Simke, L. R.; Baran, P. S. *J. Am. Chem. Soc.* **2015**, *137*, 10160.

Loach, R. P.; Fenton, O. S.; Amaike, K.; Siegel, D. S.; Ozkal, E.; Movassaghi, M. *J. Org. Chem.* **2014**, *79*, 11254.

(a) Miyaura, N. *Top. Curr. Chem.* **2002**, *219*, 11. (b) Cai, X.; Snieckus, V. *Org. Lett.* **2004**, *6*, 2293. (c) Zheng, S. L.; Reid, S.; Lin, N.; Wang, B. *Tetrahedron Lett.* **2006**, *47*, 2331. (d) Alfonsi, M.; Arcadi, A.; Chiarini, M.; Marinelli, F. *J. Org. Chem.* **2007**, *72*, 9510. (e) Klingensmith, L. M.; Bio, M. M.; Moniz, G. A. *Tetrahedron Lett.* **2007**, *48*, 8242. (f) Billingsley, K; Buchwald, S. L. *J. Am. Chem. Soc.* **2007**, *129*, 3358. (g) Fleckenstein C. A.; Plenio, H. *J. Org. Chem.* **2008**, *73*, 3236. (h) Chartoire, A.; Comoy, C.; Fort, Y. *Tetrahedron*, **2008**, *64*, 10867. (i) Steel, P. G.; Woods, T. M. *Synthesis* **2009**, 3897. (j) Molander, G. A.; Canturk, B.; Kennedy, L. E. *J. Org. Chem.* **2009**, *74*, 973. (k) Kinzel, T.; Zhang, Y.; Buchwald, S. L. *J. Am. Chem. Soc.* **2010**, *132*, 14073. (l) Del Grosso, A.; Singleton, P. J.; Muryn, C. A.; Ingleson, M. A. *Angew. Chem. Int. Ed.* **2011**, *50*, 2102. (m) De M. Muñoz, J.; Alcázar, J.; De La Hoz, A.; Díaz-Ortiz, A. *Adv. Synth. Catal.* **2012**, *354*, 3456. (n) Migliorini, A, Oliviero, C.; Gasperi, T.; Loreto, M. A. *Molecules* **2012**, *17*, 4508. (o) Berionni, G.; Morozova, V.; Heininger, M.; Mayer, P.; Knochel, P.; Mayr, H. *J. Am. Chem. Soc.* **2013**, *135*, 6317. (p) Gavara, L.; Suchaud, V.; Nauton, L.; Théry, V.; Anizon, F.; Moreau, P. *Bioorg. Med. Chem. Lett.* **2013**, *23*, 2298. (q) Shan, D.; Gao, Y.; Jia, Y. *Angew. Chem. Int. Ed.* **2013**, *52*, 4902. (r) Breazzano, S. P.; Poudel, Y. B.; Boger, D. L. *J. Am. Chem. Soc.* **2013**, *135*, 1600. (s) Zhao, Y.; Snieckus, V. *Adv. Synth. Catal.* **2014**, *356*, 1527.

Appendix
Chemical Abstracts Nomenclature (Registry Number)

N-Boc-L-Tryptophan methyl ester: L-Tryptophan, N-[(1,1-
dimethylethoxy)carbonyl]- (13139-14-5)
(1,5-Cyclooctadiene)(methoxy)iridium(I) dimer: bis[(1,2,5,6-□)-1,5-
cyclooctadiene]di-µ-methoxydi-; (12148-71-9)
4,4'-Di-*tert*-butyl-2,2'-bipyridine: 2,2'-Bipyridine, 4,4'-bis(1,1-dimethylethyl)-;
(72914-19-3)
4,4,5,5-Tetramethyl-1,3,2-dioxaborolane: 1,3,2-Dioxaborolane, 4,4,5,5-
tetramethyl-; (25015-63-8)
Palladium(II) acetate: Acetic acid, palladium(2+) salt (2:1); (3375-31-3)

Kazuma Amaike is pursuing his graduate studies in Professor Kenichiro Itami's group at Nagoya University, Nagoya, Japan. His studies focus on a range projects related to C–H activation and the synthesis of natural products. In 2013, he joined the laboratory of Professor Mohammad Movassaghi at MIT as a visiting graduate student via the National Science Foundation CCI Center for selective C–H functionalization.

Richard Loach was born in Birmingham (U.K.) and graduated from Imperial College, London in 2003 with a B.Sc. in Chemistry. In 2007 he joined the research group of Professor John Boukouvalas at Laval University in Québec (Canada), earning his Ph.D. in 2013 for his studies on the total syntheses of novel γ-hydroxybutenolide natural products. In 2014, he was granted a FRQNT fellowship to pursue his postdoctoral research in Professor Mohammad Movassaghi's group at MIT. He is currently working on the total synthesis of alkaloid natural products.

Mohammad Movassaghi carried out his undergraduate research with Professor Paul A. Bartlett at UC Berkeley, where he received his B.S. degree with Honors in chemistry in 1995. He completed his graduate studies in Professor Andrew G. Myers' group as a Roche predoctoral fellow at Harvard University. In 2001, Mo joined Professor Eric N. Jacobsen's group at Harvard University as a Damon Runyon Cancer Research Foundation postdoctoral fellow. In 2003, he joined the faculty at MIT and his research program focuses on the total synthesis of alkaloids in concert with the discovery and development of new reactions for organic synthesis.

Danilo Pereira de Sant'Ana was born in Rio de Janeiro-RJ, Brazil. He received his B.S. and M.S. degrees at the Federal University of Rio de Janeiro, Brazil (UFRJ) (2008) under the supervision of Prof. Paulo Roberto Ribeiro Costa. He got his Ph.D. in cotutelle between the State University of Campinas-SP, Brazil (UNICAMP) and National Graduate School of Chemistry, Montpellier, France (ENSCM) (2014) under the supervision of Prof Luiz Carlos Dias and Jean-Marc Campagne. He is currently a Postdoctoral Fellow with Prof. Richmond Sarpong at UC Berkeley, working on the total synthesis of prenylated indole alkaloid natural products.

Author Index Volume 92

Abbott, J. R., **92**,26, 38
Ackermann, L., **92**,131
Adolfsson, H., **92**,229
Allais, C., 26, **92**,38
Amaike, K., **92**,373
Arnold, D., **92**,277
Atmuri, N. D. P., **92**,103

Baudoin, O., **92**,76
Beaton, E., **92**,182
Bezzenine-Lafollée, S., **92**,117
Biegasiewicz, K. F., **92**,309, 320
Bitting, K. J., **92**,356
Boeckman, Jr., R. K., **92**,309, 320
Bour, C., **92**,117
Busacca, C. A., **92**,237

Campagne, J. -M., **92**,296
Carreira, E. M., **92**,1
Chen, F., **92**,213
Cho, C. -G., **92**,148
Cho, H. -K., **92**,148
Choy, P. Y., **92**,195
Curran, D. P., **92**,342

Danheiser, R. L., 13, **92**,156
Dastbaravardeh, N., **92**,58
Deng, J., **92**,13
de Figueiredo, R. M., **92**,296
Di Maso, M. J., **92**,328
Ding, Z.-Y., **92**,213

Engle, K. M., **92**,58
Eriksson, M. C., **92**,237

Fan, Q. -H., **92**,213
Fang, W., **92**,117
Fillion, E., **92**,182

Gandon, V., **92**,117
Gardner, S., **92**,342
Guérinot, A., **92**,117

Hamilton, J. Y., **92**,1
Haussener, T. J., **92**,91
He, Y. -M., **92**,213
Hendrick, C. E., **92**,356
Holder, J. C., **92**,247

Kawamoto, T., **92**,342
Kitamura, M., **92**,171
Kivijärvi, T., **92**,227
Kozhushkov, S. I., **92**,131
Krainz, T., **92**,277
Kwon, K., **92**,91
Kwong, F. Y., **92**,195

Lee, H., **92**,237
Li, Z., **92**,237
Loach, R. P., **92**,373
Looper, R. E., **92**,91
Lubell, W. D., **92**,103
Lundberg, H., **92**,227
Lygin, A. V., **92**,131

McDonald, S. L., **92**,356
Millet, A., **92**,76
Movassaghi, M., **92**,373
Murakami, K., **92**,171

Presset, M., **92**,117

Qu, B., **92**,237

Radosevich, A. T., **92**,267
Rauch, K., **92**,131

Read, J. M., **92**,156
Roush, W. R., **92**,26, 38

Sarlah, S., **92**,1
Sather, A. C., **92**,58
Senanayake, C. H., **92**,237
Shaw, J. T., **92**,328
Shimizu, H., **92**,247
Shockley, S. E., **92**,247
So, C. M., **92**,195
St. Peter, M. A., **92**,328
Stoltz, B. M., **92**,247
Suppo, J. -S., **92**,296

Thuy-Boun, P. S., **92**,58

Tinnis, F., **92**,227
Tusch, D. J., **92**,309, 320

Wang, D. -H., **92**,58
Wang, L., **92**,131
Wang, Q., **92**,356
Wang, Y.-P., **92**,13, 156
Wiesenfeldt, M. P., **92**,247
Wipf, P., **92**,277
Wong, S. M., **92**,195

Yu, J. -Q., **92**,58
Yuen, O. Y., **92**,195

Zhao, W., **92**,267